SUING THE TOBACCO AND LEAD PIGMENT INDUSTRIES

SUING THE TOBACCO AND LEAD PIGMENT INDUSTRIES

Government Litigation as Public Health Prescription

Donald G. Gifford

THE UNIVERSITY OF MICHIGAN PRESS

Ann Arbor

2013   2012   2011   2010   4   3   2   1

A CIP catalog record for this book is available from the British Library.

Library of Congress Cataloging-in-Publication Data

Gifford, Donald G., 1952–
Suing the tobacco and lead pigment industries : government
litigation as public health prescription / Donald G. Gifford.
p.   cm.
Includes bibliographical references and index.
ISBN 978-0-472-11714-7 (cloth : alk. paper)
— ISBN 978-0-472-02186-4 (e-book)
1. Products liability—United States.   2. Products liability—
Tobacco—United States.   3. Lead based paint—Law and legislation—
United States.   4. Government litigation—United States.   5. Class
actions (Civil procedure)—United States.   I. Title.

KF1296.G54   2010
346.7303'8—dc22   2009038398

ACKNOWLEDGMENTS

For more than a decade, I have been both a scholar of the law of mass products torts and a real-world advocate for stronger legislation to eliminate childhood lead poisoning. I bring both sets of experience to this assessment of litigation filed by states and municipalities against the manufacturers of products—notably cigarettes, lead pigment, and handguns—that have contributed to widespread public health problems. From 1992 through 1995, I chaired the Maryland Lead Paint Poisoning Commission, an extraordinary group of people whose efforts led to the adoption of Maryland's comparatively successful, comprehensive statutory framework for reducing the incidence of childhood lead poisoning. Since 1999, I also have been involved in efforts to adopt legislation for the prevention of childhood lead poisoning in a number of other states and in the U.S. Congress, usually working as a consultant for either the National Paint and Coatings Association (NPCA) or E. I. DuPont de Nemours and Company. While working on behalf of either the NPCA or DuPont, I have, on a few occasions, urged public officials to forgo *parens patriae* litigation against product manufacturers and instead adopt legislative proposals. Obviously, the perspective presented in this book is mine alone.

The idea for a book-length project emerged from a lively panel discussion at the 2006 annual meeting of the Law and Society Association, which included copanelists Tony Sebok, Tim Lytton, Wendy Wagner, and Richard Nagareda, as well as Larry Rosenthal. This book benefited enormously from the helpful and critical comments of a number of colleagues, both at the University of Maryland School of Law and elsewhere. Oscar Gray, Tim Lytton, and Bill Reynolds critiqued the entire manuscript for me. Michael McCann generously shared his expertise and time on the topic of the impact of the states' tobacco litigation on public opinion and the political debate. Kathleen Dachille and Rena Steinzor shared their respective expertise on various chapters. I particularly express my appreciation to dean Phoebe Haddon and former dean Karen Rothenberg, both of the University of Maryland School of Law, who were very generous in supporting my work. This book would not have been possible without the highly dedicated and competent assistance of reference librarian Maxine Grosshans and research associate Sue McCarty of the Thurgood Marshall Law Library; Greg Smith of the University of Maryland School of Law Office of Information

Technology; student research assistants Leslie Harrelson, Peter Nicewicz, Tom Prevas, Alison Silber, Rachel Simmonsen, and George Waddington; and administrative assistants Marie Schwartz and Yvonne McMorris. I thank Melody Herr and Kevin M. Rennells of the University of Michigan Press for guiding me through the publication process, and Jim Reische, formerly of the Press, for encouraging me to write the book. Most important, I thank my wife, Nancy Gifford, for her support and encouragement and for understanding the sometimes strange work habits of a scholar.

I acknowledge the permission of the American Cancer Society to republish figure 1 of this text (originally titled "Tobacco Use in the US, 1900–2004") and the permission of Thompson Reuters (West) to republish an excerpt from *Products Liability Law* (2nd ed., 2008) by David Owen. I also thank the *Boston College Law Review*, the *Washington and Lee Law Review*, and the *University of Cincinnati Law Review* for permission to republish revised excerpts from articles I previously published in their journals.

CONTENTS

Introduction   1

**PART I: THE TORT SYSTEM'S EARLY RESPONSES TO
PRODUCT-CAUSED DISEASES**

Chapter 1.   The Morning after the Consumer Century   13

Chapter 2.   Product-Caused Diseases Confront the Law of
             the Iron Horse   32

Chapter 3.   The First Wave of Challenges to the Individual
             Causation Requirement   54

**PART II: PUBLIC PRODUCTS LITIGATION AS A RESPONSE
TO REGULATORY FAILURE**

Chapter 4.   The Seeds of Government-Sponsored Litigation   83

Chapter 5.   A Failure of Democratic Processes? Legislative
             Responses to the Public Health Problems Caused
             by Tobacco and Lead Pigment   104

Chapter 6.   The Government as Plaintiff: *Parens Patriae* Actions
             against Tobacco and Gun Manufacturers   120

Chapter 7.   Judicial Rejection of Recovery for Collective Harm:
             Public Nuisance and the Rhode Island Paint Litigation   138

PART III: A CRITIQUE OF PUBLIC PRODUCTS LITIGATION

Chapter 8.   Do Litigation Remedies Cure Product-Caused
             Public Health Problems?   171

Chapter 9.   Impersonating the Legislature: State Attorneys
             General and *Parens Patriae* Products Litigation   192

Conclusion   215

Notes   231
Selected Bibliography   281
Index   299

INTRODUCTION

"The most important public health litigation ever in history"—that is how Mississippi attorney general Michael C. Moore described the lawsuit he had filed on behalf of his state against the tobacco industry in 1994. He boasted to a *New York Times* reporter, "It has the potential to save more lives than anything that's ever been done."[1] A dozen years later, when a jury in Providence, Rhode Island, returned a verdict against manufacturers of lead pigment, which causes lead poisoning when ingested or inhaled by toddlers, local childhood activist Roberta Hazen Aaronson exclaimed, "Sometimes in this not so friendly world, the Goliaths are defeated and justice triumphs." She added that the jury's verdict felt "like a home run for the families devastated by lead poisoning and for a community that has borne the cost of this industry-made public health disaster."[2]

Such effusive praise greeted a new phenomenon in American litigation—*parens patriae* litigation filed on behalf of states and municipalities against the manufacturers of products that have caused public health problems. It was not hard to understand why. The tobacco litigation represented the first significant occasion when manufacturers had been held accountable for the public health harms to which their products had causally contributed. The Rhode Island litigation was the first time that a jury had found product manufacturers responsible for similar public health problems, even though the trial court judgment would subsequently be reversed.[3] During the preceding decades, when individual victims of tobacco-related diseases or childhood lead poisoning had sued the manufacturers, they were unable to recover damages because their cases were governed by common-law doctrines (judge-made law) that had developed during the waning decades of the nineteenth century. At that time, courts had encountered a very different genre of cases—lawsuits claiming damages for the smashing of bones by railroad locomotives and other machines of the newly industrialized economy. Tobacco victims were unable to recover because judges and juries alike attributed the blame for smoking-related diseases to the victims themselves instead of to the manufacturers. Children suffering from childhood lead poisoning could not recover because a century after the walls of their homes were painted, neither their parents nor their lawyers could identify the specific manufacturer whose lead pigment was contained in the paint applied to those walls. Despite innovative legal doctrines that had emerged by the 1970s and 1980s, such as market share liability and the use of class actions against product manufacturers, smokers and children still were unable to recover by the mid-1990s.

The enthusiasm that greeted the litigation concerning tobacco and lead pigment also resulted from apparent failures of the legislative branches of federal and state governments and the administrative agencies they created, which had not addressed the distinct public health problems caused by tobacco, lead pigment, and handguns. The state attorneys general and their partners in filing such litigation, a small group of plaintiffs' attorneys specializing in mass products torts, consciously viewed these lawsuits as filling the void created by the abdication of regulatory responsibility by the political branches. John P. Coale, one of the leading private attorneys that assisted in government lawsuits against tobacco and gun manufacturers, explained, "They failed to regulate tobacco and they failed regarding guns. . . . Congress is not doing its job. . . . [L]awyers are taking up the slack."[4] The State of Rhode Island's complaint in its action against lead pigment manufacturers requested the trial court to create a statewide program to "detect and abate Lead in all residences, schools, hospitals and public and private buildings within the State accessible to children,"[5] an ambitious undertaking that administrative agencies would ordinarily undertake at the direction of the legislative branch. Thus, this new form of products litigation no longer focused primarily on compensation for victims of product-related diseases. Instead, such lawsuits emerged as tort-centered examples of what Robert A. Kagan has labeled "adversarial legalism," the uniquely American phenomenon of attempting to establish government policy through litigation.[6] The perceived failures of legislatures and regulatory agencies to address the public health problems resulting from cigarette smoking and lead paint had created what Kagan refers to as a "mismatch" between "political pressures for total justice" and "inherited legal structures and modes of government."[7]

Part I of this book discusses the challenges that diseases resulting from product exposure pose to the traditional principles of tort law—the body of American law governing liability for harm caused to others. In chapter 1, I begin by describing the important roles played in American society during the first three-quarters of the twentieth century by two consumer products, cigarettes and lead-based paint, as well as the growing understanding of the terrible health consequences caused by each of these products. In chapter 2, I analyze how the barriers to recovery inherent in traditional American tort law prevented victims of tobacco-related diseases and childhood lead poisoning from recovering from the manufacturers of these products. Not until victims of diseases resulting from exposure to another group of products, those containing asbestos, sued manufacturers in the 1970s did the law begin to eliminate some of these traditional roadblocks to recovery. In chapter 3, I focus on one specific obstacle to recovery—the requirement that a particular victim identify the manufacturer of the specific product to which he was exposed that caused his disease. In cases involving asbestos and other mass products that caused latent diseases, courts began to ease this requirement in carefully circumscribed circumstances. None of these novel means of bypassing the traditional requirement of proof of an individualized causal connection, however, helped the lead-poisoned child. When a child was poisoned in the late twentieth century, it was impossible for his legal representatives, as a practical matter, to identify the producer of the fungible lead pigment that was contained in the paint that had been applied to the walls of the child's home eighty or one hundred years earlier.

Part II of the book analyzes the emergence of *parens patriae* litigation filed by state and municipal governments against the manufacturers of tobacco products, handguns, and lead pigment manufacturers. During the 1970s and 1980s, critical seeds were planted that later flowered in this new genre of litigation. I explore these important precursors of government tort litigation against product manufacturers in chapter 4. The most important of these might be called the potential "environmentalization" of mass products tort law. Traditional products liability law viewed the harm suffered by each individual victim, even among those harmed by mass products, as discrete. Some lawyers and judges influenced by the emergence of the environmental law movement, however, increasingly (but controversially) began to view mass products torts not as an aggregation of injuries to discrete individuals but, instead, as a collective harm—an environmental harm arguably governed by a different body of law. During roughly the same period, personal injury attorneys first began to use product liability actions to address the injuries suffered by victims of automobile accidents as automobile accidents became increasingly to be perceived as a public health problem. These attorneys sued automobile manufacturers and asserted that their products were not "crashworthy" or were otherwise designed in an unsafe manner. In addition, a few public health experts began to see litigation against product manufacturers as yet another approach that the government could employ to address public health problems. Finally, the ongoing proliferation of claims against asbestos manufacturers yielded a small group of specialized mass plaintiffs' attorneys who acquired both the medical and legal expertise and the resources to tackle sophisticated litigation against manufacturers of other mass products. Together, these developments paved the way for an entirely new and different form of litigation to address public health problems resulting from product exposure.

As mentioned previously, *parens patriae* litigation against product manufacturers resulted in part from a perception shared among plaintiffs' attorneys—as well as politically ambitious state attorneys general, public health officials, and public interest advocates—that Congress, state legislatures, and administrative agencies had failed to adequately regulate products that caused disease or to address the public health consequences resulting from product use. I consider this assertion in chapter 5. It is not surprising that many lawyers and some judges, disparagingly referred to by business interests as "activist" judges, saw resort to the courts as the last best hope for public health. In the process, the principal objective of suing manufacturers of products that caused disease shifted from the compensation of victims to the regulation of the manufacturers' products or other means of preventing harm caused by such products.

States first began to sue tobacco manufacturers to seek damages caused by cancer and other illnesses in 1994. Rhode Island's lawsuit against lead pigment manufacturers followed five years later, shortly after the tobacco litigation had settled. The success of this novel form of litigation required abrupt changes in the law governing both the standing of the state to sue as *parens patriae* and the principal substantive claim of public nuisance. The capacity of the state to sue as *parens patriae*—literally meaning as "parent of the country"—originated in the ability of the English Crown to legally represent the rights of persons unable to represent their own legal interests because of mental incapacity or age.[8] By the early twentieth century, the U.S. Supreme Court had also recognized a state's ability to sue as *parens patriae* to protect its citizens from collective wrongs, such as transboundary air or water pollution.[9]

Granting the state standing to sue manufacturers for the harms inflicted by mass consumer products dramatically expanded the scope of *parens patriae* standing beyond the traditional understanding of it in

American law. In the product-focused version of *parens patriae* litigation, the state sues to collect damages it has sustained as a result of harms inflicted initially and more directly on its residents—for example, state medical assistance (Medicaid) funds already paid to the victims of tobacco-related disease for their medical expenses or the costs of abating lead-based paint hazards in tens or hundreds of thousands of private residences throughout the state. In short, the state becomes a "superplaintiff." Instead of each individual victim suing manufacturers directly, the state sues on behalf of all victims and disburses the funds to individual citizens as Medicaid benefits or lead-hazard remediation grants. In chapter 6, I analyze this new form of products litigation as it developed in lawsuits filed against tobacco and handgun manufacturers.

This innovation within the legal system is one that would make alchemists proud, because in the process of the state assuming the right to collect damages for harms inflicted more directly on its residents, manufacturers somehow lose defenses that would have prevented their liability if they had been sued by the individual victims themselves. If the individual victim of lung cancer had sued, he would not have been able to recover, because the judge and jury would have found that he either knew about the health risks of smoking and therefore had "assumed the risks" or that the dangers of cigarette smoking were "common knowledge." When the state sued to collect damages for the harms originally inflicted on individual victims, however, such defenses no longer applied. Don Barrett, a Mississippi attorney who may have been among the first to appreciate the advantages of the state proceeding as *parens patriae*, explained, "The State . . . never smoked a cigarette."[10] Similarly, when the lead-poisoned child and his parents sued in their own right, they were never able to identify the specific manufacturer of lead pigment contained in the paint that poisoned him, which had been applied to the walls of his residence thirty, sixty, or perhaps one hundred years ago. If the presence of lead on the walls of homes throughout a state is understood as a collective harm to the state, however, the need to prove that any specific manufacturer produced the harm to any particular child is avoided, and it probably is possible to identify at least some of the manufacturers whose pigment contributed to the statewide incidence of childhood lead poisoning.

States in the tobacco litigation asserted a broad variety of substantive legal theories of recovery, including unjust enrichment, indemnity, common-law misrepresentation, deceptive advertising, antitrust violations, state unfair trade practice claims, and violations of the federal Racketeer Influenced and Corrupt Organizations (RICO) Act. Mississippi, the first state to file against the manufacturers and an important leader in coordinating most of the state lawsuits, primarily rested its case, however, on an obscure common-law tort known as public nuisance. Public nuisance later became the gravamen of many *parens patriae* actions against lead pigment manufacturers, handgun manufacturers, and automobile manufacturers. Legal scholars, judges, and lawyers from an earlier generation would have been shocked to learn that public nuisance, traditionally regarded as "a species of catch-all low grade criminal offense"[11] and as part of "the great grab bag, the dust bin, of the law,"[12] had become the most conspicuous weapon in the arsenal of states and municipalities seeking to address public health problems through litigation.

At the turn of the twenty-first century, some courts, mostly trial courts and a few appellate courts, expanded the boundaries of the tort to encompass any harm or annoyance that the public should not bear, even if the product manufacturer could not have been held liable under better-established claims, such as those resting on negligence, strict products liability, or misrepresentation. Mass plaintiffs' tort attorneys, public health advocates, and some judges opined that the vantage point from which courts view mass products tort actions should be shifted from one seen through the lens of "traditional" products liability law to one categorized predominantly as environmental harms. Shortly after the parties settled the tobacco litigation, Massachusetts attorney general Scott Harshbarger predicted that *parens patriae* actions would be limited to lawsuits against tobacco manufacturers and gun manufacturers.[13] Within a decade, however, similar lawsuits were filed against manufacturers of automobiles (seeking to hold them liable for the costs of abating the public nuisance of global warming), lead pigment and paint manufacturers, and the pharmaceutical company that produced the prescription drug OxyContin, which can be addictive if misused. In the modern consumer economy, any mass-produced product that contributes to causing harm results inherently in repetitive harms, which may then be characterized as a collective public health problem and, accordingly, in the legal context, arguably as a public nuisance. I focus on tobacco and lead pigment litigation in this book in part because these major *parens patriae* litigation cycles against product manufacturers have proceeded the furthest. Only Congress's unusual legislation at the behest of the National Rifle Association and gun owners, which essentially ended all litigation claiming that gun manufacturers' marketing practices in saturating inner cities with certain types of guns constituted a public nuisance,[14] aborted yet another similar litigation cycle.

In chapter 7, I turn my attention to the litigation brought by the State of Rhode Island against the manufacturers of lead pigment, where the remedy explicitly sought by the state was broad-ranging equitable relief designed to end childhood lead poisoning. In Rhode Island's action, unlike in the tobacco litigation, there was little pretense that the primary goal of the litigation was for the state to recoup as damages the funds it had been forced to spend on Medicaid payments and other past expenditures resulting from childhood lead poisoning. When the Rhode Island Supreme Court rejected the public nuisance action brought by the state against lead pigment manufacturers in 2008, just as the New Jersey Supreme Court had done a year earlier, the prospect of using public nuisance law to overcome traditional obstacles to recovery came crashing down. Because public nuisance had become the principal claim in *parens patriae* actions brought by state and municipal governments in public health litigation against product manufacturers, these decisions and others like them inflicted a serious blow, perhaps fatal, to the entire genre of litigation.

Despite the hopes of mass plaintiffs' attorneys and public health advocates that public nuisance law would provide the magical legal basis for curing product-caused public health problems, the decisions of the supreme courts of Rhode Island and New Jersey in slamming the courthouse door shut were well reasoned. Trial court judges in Rhode Island and elsewhere, in order to craft judicial solutions to public health problems, had significantly altered a number of important legal doctrines, including the appropriate boundaries of the state's *parens patriae* standing and the requirements for liability under public nuisance claims. The common law is not and should not be frozen in time. As I have written elsewhere, "Torts can be understood as the process through which courts address the issues of compensation for injuries from accidents and from wrongs in the face of changing economic and social conditions, ideologies, and scientific understandings."[15] At the same time, the norms governing judgment within a common-law system—the craft of being a judge—impose requirements of principled "reasoned elaboration," particularly when the law is changed significantly. The few courts that had changed the law of *parens patriae* standing and of public nuisance seemed to have not fully appreciated the limits of this common-law tradition.

The newfound practice of employing civil lawsuits against product manufacturers to solve product-caused public health problems during the late 1990s and the first decade of the twenty-first century looked, at first blush, to have been promising. The decisions of the state supreme courts in New Jersey and Rhode Island suggest that such hope may have been more illusory than real, even though a similar action against pigment manufacturers remains pending in the California courts.[16] Further, even if courts were to have identified an appropriate legal basis for *parens patriae* litigation, I conclude in part III of this book that such litigation likely would be both ineffective and problematic within our constitutional structure of government.

In chapter 8, I evaluate the effectiveness of such lawsuits—whether resolved through settlement or through judicial decree—in achieving their public health objectives. Many of the public health advocates who endorsed the litigation by more than forty states against the tobacco industry were quite explicit in identifying these lawsuits as the most effective way to limit the sale and use of tobacco. A decade has passed since that litigation was settled through an agreement known as the Master Settlement Agreement (MSA). Today, many public health experts regard the MSA as a colossal failure, a capitulation that protected the tobacco industry's interests more than it did public health. The question that remains is whether the failure of this negotiated resolution was unique to the tobacco litigation or whether the characteristics of the bargaining process between government officials and their retained private attorneys, on one hand, and manufacturers, on the other hand, suggest that similar settlements are inherently likely to be unsatisfactory.

The Rhode Island experience offers no assurance that remedial decrees in cases fully litigated are more likely than negotiated resolutions to be effective in solving public health problems. Before the reversal of the Rhode Island trial court judgment in favor of the state, the trial court had begun to consider the remedial phases of the litigation. It already had become clear that the trial court's self-assigned responsibility to eliminate the conditions contributing to childhood lead poisoning in tens or hundreds of thousands of residences throughout the state was likely impossible for any trial court judge, however capable, to achieve.

In chapter 9, I consider how policy-making through *parens patriae* litigation fits within our constitutional framework for the allocation of powers among the three coordinate branches of government—the legislature, the executive, and the judiciary. In the tobacco and the lead pigment litigation, the state attorneys general, members of the executive branches of their respective states, endeavored to fundamentally alter the regulatory regimes previously enacted by Congress, the state legislature, or federal or state agencies, with ones that reflected their own visions of public welfare. Many of us would prefer a world in which fewer people smoked, childhood lead poisoning was a thing of the past, and handguns were less accessible. Yet even Robert B. Reich, former secretary of labor during President Bill Clinton's administration, is troubled by the antidemocratic aspects of product regulation through litigation initiated by attorneys general.

> The biggest problem is that these lawsuits are end runs around the democratic process. We used to be a nation of laws, but this new strategy presents novel means of legislating—within settlement negotiations of large civil lawsuits initiated by the executive branch. This is faux legislation, which sacrifices democracy to the discretion of . . . officials operating in secrecy.[17]

The state attorney general clearly has the authority to file claims on behalf of the state when the state government suffers a direct loss as a result of a defendant's conduct that violates established statutory or common-law norms. When the attorney general sues to supersede a product-regulatory structure already in place, however, he dramatically changes the traditional allocation of powers among the three coordinate branches of state government.

There can be no doubt that tobacco-related illnesses and childhood lead poisoning constitute serious public health problems. All too often in the past, Congress and state legislatures, influenced by lobbying and campaign contributions from tobacco companies and the owners of residential property, have failed in their missions to prevent or to remediate these public health problems. Resorting to the courts, sometimes a reflexive response among those of us educated in the years following the dramatic litigation successes of the civil rights and environmental law movements, seemed the logical response and the last best hope. Because the legislative process inherently involves lobbying and compromise, it appears to many scholars to be less principled and less elegant than public interest litigation. Yet legislatures and administrative agencies are the appropriate bodies within our constitutional systems of government to engage in *ex ante* macroregulation of products, and they are the only ones equipped to enact effective regulatory and financing measures to address widespread public health problems. In the concluding chapter of this book, I describe the essential features of legislatively enacted solutions to tobacco-related illnesses and childhood lead poisoning. Solutions to tobacco-related illnesses and childhood lead poisoning do exist. But when it comes to finding these solutions, when we turn to the courts, we are looking in all the wrong places.

PART I

THE TORT SYSTEM'S EARLY RESPONSES TO PRODUCT-CAUSED DISEASES

CHAPTER 1

The Morning after the Consumer Century

The 1893 World's Columbian Exposition, held in Chicago, was America's way of letting the world know that it had left behind its agrarian childhood and was emerging as a world leader in a new era of industrialized consumerism. Here, the Ferris wheel, Juicy Fruit gum, Shredded Wheat, Quaker Oats, and Pabst Blue Ribbon beer were introduced to the world for the first time. The stage for the exposition was known as the "White City," made all the whiter by its having been painted with more than fifty thousand pounds of pure white lead.[1] Visited by poet Katharine Lee Bates, the "White City" inspired the line "Thine alabaster cities gleam" in the popular patriotic poem "America the Beautiful," written by Bates and later set to music.

Nearly fifty years later, at the 1939 "World of Tomorrow" World's Fair in New York, Westinghouse displayed "Elektro the Moto-Man," a seven-foot-tall robot who, like hundreds of thousands of visitors to the fair, smoked cigarettes.[2] Nothing could have captured the spirit of the age better than the smoking technological marvel. Per capita cigarette consumption in the United States was nearly twenty-eight times as great as it had been at the time of the 1893 exposition less than a half century earlier.[3]

White lead paint and cigarettes were among the products chosen at the dawn of the twentieth century to symbolize the bright future of America in the newly emerging technological era. By the end of the century, however, an alternate reality prevailed. Millions of Americans, unlike the robot Elektro, had died of lung cancer. Potentially millions of others faced the perils of diminished intellectual capabilities as a result of ingestion or inhalation of lead during their childhoods, just as a more complex, technological world required higher aptitudes.

The twentieth century was a time of unparalleled prosperity and progress for many Americans, but it also marked the onset, in dramatic fashion, of diseases caused by exposure to mass-produced products. Many products, particularly asbestos products, but also those ranging from prescription drugs to pesticides, were to cause serious detrimental health consequences. Yet it was cigarettes and lead pigment that would pose the most fundamental challenges to both the foundational prerequisites of the law of injury compensation in the American courts and, even more significantly, the constitutional allocation of powers among the judicial, executive, and legislative branches. In this chapter, I describe how these products offered Americans a brighter future during the first half of the century, only to be uncovered during succeeding decades as a source of death or disability.

CIGARETTES: THE HABIT OF THE CONSUMER CENTURY

The glorious reign of cigarettes during the early twentieth century gave few hints of the public health crisis they would later create. Perhaps no other consumer product, with the possible exception of the automobile, achieved such iconic status within American culture. General John J. ("Black Jack") Pershing, commander of the American Expeditionary Forces during the First World War, asked Americans back home to send cigarettes to his soldiers: "You asked me what we need to win this war. I answer tobacco as much as bullets."[4] Shortly after the start of the next World War, a New York City billboard displaying the Camel Man began blowing perfect smoke rings across Times Square, and he would not kick his habit for more than two decades.[5] Through persistent advertising, tobacco companies conveyed the message that cigarettes defined the smoker. The Marlboro Man of the late twentieth century was but the latest in a series of virile male role models featured by tobacco companies to promote their products. For women, smoking became an important symbol of gender equality and independence as early as the 1920s.

In short, America became a smoking society during the twentieth century. By 1965, 42 percent of all U.S. adults were smokers, including 52 percent of all men and 34 percent of all women.[6] The number of cigarettes consumed in the United States per capita—that is, for every woman, man, and child—had increased from 54 in 1900 to 4,345 in 1963. By 1981, Americans annually consumed 640 billion cigarettes. Only three decades later, it seems difficult to imagine a time when nearly half the population smoked in restaurants, workplaces, and hospitals and on planes and trains.

The cigarette habit was largely a phenomenon of the twentieth-century age of mass production and mass marketing. Even before the twentieth century, tobacco, in various forms, had played a key role throughout American history.[7] Cigarettes themselves, however, were not widely available until after the Civil War. In the early decades of the Republic, Americans typically smoked a pipe, snorted or pinched snuff, or chewed tobacco.[8] Of course, none of these means of enjoying tobacco delivered significant amounts of smoke directly to the lungs. Meanwhile, cigarettes remained a novelty, in large part because of the labor-intensive necessity of hand rolling them.

Following the Civil War, a new variety of tobacco, white burley, grew in popularity. It was easier to harvest and readily absorbed the sweeteners and flavorings that purveyors added.[9] In addition, a new form of technology, *flue curing*, produced a milder, sweeter, more consistent tobacco. These innovations made the deep inhalation of cigarette smoke into the lungs far more palatable than the inhalation of smoke produced by pipes and cigars, which typically was held only in the smoker's mouth. Once within the lungs, nicotine, the addictive ingredient in tobacco, entered the bloodstream quickly and reached the brain within seven seconds. Allan M. Brandt, professor of the history of medicine at Harvard University, has remarked, "Nicotine addiction was born in the serendipitous marriage of bright tobacco and flue-curing."[10]

Of course, the massive public health crisis created by the smoking of cigarettes in the middle and late parts of the twentieth century required the mass production of these nicotine delivery systems. Skilled nineteenth-century hand rollers could produce only about two hundred cigarettes per hour.[11] In 1881, however, Virginia inventor James Bonsack introduced a rolling machine that produced over two hundred cigarettes a minute. When combined with the new, tastier forms of tobacco, this latest innovation facilitated the mass consumption of the cigarette and paved the way for a major public health crisis decades later.

Once the tobacco manufacturers had invested in the costly Bonsack rolling machines, they ran the risk that the demand for cigarettes would not keep up with the greatly increased supply. Therefore, it should come as no surprise that during the early twentieth century, tobacco manufacturers virtually created the modern advertising and marketing industry as it is known today.[12] Cigarettes were portrayed as the ultimate product of the modern age, offering instant pleasure and relief from the stresses of everyday hectic life. Advertising showing glamorous actresses and actors, as well as other members of the rich and famous, smoking cigarettes conveyed an egalitarian message: for ten or fifteen cents a pack, even the ordinary factory worker could share the pleasures of the elite. From the beginning, cigarette manufacturers' advertising targeted youth. Some manufacturers included colorful collecting cards illustrating baseball players or scantily clad actresses.[13]

The combination of mass-produced cigarettes and their relentless promotion transformed American life and culture. By midcentury, the purchase of cigarettes represented 1.4 percent of the gross national product and 3.5 percent of spending on nondurable goods.[14] Pulitzer Prize–winning author Richard Kluger has remarked about smoking during the Depression, "Americans were in love with smoking at a time when their collective life was low on other consolations. If the habit was bad for their health, so was hunger, and the latter seemed the more pressing peril by far."[15]

LEAD PAINT: "THINE ALABASTER CITIES GLEAM"

Since ancient times, humans have been familiar with lead, in no small part because it so often is present alongside silver in the earth's crust.[16] Lead pipes carried water in both Ur and Rome. Lead coffins and lead projectiles fired from weapons similarly date from ancient times. Many cultures used lead salves and potions for medicinal purposes. Dentists filled cavities with lead. Perhaps most important for the advancement of civilization, because alloys of lead and tin were so easy to cast and recast, they were formed into the movable type of printing presses following Gutenberg's invention.

Lead became an even more ubiquitous metal in the newly emerged industrialized consumer society of twentieth-century America. Builders used it in roofing, electrical conduits, cable coverings, tanks, and water and sewer pipes. Lead-acid batteries started automobiles. Lead solder connected the wires of the telecommunications system and early televisions and radios. Small arms ammunition containing lead killed millions in the two world wars. Lead vests protected medical and dental personnel from the effects of X-ray exposure. Lead was also an important ingredient in pesticides used pervasively during the middle decades of the twentieth century.

From the perspective of public health problems, the two most important uses of lead in the twentieth century were ones not obviously identified with lead as a metal. First, tetraethyl lead was added to gasoline to make automobile engines run more smoothly and efficiently by preventing "knocking," the faulty combustion of the fuel-air mixture in the engine cylinders. The sudden emergence of the automobile in American society during the early decades of the twentieth century created huge demand for automobile fuel.

Gasoline produced from low-grade crude petroleum caused automobile engines to knock, resulting in poor performance and limited mileage.[17] After trying a variety of gasoline additives, scientists discovered that tetraethyl lead, a lead compound, prevented knocking, thereby doubling gasoline mileage and enabling the use of larger and more powerful engines. Similarly, high-octane aviation fuel containing lead additives powered American aircraft during the Second World War. From the beginning, however, health questions surrounded the use of tetraethyl lead (TEL). In the 1920s, hundreds of workers at a TEL-production facility became ill, and several died. Less clear at the time were the possible negative health consequences that could result among members of the general public from atmospheric lead exposure caused by automobile exhaust emissions.[18]

With the development of new, higher-performance engines during the 1950s, American automobiles consumed increasing amounts of tetraethyl lead. By 1960, the consumption of tetraethyl lead had increased 70 percent in only a decade.[19] However, during the 1960s and 1970s, the American public became increasingly concerned about the harmful effects of air pollution. Public health scientists and lead industry officials heatedly debated whether lead particles in the atmosphere posed a health hazard.

As it turned out, leaded gasoline also poisoned the catalytic converter, a device now required in all American automobiles to reduce automobile emissions. Motivated by lawsuits filed by the Natural Resources Defense Council, the Environmental Protection Agency required the reduction of lead in gasoline. As changes in public health go, the results were sudden and dramatic. Blood lead levels dropped nearly 40 percent within a four-year period.[20] The 1990 amendments to the Clean Air Act finally required the total elimination of leaded gasoline for cars and trucks.[21]

The second new important source of lead exposure to emerge during the late nineteenth and early twentieth centuries was the use of lead carbonates as an ingredient in paint to decrease the time required for paint to dry, increase its durability, and help it resist moisture and retain its fresh appearance. As both the population of the United States and its economy expanded between 1910 and 1977, the American housing industry boomed. Over four thousand tons of lead pigments were applied to American homes and products during this period.[22] Physicians recommended using lead paint to repaint surfaces to cover walls that might be contaminated by "disease germs."[23] Following the Spanish-American War, lead industry promotional materials claimed that "Uncle Sam" had eliminated yellow fever in Havana and Santiago by introducing "cleanliness" and that "for cleanliness there is nothing like paint—the *best* paint—Pure White Lead."[24] Decades later, government publications during the Depression extolled the virtues of public housing brightly painted with "white lead," as elevating the spirits of African Americans living in the rural South.

Professional painters found that surfaces previously painted with lead paint were easier to repaint, because of the natural chalking process of white lead. At the urging of the lead industry, the painters proclaimed themselves to be "white-leaders."[25] Government regulation of paint during the first several decades of the twentieth century did not focus on the potential health effects of lead; rather, regulators debated whether paint that was not "pure white lead" should be regarded as "adulterated" and forced to include warning labels to this effect.[26] Ignored in all of this, of course, was the fact that, as Christian Warren reports, "anyone with passing knowledge of the paint industry knew that making and using lead pigments was potentially deadly."[27]

THE CHANGING FOCUS OF PUBLIC HEALTH: VICTORY OVER BACTERIA AND THE NEW FOCUS ON TOXIC THREATS

For the health risks caused by exposure to cigarette smoking and walls painted with lead paint to be recognized, the focus of public health itself needed to change. At the beginning of the twentieth century, infectious illnesses—such as tuberculosis, influenza, poliomyelitis, smallpox, diphtheria, and yellow fever—were far more visible than latent diseases resulting from product exposure. Infectious illnesses accounted for 56 percent of all deaths in 1900.[28] During the next forty years, life expectancy would increase from 47.3 years to 62.9 years.[29] By the Second World War, better public sanitation, penicillin and other antibiotics, and vaccinations brought many of these traditional health threats under control for most Americans. Poliomyelitis, a disease frequently causing death or paralysis, haunted children even during the postwar period, but a vaccine was discovered for this dreaded disease in 1954, and most American children would be vaccinated within less than a decade.

Against the background of the rosy picture created by the dramatic success of drugs in wiping out disease, the publication of Rachel Carson's *Silent Spring*[30] in 1962 sounded a discordant note. Like lead paint and cigarettes, DDT (dichlorodiphenyltrichloroethane) was one of the technological marvels supposedly destined to make "the good life" possible in the twentieth century. DDT was a powerful pesticide used, among other things, to destroy insects that carried viruses and bacteria that caused a variety of serious illnesses. However, Carson, a marine biologist, wrote a text that envisioned a world in which, as a result of the toxic effects of DDT exposure, birds no longer sang, frogs no longer chirped, and humans died as a result of exposure to pesticides. One of the chapters, entitled "One in Every Four," analyzed likely cancer rates among humans as a result of exposure to carcinogenic substances. Silent Spring has been identified repeatedly as one of the most important books of the twentieth century[31] and is often credited with inspiring the environmental movement.[32]

The roots of Silent Spring lay deep within the fields of occupational health and industrial hygiene. While the remainder of the medical and public health communities had focused largely on infectious illness during the preceding decades, occupational health specialists increasingly had identified and cataloged the harms resulting from exposure to a variety of substances during the manufacturing processes, including, notably for the purposes of the present discussion, lead. Not much attention had been paid to occupational illness at the beginning of the twentieth century. The most influential medical textbook of the time, William Osler's The Principles and Practices of Medicine, published in 1915, covered illnesses resulting from occupational exposure in only 7 of 1,225 pages.[33] Firms in dangerous industries, such as mining and lumber, employed "company" physicians who predictably took the view that health and safety at the workplace were the responsibilities of the individual worker.[34]

In 1910, Dr. Alice Hamilton, generally credited as the founder of the occupational health field in the United States, began to study the health consequences to those who worked regularly with phosphorus, lead, radium, and other substances deemed dangerous.[35] Originally working with other socially conscious physicians, labor unions, and other progressive groups, Hamilton later sought to establish "a more self-consciously objective science" and promoted the idea of safe levels of exposure to hazardous substances.[36] Hamilton's critical work regarding occupational lead poisoning will be considered more fully later in this chapter.

The genesis of environmental science and policy can largely be traced to the occupational health movement. But the contexts of toxic exposure in the workplace, on one hand, and in the environment or through product exposure, on the other hand, differ. The worker employed in the often poorly ventilated factory was constantly exposed to hazardous substances at a level higher than those typically experienced by consumers of products or members of the general public.

THE DISCOVERY OF HARMFUL CONSEQUENCES FROM PRODUCT EXPOSURE

The changing focus within medical science and public health paved the way for a fuller appreciation of the health consequences that product users and consumers faced as a result of exposure to cigarettes and lead paint. In the case of diseases resulting from exposure to tobacco, the long period of latency between product exposure and disease manifestation, generally fifteen to forty years, delayed the recognition of the causal connection between product and illness. Mass production of cigarettes did not begin until the late nineteenth century; consequently, the widespread evidence of diseases resulting from product exposure was not easily observed until decades into the twentieth century. In the remainder of this chapter, I trace the discovery of the harmful consequences resulting from exposure to tobacco products and lead paint.

The Health Consequences of Cigarette Smoking

The earliest opposition to cigarette smoking intertwined unsubstantiated prognostications of damage to the smoker's health with moral concerns. As early as 1604, King James I of England warned that smoking tobacco was both "barbarous and . . . godlesse" and "lothsome to the eye, hatefull to the nose, harmefull to the braine, dangerous to the Lungs."[37] In 1859, even before cigarette smoking became popular, antitobacco reformer George Trask warned that smoking cost twenty thousand lives a year, but his admonition lacked any empirical foundation.[38] Health concerns usually were combined with warnings of moral turpitude during the Victorian era, when anything that provided pleasure was regarded as evil.[39] By 1901, the National Anti-Cigarette League claimed three hundred thousand members. In 1916, as Model Ts rolled off the Ford assembly line, Henry Ford vowed not to hire smokers, because they were "loose in their morals."[40] Even baseball star Ty Cobb, an unlikely guru of morality, warned, "Cigarette smoking stupefies the brain, saps vitality, undermines one's health, and lessens the moral fiber of the man."[41]

In the decades between the two world wars, medical opinion about the possible health effects of smoking was sharply divided. Perhaps the most common opinion during the 1920s and 1930s was that smoking in moderation did one no harm. In 1936, the editors of *Scientific American* proclaimed, "Most smokers—probably all smokers—are doubtless harmed to some extent, usually not great, by smoking."[42] The article went on to comment that climbing stepladders, playing football, crossing the street, and even just living and breathing have risks that we willingly accept.

During these decades, three factors impeded the discovery of the link between smoking and cancer. First, medical science and public health were in the midst of the bacteriological revolution. The assumption was that a disease like cancer must have a bacteriological or viral cause. Second, cancer typically occurs only after a latency period following the victim's commencement of smoking, which averages fifty years.[43] Thus, the dramatic increase in the incidence of lung cancer was not statistically visible until the 1940s, because Americans had not begun to smoke in massive numbers until the First World War. Finally, the causal relationship between smoking and cancer was difficult to establish because smoking is not the only causal variable in producing cancer: not everyone who smokes gets cancer, and conversely, not everyone who gets cancer is a smoker.

Eventually, however, the mass production of cigarettes resulted in a lung cancer epidemic in twentieth-century America. In 1926, the American health care system included eight hundred clinics for venereal disease and six hundred tuberculosis sanatoriums, but there was only one cancer center.[44] As the number of cases of lung cancer in the United States rose from 371 in 1914 to 7,121 in 1950, the possible correlation with America's new addiction to tobacco became impossible to ignore. In 1939, Franz Hermann Müller of Germany published a paper noting the relationship between increased smoking and the increasing incidence of lung cancer.[45] Müller conducted a study comparing the case histories of eighty-six men with lung cancer with a comparably sized population of those not afflicted with lung cancer. Although there was not a perfect correlation between smoking and cancer, smoking dramatically increased the likelihood of developing the dreaded disease.

In the face of these and similar studies, tobacco companies responded. Physicians at the 1947 American Medical Association (AMA) convention stood in line to receive free cigarettes from Philip Morris, which claimed that it produced the most healthful cigarette.[46] Its competitor, R. J. Reynolds, advertised that doctors smoked more Camels than any other brand.

According to Richard Kluger, several studies linking cigarette smoking and lung cancer released in the early 1950s "marked the end of the age of innocence about the blithe charms of the cigarette."[47] A study by Ernst L. Wynder and Evarts Graham, published in the *Journal of the American Medical Association* in 1950, showed that within a large sample population,

96.5 percent of lung cancer victims were moderate to heavy chain-smokers, and most of these victims had smoked for twenty years or more.[48] The results of the Wynder and Graham study created a firestorm when, in 1952, they were summarized in a brief article in *Reader's Digest*,[49] the most widely circulated periodical of the era. At about the same time, A. Bradford Hill, a pioneer of the use of medical statistics, and his colleague Richard Doll, began to study the link between smoking and cancer.[50] Doll approached the research hypothesis as a skeptic, initially believing that automobile exhaust and road tar were more important causal factors in producing lung cancer. Yet Hill and Doll concluded that the possibility that the correlation between smoking and lung cancer was merely a matter of chance was less than one in a million and that heavy smokers were more than fifty times as likely as nonsmokers to develop lung cancer.[51] Finally, at the 1954 AMA convention, Cuyler Hammond and Daniel Horn, epidemiologists for the American Cancer Society, presented the findings of their prospective study of 187,766 white males between the ages of fifty and sixty-nine.[52] Among this group, smokers died of lung cancer at a rate five to sixteen times higher than nonsmokers.

For the first time in decades, smoking rates began to decline following the release of these studies in the early 1950s.[53] Predictably, the tobacco industry again responded. Cigarette companies promoted cigarettes that were "milder," "mentholated," or filtered. They competed in the "tar derby," claiming that their particular brands of cigarettes had less tar and nicotine than competing brands and therefore were less hazardous to smokers' health. Their advertisements became so egregiously inaccurate that the Federal Trade Commission stepped in during the late 1950s to prohibit many of the competing claims. In a tragic and maleficent irony, Lorillard, one of the manufacturers, responded to health concerns by designing and marketing Kent, a cigarette that originally included the Micronite filter made from asbestos; the filter was highly promoted and described in advertisements as made from "completely harmless material."[54] In fact, research conducted by two independent laboratories for Lorillard already had determined that these asbestos fibers, known to cause the horrible diseases of asbestosis, lung cancer, and mesothelioma, became coated with carcinogens from the tobacco itself and then imbedded themselves in smokers' lungs.

In 1954, the tobacco industry, working with the prestigious public relations firm of Hill & Knowlton, released "A Frank Statement to Cigarette Smokers." Claiming, "We believe the products we make are not injurious to health," the companies said that they "always have and always will cooperate closely with those whose task it is to safeguard the public health."[55] The statement then announced the creation of the Tobacco Industry Research Committee (TIRC). Confidential reports of the TIRC meetings, subsequently made available to the public, make it clear that the purpose of the TIRC was "to build a foundation of research sufficiently strong to arrest continuing or future attacks" on the tobacco industry.[56] As Brandt concludes, during the late 1950s, the tobacco companies succeeded in puncturing the scientific consensus, which had emerged by 1953, that cigarette smoking caused lung cancer. They created a "cigarette controversy."[57] In fact, internal documents show that scientists working for both R. J. Reynolds[58] and Philip Morris[59] had concluded by the late 1950s and early 1960s that cigarette smoke included carcinogens, and that cigarette companies were seeking ways to reduce or eliminate these substances.

In 1959, U.S. surgeon general Leroy Burney issued a statement on behalf of the Public Health Service that associated cigarette smoking with cancer. The statement concluded, "The weight of evidence at present implicates smoking as the primary etiological factor in the increased incidence of lung cancer."[60] Shortly thereafter, however, the *Journal of the American Medical Association* editorialized, "A number of authorities who have examined the same evidence cited by Dr. Burney do not agree with his conclusions."[61] In 1962, only 38 percent of Americans believed that cigarette smoking caused cancer.[62]

In June 1962, President Kennedy appointed the Surgeon General's Advisory Committee on Smoking and Health.[63] Two years later, this distinguished group of scientists with disparate medical and public health specialties—and different perspectives, at least initially, regarding the possible causal connection between smoking and lung cancer—issued a unanimous 387-page report. Its most important conclusion read, "Cigarette smoking is causally related to lung cancer in men; the magnitude of the effect of cigarette smoking far outweighs all other factors. The data for women, though less extensive, point in the same direction."[64]

In the decades to come, it would become clear that cigarette smoking had a variety of other harmful health consequences. Dr. C. Everett Koop, surgeon general during the administration of President Ronald Reagan, issued a report in 1983 that strongly suggested that cigarette smoking was a serious risk factor for heart disease.[65] The report stated that smokers were two to three times more likely to have cardiac complications than nonsmokers. In the following year, Koop issued a report that found that cigarette smoking was the primary cause of emphysema and other chronic obstructive lung diseases.[66] Finally, in 1988, Koop issued a further surgeon general's report that declared that cigarettes were addictive because of the presence of an addictive chemical, nicotine, in much the same way that heroin and cocaine were addictive.[67] Within a few years, the release of internal tobacco industry documents from the early 1960s revealed that the industry had long been aware of the addictive properties of its product. For example, in 1963, corporate counsel for one of the major cigarette companies had written, "We are, then, in the business of selling nicotine, an addictive drug effective in the release of stress mechanisms."[68] Other industry documents revealed that chemical additives to cigarettes, notably ammonia, quickened the rate at which nicotine was absorbed into the smoker's body, increasing this addictive effect.[69]

Despite the overwhelming body of scientific research establishing that cigarette smoking both causes cancer and is addictive, the chief executives of major tobacco companies continued to deny these facts. As recently as 1994, Lorillard's CEO Andrew Tisch testified under oath that he did not believe that cigarettes caused cancer.[70] That same year, R. J. Reynolds CEO James W. Johnston compared the addictive qualities of cigarettes to those of coffee, chocolates, and Twinkies.[71]

In a society in which nearly half of all people had smoked during the mid-twentieth century, cigarette smoke pervaded not only homes but restaurants, workplaces, colleges and universities, airplanes, and buses. Within a few years after the 1964 surgeon general's report, distinguished scientists began questioning whether secondhand smoke, or "passive smoke," was an environmental hazard. Two studies released in the early 1980s demonstrated that wives living with smoking husbands were at significantly greater risk of developing lung cancer.[72] Another study at the time showed that residual particles from cigarette smoking were present in restaurants, bars, and other commercial establishments at levels likely sufficient to pose a health hazard.[73]

The risks attributable to exposure to secondary smoke posed a challenge to the cigarette industry's twenty-year-old response to the 1964 surgeon general's report. For years, the industry had reinforced the inherently American opinion that each individual should be able to choose whether or not to smoke and expose himself to the risks caused by smoking. However, now it was becoming clear that the smoker's decision affected the health of others. Even in the face of knowledge of the risks of secondhand smoke, Philip Morris wrapped itself in traditional American freedoms. For example, it sponsored an essay contest on free speech.[74] Walker Merryman of the Tobacco Institute compared the tobacco companies' protection of the rights of smokers to the civil rights movement, suggesting that it is danger-

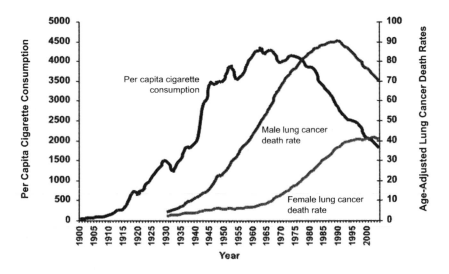

Fig. 1. Twentieth-century cigarette consumption and lung cancer death rates. (Data from American Cancer Society, "Tobacco Use in the U.S., 1900–2004," *Cancer Statistics 2008*, PowerPoint no. 396, slide 27, http://www.cancer.org/downloads/STT/Cancer_Statistics_2008.ppt. Copyright 2008 American Cancer Society, Inc. Reprinted with permission from www.cancer.org. All rights reserved.)

ous to treat smokers as "social pariahs" and "second class citizens" who "already sit in the back of the plane and the bus."[75] Philip Morris proffered a "Smoker's Bill of Rights" that included the "inalienable" rights "to choose to smoke," "to accommodation in public places," and "to freedom from unnecessary government intrusion."[76]

Unlike lead paint, the sale and distribution of cigarettes, at least to adults, has remained legal and essentially unregulated—although the government has restricted promotion of the product. According to the Centers for Disease Control and Prevention, in 2005, 20.9 percent of all American adults were smokers,[77] slightly less than half the percentage at the time of the release of the 1964 surgeon general's report. Perhaps more troubling, in 2003, 21.9 percent of high school students smoked.[78]

Childhood Lead Poisoning

It has been known for thousands of years that exposure to lead, in sufficiently high concentrations, results in harmful health consequences. Hippocrates, in 370 BCE, diagnosed colic in a miner as resulting from lead.[79] In the eighteenth century, Bernadino Ramazzini, an Italian physician and professor, observed similar symptoms among those who used lead to glaze pottery. By the late eighteenth and the nineteenth century, colic attributed to lead poisoning was noted among those who consumed ciders or wines that had been stored in lead containers.

In the modern era, as previously noted, exposure to lead occurs largely through three distinct pathways: occupational exposure, atmospheric exposure resulting from breathing air contaminated by automobiles using gasoline containing tetraethyl lead, and exposure to lead paint.[80] Most early reports of the harmful effects of lead exposure focused on occupational poisoning, that is, exposure to very high doses of lead on an ongoing basis resulting in severe and acute lead poisoning of adult workers. The "pure" white lead that early twentieth-century painters and consumers preferred was made by corroding heavy ingots of metallic lead with acid, resulting in a crusting of white lead.[81] Similarly, foundries producing lead type, so critical to the printing process before the computer age, resulted in rooms "always filled with a fine lead dust."[82] The constant breathing of dust from paint plants and type plants sometimes resulted in acute lead poisoning—colic, gastrointestinal problems, and often death. Painters, particularly those working in interiors or other confined spaces, experienced problems similar to those of factory workers. In October 1924, several Standard Oil laboratory employees who were working with tetraethyl lead became severely paranoid, hallucinated, convulsed, and screamed as a result of lead exposure.[83] Five workers died, and another thirty-five suffered severe neurological problems. However, within a few decades, industrial lead poisoning was to become a prominently featured success story of the emerging field of occupational health.[84]

While lead poisoning among adults has become largely a thing of the past, children suffer a variety of harmful effects at much lower levels of exposure than do adults. Pathways to lead exposure that pose little or no risk for adults are extremely dangerous to children. Though children were occasionally poisoned as a result of helping their parents paint or working in factories with heavy concentrations of lead, the two most significant sources of lead exposure for children during the twentieth century were lead compounds in the air, resulting from the use of "leaded" gasoline, and lead paint on the walls of homes and schools.

For much of the twentieth century, children breathed air containing lead particles resulting from automobile exhaust of cars using tetraethyl lead (TEL) in amounts sufficiently great to cause neurological damage to children (but not to adults).[85] As previously described, the regulatory process worked, albeit belatedly, and TEL was removed from gasoline; this action resulted in almost immediate and dramatic reductions in the blood lead levels of American children. Accordingly, TEL has not played a major role in litigation against product manufacturers brought by individuals, representatives of class actions, states, or municipalities. For our purposes, all that needs to be said about TEL is that from the mid-1920s through the 1980s, atmospheric exposure to lead was an important contributing factor in causing childhood lead poisoning.

Today, soil that was soaked with leaded gasoline continues to play a role in a relatively small number of cases of childhood lead poisonings. But it is lead paint that poses the predominant risk. In a minority of cases, jewelry or other toys containing lead, some pottery glazed with lead compounds, and native remedies also play a role—in some instances, the predominant role.

Remarkably few cases of childhood lead poisoning were diagnosed in the United States during the first three decades of the twentieth century.[86] According to Peter English, through 1925, there were only eight diagnosed cases of childhood lead poisoning in the United States that were caused by lead paint on cribs, toys, windowsills, or railings.[87] Physicians of the era were far more concerned with the high infant and childhood mortality rates resulting from infectious illnesses. They looked for bacteriological causes for illness and found them, even when they did not exist. Retrospective examination of hospital records during this period reveals numerous cases describing gastrointestinal disorders followed by convulsions, sometimes followed by paralysis, all of which are classic symptoms of severe childhood lead poisoning. The terminal phases of lead poisoning were often misdiagnosed as tubercular meningitis.

Childhood lead poisoning was diagnosed during the first several decades of the twentieth century only when a child manifested acute symptoms, such as seizures, coma, and severe abdominal pain. These symptoms required a very high concentration of lead in the blood, and consequently could only result from constantly chewing objects painted with lead. At the time, pediatric texts referred to this gnawing or chewing behavior as *pica,* understood to be an inordinate desire to eat inanimate objects.[88] Even as late as the period from 1959 through 1961, of the 103 lead-poisoned children treated at the Philadelphia Children's Hospital's outpatient clinic, 83 were found to have had a history of pica.[89] Pica was to assume enormous importance for decades in debates regarding responsibility for childhood lead poisoning. On the one hand, advocates for the lead paint industry would claim that because the causes of childhood lead poisoning so often included a child's gnawing on painted substances, some of the causal responsibility could be attributed to the unusual behavior of the child or the failure of his parents, often lower-income parents, to adequately supervise him. On the other hand, public health advocates and pediatricians noted that children frequently put objects in their mouths and that in these circumstances, toxic substances should not be readily available in a child's environment.[90]

In the early decades of the twentieth century, France and several other European countries already had outlawed the use of lead paint in residential interiors. Dr. Alice Hamilton, mentioned previously in this chapter as an early leader in the field of occupational health, traveled to Europe at the beginning of the twentieth century to study this issue. Her writings and speeches make it clear that the purpose of the European ban on lead paint was to protect painters who were working in confined interiors, with no mention of risks to children living in the painted residences.[91] Australian studies published between 1885 and 1904 described children in Queensland who had contracted childhood lead poisoning from porches painted with lead paint, but contemporary American observers dismissed these reports as irrelevant to the American experience.[92] Climatic conditions were extreme in Queensland, with average high temperatures running as high as 108.9 degrees during the Australian summer, and the symptoms observed in the Australian children, though similar to those displayed by victims of occupational workplace poisoning, were different from those experienced among the handful of children poisoned in the United States. In short, the reports were regarded as just too exotic to have any impact in a world not yet conceived of as a global economy.

During the 1930s, the lead paint industry took modest steps to try to prevent its products from being used on cribs and toys, particularly blocks,[93] but it did little to warn the public that such objects should not be repainted with lead paint. The industry also did not address researchers' suggestion that lead paint not be used on woodwork and railings that were accessible to small children. Increasingly, childhood lead poisoning was portrayed as a disease of poverty and ignorance. Prevention efforts during this decade, even those recommended by the most effective and enlightened public health experts of the time, focused on better parenting. For example, the Baltimore Health Department's early prevention efforts are exemplified by the title of its 1937 radio address "Children Who Eat Paint,"[94] not so subtly sending the message that children would only become poisoned by lead if their irresponsible parents allowed them to eat paint. The city continued to apply lead paint to the walls of hospitals constructed at the time. Meanwhile, the federal government required that the paints it purchased contain between 45 percent and 70 percent white lead, an unusually high lead content, even for that time.[95]

During the 1940s and 1950s, childhood lead poisoning would be redefined, though not for the last time. In 1943, a research paper published by Randolph Byers (posthumously) and Elizabeth Lord revealed that a group of children displayed learning disabilities, behavioral problems, and attention deficit problems long after their exposure to lead.[96] The impact of children's exposure to low levels of lead on their mental development later became the focus of public health concern about childhood lead poisoning in the waning decades of the twentieth century.

In the meantime, more systematic blood testing in Baltimore of children with symptoms suggesting possible lead poisoning revealed that such poisoning was far more prevalent than had been previously imagined.[97] Further, the cause of lead poisoning was found to include not only the chewing of toys and cribs but also the chewing of woodwork and the eating of fallen paint chips and flakes.[98] According to Baltimore city health commissioner Huntington Williams, a pioneer in the public health aspects of childhood lead poisoning, the problem was concentrated in areas of Baltimore with a high rate of deteriorated properties, that is, the slums. Baltimore public health officials who visited the homes of children poisoned in the late 1940s and early 1950s described lead paint peeling off the walls and ceilings in sheets, even hanging from the ceiling "like stalactites."[99]

Perhaps ironically, most residences of lead-poisoned children were apartments that had been created by subdividing the former homes of affluent owners. These homes had been painted with the "best" paint available at the time of their construction and had been repainted several times to create a build-up of thick layers of potentially dangerous leaded paint. Most often, the affected toddler did not chew on painted walls; rather, the toddler ingested deteriorated paint produced flakes and chips that were within easy reach. Lead paint was a naturally sweet substance and so toddlers understandably placed these flakes and chips in their mouths. More important, the natural chalking characteristics of uncovered lead paint, valued highly in earlier generations because it maintained a fresh appearance and facilitated repainting, now yielded a fine lead dust. A toddler crawling around on floors covered with such lead dust ended up with lead dust on his hands and then, as toddlers are apt to do, frequently put his fingers in his mouth. The result, unfortunately, often was a child with elevated levels of lead in his blood.

Asymptomatic childhood lead poisoning emerged as a "silent epidemic" during the 1950s. In 1956, the Baltimore City Health Department took blood samples from 148 asymptomatic babies and found that 40 percent of them had blood lead levels in excess of 50 micrograms of lead per deciliter of blood (μg/dL). In comparison, the Centers for Disease Control and Prevention today "recommends that public health actions be initiated" for children with blood lead levels in excess of 10 μg/dL.[100] A child with an elevated blood lead level of 50 μg/dL would be regarded as seriously poisoned and at extreme risk.

The role of the lead industry in investigating the dangers posed by its products, if not in all ways commendable by modern standards, probably compares favorably with the response of the tobacco industry in parallel circumstances.[101] From the 1920s to the mid-1990s, lead manufacturers and their trade association, the Lead Industries Association (LIA), funded research about the health consequences of lead exposure—at institutions such as Harvard University, Johns Hopkins University, and the Baltimore City Department of Health, among others.[102] Peter English concludes that although "the funding was motivated by self-interest," it "was noticeably independent of industry control." For example, all the research findings were reported in peer-reviewed medical journals, and English concludes that there was no evidence of censorship.[103]

By the 1950s, it was clear that childhood lead poisoning typically could not be attributed exclusively to children gnawing on toys or cribs. The presence of lead paint in residential dwellings, at least when the paint was allowed to deteriorate, posed a substantial risk to children. In 1951, Baltimore became the first major American city to outlaw the use of lead paint in residential interiors.[104] New York City followed Baltimore with a similar ordinance in 1960.[105] In 1955, the paint industry endorsed a standard reducing the amount of lead in paint to no more than 1 percent by weight.[106] In contrast, some paints in the 1920s and earlier advertised that they included 70 percent or more lead by weight. This development was not as self-sacrificing as it might seem. During the intervening decades, paint manufacturers had developed a variety of new pigments, notably titanium dioxide, to replace lead. Not until twenty years later, on September 1, 1977, did the federal Consumer Product Safety Commission ban the consumer use of lead paint.[107]

In recent years, public health officials have persistently lowered their estimates of the level of lead exposure understood to detrimentally affect children. The children first diagnosed with severe acute lead poisoning in the early decades of the twentieth century likely had elevated blood lead levels of at least 70 μg/dL, a level now known to cause severe lead toxicity.[108] In the 1960s, a child was diagnosed with an elevated blood lead level if the concentration of lead in his blood exceeded 60 μg/dL.[109] By the late 1970s, the research of Herbert Needleman and others demonstrated that lower levels of lead in a child's blood could reduce his intelligence and result in behavior problems. The U.S. Public Health Service reduced the blood lead level that it defined as potentially harmful to 40 μg/dL in1971, 30 μg/dL in 1978, and 25 μg/dL in 1985. In 1991, the Centers for Disease Control reduced the level to 10 μg/dL, a level where it remains today.[110] Research since 1991 strongly suggests that lead levels even lower than 5 μg/dL affect cognitive development and academic success.[111]

Largely as a result of the end of the leaded gas era, but also as a result of progress being made in reduced lead poisoning from lead-based paint hazards, childhood lead poisoning in America has declined precipitously during the past generation. In 1975–76, the median blood lead level in children under the age of six in the United States was 16.5 μg/dL, a level significantly in excess of the CDC's current standard of concern.[112] The percentage of children tested at a level of 10 μg/dL or above declined from 77.8 percent of all children in the mid-1970s to 4.4 percent of all children by the mid-1990s.[113] By 2001–2, the median blood lead level had declined to 1.7 μg/dL.[114] Nevertheless, it is estimated today that approximately 310,000 American children, most of them asymptomatic and many of them unknown, have blood lead levels at or above the CDC's level of concern. In 2002, approximately 38 million American homes still included old lead paint.[115]

CONCLUSION

The American prosperity and the technological advances of the late twentieth century would have awed even the promoters of the 1893 World's Columbian Exposition in Chicago and the 1939 World's Fair in New York. Despite experiencing significant social problems and the horrors of war, mid-twentieth-century Americans generally shared a fundamental, probably naive optimism, perhaps best captured by President Ronald Reagan's 1984 campaign slogan, "It's morning in America again." Yet the American consumer society had a darker side. For the millions of people suffering from diseases and lost mental acuity caused by product exposure, their morning in America was not the one described by President Reagan but the one prophesied by Rachel Carson in *Silent Spring*—a morning in which toxic exposures meant that birds no longer sang.[116]

CHAPTER 2

Product-Caused Diseases Confront the Law of the Iron Horse

The common law, or judge-made law, typically governs claims filed by victims against those they believe should be held legally responsible for causing their injuries. It relies on basic principles of law established by judges in previous cases, or precedents. The operating premise that lies behind this precedent-based system is that a certain degree of fairness is assured when the court, at least in the absence of articulated, compelling contrary justification, applies the same principles to the case it considers in the present to similar cases in the past. The judge is not able to pick and choose which principles of law apply to a particular case according to his biases for or against any of the parties appearing before the court. Philip Soper expresses this fundamental notion succinctly: "To judge in general means to decide by reference to pre-existing standards rather than personal whim."[1] Gregory C. Keating echoes this principle when he states, "As far as legal decision is concerned, the minimum implication of the ideal of the rule of law is that courts should decide cases in accordance with general, public pre-existing laws—and that they should do so impartially and fairly—'without zeal or bias.'"[2]

At the same time, the common law is not frozen in time. It changes, hopefully in a principled manner, to address new realities inherent in the cases before the courts. Product-caused diseases, a phenomenon new in the twentieth century, presented just such a challenge to the common law governing accidental harms. As this chapter illustrates, actions filed by victims of these diseases created tension between the expectation that judges would follow the basic principles of the preexisting legal regime, on the one hand, and the goal of achieving just results in new contexts, on the other.

When victims of diseases resulting from exposure to cigarettes and lead pigment first went to court seeking compensation, they encountered a legal regime that made victory on their part impossible. The law governing when someone injured by a business could recover damages had been developed by judges considering cases that arose in very different factual contexts in the late nineteenth century. As American legal historian Lawrence M. Friedman describes it, the "new machines" of that era "had a marvelous capacity for smashing the human body."[3]

The harms suffered by victims of lung cancer or childhood lead poisoning were fundamentally different. The claimant was not the victim of a traumatic "smashing" of the human body that occurred instantaneously upon impact with the "iron horse," the railroad locomotive. Instead, the victim's harm was an insidious disease that developed imperceptibly, sometimes over a period of half a century or more. Under the "law of the iron horse," as I refer to traditional American tort law that emerged during the heyday of railroads, the identity of the tortfeasor, the party whose allegedly wrongful conduct caused the harm, usually was not in doubt. The owner of the railroad, the owner of the machinery that injured the factory worker, or the driver at fault in an automobile accident most often could be readily identified. The victim of latent disease, however, probably came into conduct with fungible products manufactured by several different manufacturers. There was never any doubt that the broken bones of the victim of the earlier era resulted from the collision, but the cigarette manufacturer might claim that the cancer victim's disease resulted from industrial exposure to hazardous chemicals or from a genetic predisposition.

Classical American tort law also almost always required that the injured party, the plaintiff, prove both that his injury was caused by negligence or fault on the part of the defendant, the alleged injurer, and that the plaintiff's own lack of reasonable care did not in any way causally contribute to his harm. Oliver Wendell Holmes, perhaps the most important spokesperson justifying the negligence standard, wrote that as a general principle in our legal system, "loss from accident must lie where it falls," even if "a human being is the instrument of misfortune,"[4] unless the victim can show that the person who caused the injury acted with fault. Except when the tortfeasor acted with fault and the victim without fault, Holmes argued, there was no justification for setting the "cumbrous and expensive machinery" of the state's litigation process in motion.[5] In the late nineteenth century, imposing strict liability on businesses was seen as a surefire way to bankrupt emerging industries. Rejecting the application of a strict liability standard, the New York Court of Appeals stated, "We must have factories, machinery, dams, canals and railroads. They . . . lay at the basis of all our civilization."[6] By 1911, fault as a requirement of liability was so firmly embedded in American law that the same court held that an early workers' compensation statute enacted by the legislature and granting employees compensation for workplace injuries through insurance (in effect, creating liability even in the absence of negligence on the employer's part) constituted a violation of the due process clauses of the federal and state constitutions.[7]

The traditional requirement of proof of negligence on the part of the tortfeasor was particularly burdensome to those injured by products. It is one thing for a victim injured by a railroad locomotive or an automobile to prove that the defendant or its employees acted carelessly. It typically is far more difficult to prove such lack of due care when someone is injured by a product that was designed and manufactured by unknown persons in a factory far away. Further, if a product poses an unreasonable danger to a consumer or a bystander and in fact injures him, even the absence of negligence on the part of the manufacturer arguably should not preclude liability.[8]

In early litigation against product manufacturers, the victim faced an additional and intertwined roadblock to recovery. During the nineteenth century, the general rule was that someone injured by a product could not recover from the manufacturer unless he had purchased the product directly from it; that is, to use the legal term of art, he had to be in "privity" with the manufacturer.[9] To recover compensation, the victim was required to prove both privity with the manufacturer and fault on its part. In his famous 1916 opinion in *MacPherson v. Buick Motor Co.*,[10] Judge Benjamin Cardozo eliminated the privity requirement in those cases where the victim could prove that the manufacturer's conduct was negligent. For decades thereafter, however, courts continued to deny recovery to victims of product-caused harms who were neither in privity with the manufacturer nor able to prove negligence on the manufacturer's part. A victim who was in privity with the manufacturer might be able to recover under an implied warranty theory without proof of fault.

In this chapter, I next address how doctrines central to the law of the iron horse prevented recovery in early litigation against tobacco manufacturers. I then turn my attention to litigation against manufacturers of lead pigment, where entirely separate doctrines from the same legal regime precluded recovery. Finally, I consider litigation against asbestos manufacturers, as this litigation provided victims of product-caused diseases with their first opportunity to recover and, in the process, began to transform American tort law.

TOBACCO LITIGATION AS AN "INTERPERSONAL MORALITY PLAY"

The Setting: The Dawn of Strict Products Liability

The first significant wave of litigation seeking to hold tobacco manufacturers liable occurred during the early 1960s, at about the same time as the emergence of the modern law of strict products liability.[11] In 1965, the American Law Institute (ALI), a select group of distinguished lawyers, judges, and law professors with the self-delegated responsibility to "restate" various fields of American law and make recommendations to courts and legislatures, promulgated what has become its single most influential restatement provision, section 402A of the *Restatement (Second) of Torts*. Section 402A provides that a product manufacturer or other product seller should be held strictly liable, without a requirement that the plaintiff prove either negligence on the part of the manufacturer or privity between the manufacturer and the injured party, if the product was "in a defective condition unreasonably dangerous to the user" and caused injury to the "ultimate consumer or user."[12] American common-law courts had been moving toward eliminating the barriers of both fault and privity in products litigation for more than half a century. That trend culminated in a series of cases in the early 1960s in several different jurisdictions, holding manufacturers strictly liable in product cases. The adoption of strict products liability by the ALI and the courts has been called "the most radical and spectacular" development in American tort law during the twentieth century.[13]

Justice Roger Traynor of the California Supreme Court had explained the justifications for strict liability for defective products most eloquently in his concurring opinion nearly three decades earlier in *Escola v. Coca Cola Bottling Co.* His words failed to convince a majority of his colleagues on the California Supreme Court at that time, but his perspective would prevail by 1963.[14] Traynor had argued, "The manufacturer can anticipate some hazards and guard against the recurrence of others, as the public cannot." Additionally, strict products liability furthered the goal of loss distribution. Traynor explained, "The cost of an injury and the loss of time or health may be an overwhelming misfortune to the person injured, and a needless one, for the risk of injury can be insured by the manufacturer and distributed among the public as a cost of doing business."[15]

"Good tobacco is not unreasonably dangerous"

Surprisingly, litigation against tobacco manufacturers that occurred at the same time as the shift toward strict products liability was left largely unaffected by the transformation. The claims in the tobacco cases were not based primarily on strict product liability. Instead, they rested on the law of negligence, or on the law of implied warranty, which provided that a seller's product came with a guarantee implied by law that the product was reasonably fit for the purpose for which it was sold[16]—that is, in the case of cigarettes, that they were fit for human consumption. Robert L. Rabin has characterized litigation brought by victims of cancer and other tobacco-related diseases against cigarette companies over a period of decades as "a last vestige of a perhaps idealized vision of nineteenth-century tort law as an interpersonal morality play."[17]

As was to become customary in litigation against manufacturers of other products causing disease, tobacco companies argued for years that they should not be held liable because they could not have foreseen the harmful health consequences of smoking. Under a legal theory based on implied warranty, one would have expected courts to regard the manufacturer's foreseeability of harm as irrelevant.[18] Instead, the lack of foreseeability of harm or fault on the part of the tobacco companies often defeated victims' claims even under implied warranty theories that supposedly did not require fault on the part of the manufacturer.

Typical among the early cases was *Ross v. Philip Morris & Co.*[19] The plaintiff was a chain-smoker who had smoked Philip Morris cigarettes almost exclusively for eighteen years. He developed cancer of the throat and submitted to an operation that included a laryngectomy, a neck dissection, and a tracheotomy. At trial, his attorney requested a jury instruction that would have allowed the jury to find Philip Morris liable for violation of an implied warranty, but the trial court refused to instruct the jury on this basis, and the plaintiff appealed. On appeal, the federal Court of Appeals for the Eighth Circuit found an implied warranty cause of action to be applicable to claims against cigarette manufacturers, just as it was to claims against purveyors of food and beverages. Further and more important, the court held that the victim's lack of privity with the manufacturer did not preclude him from recovering on an implied warranty theory. The court then held, however, that cigarettes conforming to the standard of the industry did not violate the implied warranty. Instead, a violation of such a warranty extended only to "flagrant deviations from what a consumer expects when he buys a product and what the manufacturer intends his product to contain."[20] The court noted that implied warranty precedents usually involved things like "a mouse in a bottled soft drink"[21] or "a fly in a can of salmon."[22]

The *Ross* court also stressed the importance of foreseeability in rejecting the notion that the manufacturer should be held liable "even though no developed human skill or foresight could afford knowledge of the cancer-smoking relationship."[23] Similarly, in rejecting both negligence and implied warranty claims against another tobacco manufacturer in *Lartigue v. R. J. Reynolds Tobacco Co.*,[24] the federal Court of Appeals for the Fifth Circuit held that under Louisiana law, "there must be foreseeability of harm."[25] The courts viewed this issue from the perspective of the time when the cigarettes had been sold. They accepted the manufacturers' argument that the risks of cigarette smoking were not known to them until the 1950s. Notably, in *Lartigue*, the court unconvincingly attempted to distinguish the earlier opinion of the Florida Supreme Court in *Green v. American Tobacco Co.*, in which the court had held that "implied warranty liability is not limited by the foreseeability doctrine."[26] The opinions in *Ross* and *Lartigue*, critical victories for tobacco manufacturers in that era, appear to be in conflict not only with the decision in *Green* but also with the general rule under the law of implied warranties that the manufacturer's foreseeability of harm was irrelevant.

Viewed from today's perspective, what is most striking about these cases against tobacco manufacturers and others of the same vintage is the extent to which, at least implicitly, the courts placed the obligation to assure that the use of a product would be safe as much on the consumer as it did on the manufacturer. Perhaps reverting to an era when smoking was regarded as an immoral act, the courts compared the smoker who develops cancer to "a man [who] buys whiskey and drinks too much of it and gets some liver trouble."[27] Justice Traynor's prescient observation in 1944 that "the manufacturer can anticipate some hazards and guard against the recurrence of others, as the public cannot,"[28] had not yet percolated into claims against cigarette manufacturers. In these cases, the courts declined to impose any special obligation on the manufacturer to test for product risks or to make itself aware of risks known to independent researchers. This omission is all the more surprising because many courts, in cases involving products other than cigarettes, already had held manufacturers to the level of an "expert" regarding the risks posed by their products.[29] This anomaly probably can best be explained by the notion that smokers often appreciated that smoking was not good for them and that it caused a shortness of breath. Courts failed to differentiate this widespread but vague public knowledge that "smoking is bad for you" from the growing body of scientific evidence establishing the link between smoking and cancer, of which manufacturers either were or should have been aware.

Further, for all intents and purposes, the first wave of products liability cases in the early 1960s largely focused on liability when the product that injured the consumer deviated from the manufacturer's intended design. In other words, in modern parlance, the law of the early 1960s covered what is now known as "manufacturing defects."[30] For the most part, courts had not yet held manufacturers liable for products that were inherently designed in an unsafe or unhealthy manner or that failed to carry adequate warnings. The concept that manufacturers should not be liable for manufacturing products that were inherently dangerous and generally known to be such was soon consolidated in a comment to section 402A of the *Restatement (Second) of Torts*.

> The article sold must be dangerous to an extent beyond that which would be contemplated by the ordinary consumer who purchases it, with the ordinary knowledge common to the community as to its characteristics. . . . Good tobacco is not unreasonably dangerous merely because the effects of smoking may be harmful; but tobacco containing something like marijuana may be unreasonably dangerous.[31]

This principle was one of several factors that stalled meaningful attempts to hold tobacco companies liable for the next several decades. Another was federal legislation enacted after the surgeon general's 1964 report that had publicized the causal link between smoking and lung cancer. The Federal Cigarette Labeling and Advertising Act of 1965 and its successor, the Public Health Cigarette Smoking Act of 1969,[32] required specific health warning labels on cigarette packages. In *Cipollone v. Liggett Group*,[33] the U.S. Supreme Court held that the 1969 act provided that these federal labeling requirements preempted any obligations to warn required by state law, including those resulting from state common-law damage claims holding manufacturers liable for failure to provide adequate warnings of smoking consequences or for negligence in the manner in which cigarettes were tested, researched, sold, promoted, or advertised.[34] The final factor that doomed individual litigants who sought to recover for their tobacco-related diseases during the 1960s was that tobacco companies adopted a "scorched-earth" litigation strategy, contesting all issues as vigorously as possible and burying often underresourced plaintiffs' attorneys with an exhaustive stream of motions and discovery requests.[35]

Personal Choice or Blaming the Victim?

Ironically, the developments that defeated the first wave of tobacco litigation—principally the enactments of the federal cigarette warning legislation and the American Law Institute's *Restatement (Second) of Torts*—occurred simultaneously with the growing public awareness of the harmful health consequences of smoking. In a sense, therefore, the surgeon general's 1964 report and the federally mandated warning labels set the stage for the second act of the morality play. Another pillar of the traditional American fault-based system—that a plaintiff whose injury was caused in part by his own fault could not recover from the defendant—defeated victims' next wave of claims during the 1980s. Looking back to the period when the fundamental principles of American tort law first developed, legal historian John Fabian Witt has found that courts considered issues of contributory negligence—one of the traditional doctrines denying plaintiffs recovery in these circumstances—in more than two-thirds of all cases during the early decades of the negligence regime between1860 and 1880.[36]

A close sibling of contributory negligence at common law—indeed, one that frequently overlapped with contributory negligence—was assumption of risk, a doctrine that was often said to provide that a plaintiff could not recover if he had knowingly exposed himself to the risk. Today, the nineteenth-century origins of the assumption of risk defense are viewed with particular disdain.[37] During that era, even if an injured factory worker could prove that his employer failed to provide a reasonably safe workplace, the employer might still escape liability by arguing that the worker assumed the inherent risks of continuing to work.

At common law, the plaintiff's contributory negligence or assumption of risk totally barred the plaintiff from recovery, even if the defendant's conduct was far more egregious or substantial in causing the plaintiff's injury. However, by the 1980s, many states had adopted a modification of contributory negligence known as *comparative fault*, which allowed the plaintiff to recover but reduced his recovery in proportion to the ratio that his fault bore to that of the defendant.

Representative of the second wave of tobacco litigation brought by individual victims of tobacco-related diseases and their survivors, in which tobacco manufacturers prevailed by pinning the blame for continuing to smoke on the cancer victim, is *Horton v. American Tobacco Co.*[38] In *Horton*, the plaintiffs' decedent, who died of cancer prior to trial, had begun to smoke cigarettes manufactured by the defendant when he was eighteen. During his deposition, he admitted that "he had been told that smoking was bad for him" and that he "himself referred to cigarettes as 'coffin sticks' and 'cancer sticks' through the years."[39] He also admitted that he had read the federally required package warnings but continued to smoke. The trial court judge submitted the case to the jury on a strict liability basis but also gave the jury the opportunity to decide for the defendant on the basis of assumption of risk or to reduce the plaintiff's recovery because of comparative negligence. The plaintiffs' medical expert had testified that between 95 and 97 percent of all lung cancers occurred in smokers and that cigarette smoking was highly addictive. The former president and CEO of the American Tobacco Company, the defendant-manufacturer, had testified through video deposition that his company was " well aware of scientific reports that linked smoking to lung disease" and that his company "wanted the public to believe that cigarette smoking was not dangerous to their health."[40] In the face of this testimony, the jury held the American Tobacco Company liable on a negligence or strict liability basis, yet it awarded zero dollars in damages, finding that Horton's own conduct, whether termed contributory negligence or assumption of risk, was 100 percent responsible for his harm.

In other instances, tobacco companies defeated claims by smokers or their survivors by arguing that because the dangers of smoking were "common knowledge," cigarettes were not defective in the first place.[41] The genesis of this argument lies in comment g to section 402A of the *Restatement (Second) of Torts*, which provides that a product is not to be regarded as "unreasonably dangerous" and therefore is not "defective" unless it is "dangerous to an extent beyond that which would be contemplated by the ordinary consumer who purchases it, with the ordinary knowledge common to the community as to its characteristics."[42] A number of state legislatures, at the behest of tobacco lobbyists, later passed statutes, grounded in the *Restatement* comment, explicitly providing that where common knowledge of a product's harmful consequences existed, the product manufacturer could not be held liable.[43]

By the mid-1990s, the tobacco industry was undefeated in actions brought by victims of lung cancer and other related illnesses and their survivors. Despite the revolution brought about by strict products liability in the 1960s, the industry continued to prevail in litigation, because of the assessments of both judges and juries that injury was the fault of smokers themselves and not that of the manufacturers of products that people bought and consumed. The extent to which tobacco companies had concealed and misrepresented the harmful health consequences and addictive qualities of cigarettes had not yet been exposed, and the tobacco companies' public relations campaign promoting the idea that smoking was a matter of personal choice was succeeding in the jury room, just as it was in legislative chambers and in the marketplace.

In short, during the early years of the tobacco litigation, juries found that tobacco companies had not been "at fault": they were not negligent, nor had they violated implied warranties, because the risks posed by their products were supposedly unforeseeable ones. Later, courts concluded not only that smokers had either assumed the risk or been contributorily negligent but also that this conduct on the victims' part trumped any tortious conduct of the manufacturers, even in states that had adopted comparative fault as a replacement for the total bar of contributory negligence. In tobacco litigation, the classical fault-based tort law developed during the railroad litigation of the late nineteenth century remained intact.

CHILDHOOD LEAD POISONING: IDENTIFYING
THE INJURER A CENTURY LATER

Causation

With or without a requirement that, in order to recover, a plaintiff prove that the injurer acted with fault, tort law traditionally accepted the idea that a particular plaintiff must prove that a particular defendant's acts caused the plaintiff's injuries.[44] William Prosser described this as "the simplest and most obvious" aspect of determining tort liability.[45] In 1938, the New York Court of Appeals held, "Where the facts proven show that there are several possible causes of an injury, for one or more of which the defendant was not responsible, . . . plaintiff cannot have a recovery."[46] Nearly a half century later, in *Payton v. Abbott Labs*,[47] the Massachusetts Supreme Judicial Court noted that "identification of the party responsible for causing injury to another is a long-standing prerequisite" for liability. The court reasoned that the requirement "separates wrongdoers from innocent actors, and also ensures that wrongdoers are held liable only for the harm that they have caused."[48] David Rosenberg aptly characterized the traditional common law as "starkly individualistic" in his classic 1984 article "The Causal Connection in Mass Exposure Cases: A 'Public Law' Vision of the Tort System."[49]

Frequently, the victim of a latent disease caused by exposure to products that are fungible or nearly fungible is not able to identify the particular tortfeasor that manufactured the product that caused his harm. This is particularly true when—as is often the case—a substantial period of time, often several decades, has passed between the time that the product was manufactured and the time of diagnosis of the plaintiff's illness. Rosenberg concludes, "The difficulties presented by mass exposure tort litigation typify the recumbent problem of legal rules that perform fairly well in common-place settings but, like Newton's laws of physics, lose their ordering power under extraordinary conditions." He characterizes the "demand for 'particularistic' evidence in mass exposure cases" as "quixotic" and capable of resulting in "calamity."[50]

Some victims of tobacco-related diseases are "lucky" enough, for litigation purposes, to have smoked only one brand of cigarette, but others are not. For all intents and purposes, the victim of childhood lead poisoning was never able to identify the specific manufacturer that produced the harm-causing product.[51] The reasons why are illustrated by the facts in *Skipworth v. Lead Industries Ass'n*,[52] in which the guardian of a child suffering from lead poisoning sued substantially all the manufacturers of lead pigment, the toxic ingredient in the paint applied to the house in which the child lived. The court found that the house had been painted many times between its initial construction in 1870 and 1977, when, according to the court, the manufacture of paint containing lead pigment for use in residential interiors ceased. No records were available to determine when the house had been painted, which paint manufacturers' products were used, or which pigment manufacturers' products were contained in any given paint. There was also no chemical signature that could identify any particular manufacturer's paint or pigment.

In the decades preceding the 1980s, the inability of the victim of childhood lead poisoning to prove the identity of the manufacturer of the products causing his harm probably was the most important, though not the only, factor preventing his recovery. Cases brought by victims of tobacco-related illness seldom, if ever, reached this question, both because other issues prevented plaintiffs from recovering and because those victims who did choose to pursue lawsuits often were able to testify to their exclusive use of a single product brand (e.g., "I never smoked anything but 'Luckies'" (Lucky Strike cigarettes)).

Conduct of Third Parties

The product manufacturer, facing liability for the costs incurred as a result of diseases caused by product exposure, sometimes asserts that someone else should bear that expense. Blaming the victim of the disease, as in tobacco-related illness, obviously is one of many ways to shift responsibility. In other instances, notably childhood lead poisoning, manufacturers have asserted that third parties are responsible.

Several actors in addition to the manufacturers of lead paint or lead pigment may be considered factual causes in a child's lead poisoning. Because most cases of such poisoning occur in poorly maintained, older housing units where lead-based paint has been allowed to deteriorate, the landlord or other property owner in such cases causally contributes to the child's lead poisoning.[53] Additionally, in the 1950s and 1960s, it was popular to blame parents, particularly low-income parents, for allowing their children to eat paint chips, neglecting to maintain a clean environment, and failing to provide proper nutrition for their children (which increases a child's vulnerability to the harms caused by lead poisoning).[54]

At common law, the mere fact that other parties contributed causally to a harm did not prevent a defendant whose tortious conduct also was a cause-in-fact of the plaintiff's harm from being held liable for the entire amount of the damages. However, the conduct of a third party that causally contributed to the harm and that occurred after the original defendant's conduct is sometimes considered to "break the chain" of proximate causation, particularly if the third party's conduct was intentional or unforeseeable. In these cases, the third party's conduct then represents a "superseding cause" and relieves the original defendant of liability for the harm. During recent decades, even if the third party's conduct does not break the chain of causation, the defendant who has paid more than its fair share of the plaintiff's damages has usually been able to then sue the other tortfeasors for contribution, assuming that they are solvent, within the jurisdiction of the court, and not legally immune from liability.

In *City of Chicago v. American Cyanamid Co.*,[55] the Illinois Appellate Court held that defendant-manufacturers of lead-based paint and lead pigment had not proximately caused damages resulting from childhood lead poisoning. The court concluded that "the conduct of defendants in promoting and lawfully selling lead-containing pigments decades ago, which was subsequently lawfully used by others, cannot be a legal cause of plaintiff's complained-of injury, where the hazard only exists because Chicago landowners continue to violate laws that require them to remove *deteriorated* paint."[56] In other words, Chicago landowners' violations of law were a superseding cause. Though not explicitly based on proximate causation, an influential 2007 opinion of the New Jersey Supreme Court justified its dismissal of claims against the manufacturers of lead paint and lead pigment on the grounds that legislation passed by the state legislature had allocated to landlords the responsibility for remedying the conditions resulting in childhood lead poisoning.[57] In the political arena, as well as in courtrooms, advocates acting on behalf of lead-poisoned children and lobbyists representing paint companies sometimes joined forces to enact legislation requiring landlords to take steps to significantly reduce the incidence of childhood lead poisoning—in effect placing the responsibility for preventing childhood lead poisoning on the landlords instead of on the paint manufacturers.[58]

In rare instances, defendants held liable to lead-poisoned children have been able to seek contribution from another set of parties—the parents of the children. In *Ankiewicz v. Kinder*,[59] the Massachusetts Supreme Judicial Court allowed landlords to implead and seek contribution from parents whose allegedly negligent acts arguably contributed to the lead poisoning of their children.

The Timeliness of Legal Actions

Another basic principle of the traditional tort system required an action against an alleged tortfeasor to be brought within a limited period of time, specified by the legislature, following the time that the plaintiff's cause of action had "accrued." Several reasons support the adoption of these statutes of limitations. First, defendants should be assured that after a specified period of time they are not subject to liability for their actions, so that they can plan and proceed with their businesses and lives. Second, after an extended period of time, it often may be difficult to locate witnesses, their memories may fade, and documents or other evidence may be lost or destroyed.

Tobacco-related diseases and childhood lead poisoning provide unique challenges for statutes of limitations. The manufacture and sale of the product allegedly causing the disease usually occurred twenty-five, fifty, or even one hundred years before the victim became aware of his injury or, in some cases, even was injured. In an early case concerning occupational disease, workers had filed suits alleging that negligence and statutory violations committed by their employers had resulted in their contracting respiratory illnesses as a result of exposure to dust. In its 1937 opinion, the New York Court of Appeals held that the cause of action "accrues at the time when through lack of care by an employer, deleterious substances enter the lungs of an employee though the development of consequential damages may be long delayed."[60] In other words, the statute of limitations may run—the period of time during which a claim must be filed may expire—before the employee has been harmed or realizes that he has been harmed. The court recognized that when applied in this manner, the statute of limitations may "bar the assertion of a just claim" and "cause hardship"; however, the court interpreted the enactment of the statute as expressing the legislature's intent "that such occasional hardship is outweighed by the advantage of outlawing stale claims."[61]

THE ASBESTOS BREAKTHROUGH: *BOREL V. FIBREBOARD*
PAPER PRODUCTS CORP.

For victims of tobacco-related diseases or childhood lead poisoning, any attempt to sue the manufacturer whose product resulted in their harm was utterly hopeless prior to the mid-1990s. Legal doctrines and jury reactions to a smoker's own conduct defeated his attempt to be compensated for his tobacco-related illness. Victims of childhood lead poisoning could not even prove which manufacturer's product caused their harm. In addition, both courts and legislatures often found other parties, notably landlords, to be more responsible for causing lead poisoning.

Litigation against manufacturers of asbestos products was among the first and probably the most significant breakthrough in establishing the liability of manufacturers for product-caused public health crises in actions brought by individual victims against product manufacturers. No other single case has been as important to the law governing actions by victims of latent diseases resulting from product exposure as *Borel v. Fibreboard Paper Products Corp.*[62] In two opinions in that case, Judge John Minor Wisdom of the federal Court of Appeals for the Fifth Circuit became one of the first appellate judges to apply the strict products liability doctrines of section 402A of the *Restatement (Second) of Torts* to actions alleging product-caused diseases. At the same time, he struck down many of the traditional doctrines that arose from the "law of the iron horse" and that had precluded recovery by victims of tobacco-related diseases for more than a decade.

Clarence Borel developed asbestosis and later mesothelioma (from which he died prior to trial) as a result of exposure to asbestos during the thirty-three years he worked with asbestos insulation, including stints installing asbestos insulation in destroyers at a shipyard and at oil refineries.[63] In his pretrial deposition, Borel testified "that at the end of a day working with insulation material containing asbestos his clothes were usually so dusty he could 'just barely pick them up without shaking them.'" He continued, "You just move them just a little and there is going to be dust, and I blowed [sic] this dust out of my nostrils by handfuls at the end of the day."[64] Borel's attorney, Ward Stephenson, filed a complaint against eleven manufacturers of asbestos insulation products, alleging causes of action in both negligence and in strict liability or warranty. The complaint alleged that the defendants had been negligent in their failure to use reasonable care to

1. warn Borel of the risks of exposure to its products,
2. provide Borel with instructions regarding the wearing of masks

and other protective gear and otherwise how to handle the product safely,

3. adequately test the product to ascertain product risks, and
4. withdraw the product from the market once it was aware of the health risks caused by asbestos exposure.

The litigation brought by Borel and others in subsequent years in many ways paralleled the morality play that took place in the early tobacco litigation. At issue was whether asbestos product manufacturers or the victims themselves were responsible for their illnesses. This time, the outcome would be different.

Special Obligations of the Manufacturer

In *Borel*, Judge Wisdom became the first judge to recognize the full impact of the principles outlined in section 402A of the *Restatement (Second) of Torts* on litigation claiming damages for product-caused diseases. The complaint in *Borel* alleged that the asbestos insulation materials manufactured by the defendants were unreasonably dangerous and therefore defective because the defendants failed to adequately warn users of "the known or knowable dangers involved."[65] On appeal, Judge Wisdom, writing for a three-judge panel, explained that asbestos insulation materials, like new drugs used to treat diseases, were "unavoidably unsafe products" as described in comment k to section 402A. Such products are those that are incapable of being made safe given "the present state of human knowledge," but for which "strict liability may not always be appropriate . . . because of the important benefits derived from the use of the product."[66] Nevertheless, as Judge Wisdom explained, "the seller still has a responsibility to inform the user or consumer of the risk of harm."[67] When the consumer or user is adequately warned of the risks of using the product, he can make an informed decision about whether to expose himself to the risk. The failure to warn of product risks, however, makes the product unreasonably dangerous and imposes liability on the manufacturer. Applying these principles to the *Borel* case, Judge Wisdom reasoned,

> The utility of an insulation product containing asbestos may outweigh the known or foreseeable risk to the insulation workers and thus justify its marketing. The product could still be unreasonably dangerous, however, if unaccompanied by adequate warnings. An insulation worker, no less than any other product user, has a right to decide whether to expose himself to the risk.[68]

Under the new paradigm for failure to warn outlined in section 402A, the question then became, of which risks must the manufacturer warn? In answering this question, Wisdom held that the manufacturer was held to the level of knowledge and skill of an expert, meaning that the manufacturer "must keep abreast of scientific knowledge, discoveries, and advances" and "even more importantly . . . has a duty to test and inspect his product."[69] At trial, the jury had heard testimony that the health risks caused by exposure to asbestos, specifically asbestosis, had "been recognized as a disease for well over fifty years."[70] Wisdom therefore had no difficulty in upholding the jury's verdict that the defendants were liable on the strict liability count.

Borel's Own Conduct in Working with Asbestos Insulation

As in the tobacco litigation, the manufacturer's knowledge and conduct was only half the story in Borel's lawsuit against the asbestos industry. Intertwined with the court's finding that the manufacturers had behaved tortiously and ought to be held liable was its evaluation of Borel's own knowledge and conduct. During his deposition, Borel testified "that he had known for years that inhaling asbestos dust 'was bad for me,'" but "that he and his fellow insulation workers thought that the dust 'dissolves as it hits your lungs.'" Further, he testified that in his later years of working, respirators were made available for those on the job, but he had declined to use them because "you can't breathe," and they were uncomfortable and could not be worn in hot weather.[71] As the defendants pointed out, between 1964 and 1966, several years before Borel stopped working, their insulation products contained warning labels advising that "inhalation of asbestos in excessive quantities over longer periods of time may be harmful" and that users should "avoid breathing the dust" and wear respirators.[72] Finally, in 1964, Borel's physician had told him that "x-rays of his lungs were cloudy" and "advised him to avoid asbestos dust as much as he possibly could."[73]

Despite the emergence of strict products liability, Borel became a morality contest that rivaled the tobacco litigation in intensity. Should the manufacturer or the worker, who admittedly had some knowledge that working with asbestos products could be dangerous, bear the financial consequences of asbestos-related diseases? That basic question played itself out in the Borel decision in several different doctrinal pigeonholes, some addressing the manufacturers' prima facie liability and others focused on affirmative defenses arising from the plaintiff's own conduct.

On the prima facie liability side of the ledger, manufacturers, simultaneously with arguing that the risks of exposure to asbestos products were not foreseeable, argued that the public health hazards of their products were "obvious," thus precluding any duty to warn on the manufacturers' part. Judge Wisdom responded, in essence, that although Borel may have understood that asbestos insulation posed some health risk, he had no idea that it could cause serious illness. Therefore, it was the manufacturers' responsibility to warn him. In a second, separate opinion issued after a second round of arguments in the case, Wisdom concluded that the warnings that the manufacturers first placed on their insulation products in the mid-1960s did not adequately communicate the extent of the dangers posed.[74]

On the side of the ledger analyzing affirmative defenses arising from the plaintiff's own conduct, Judge Wisdom essentially adopted the position of the *Restatement (Second) of Torts* on the effect of the plaintiff's conduct in a strict products liability action,[75] an interpretation common among American courts during the twenty-year period following the decision in *Borel*. He held that the plaintiff's conduct constituted contributory negligence and barred the liability action only when the plaintiff unreasonably, voluntarily, and knowingly encountered a product risk. Again, Wisdom concluded that the evidence showed that Borel "never actually knew or appreciated the extent of the danger involved."[76]

Finally, the asbestos manufacturers argued a separate and distinct defense based on Borel's own conduct: namely, that he had "misused" the asbestos insulation by applying it without wearing a respirator. Product misuse was to become one of the most misunderstood defenses to product liability actions.[77] At this early point in the development of the law of strict products liability, Judge Wisdom simply understood misuse as a form of contributory negligence consisting of the plaintiff's failure to comply with the manufacturer's product instructions and warnings.[78] He noted that there really were not any instructions requiring the use of respirators and, as described previously, that Borel had not been contributorily negligent.

Thus, the *Borel* decision clearly places the principal responsibility for avoiding latent diseases arising from product exposure on the manufacturer and not on the ultimate consumer or user. This result contrasts with the conclusions that the courts were reaching less than a decade earlier in the tobacco litigation. State courts' widespread adoption of section 402A of the *Restatement (Second) of Torts* during the intervening years helps account for the differences in the judicial handling of the two litigation cycles. However, the shift toward strict products liability is only part of the explanation for why victims of tobacco-related diseases lost while victims of asbestos-related diseases prevailed. There were three other key factors that contributed to the success of the plaintiffs in asbestos litigation. First, even by the early 1970s, it was becoming clear that asbestos manufacturers had actively concealed the health risks of their products. Similar perfidy among cigarette manufacturers would be not revealed until the 1990s. Second and more important, asbestos products were somehow more exotic, their risks more alien to the consumer or user, than were the harmful health consequences of smoking, about which religious and moral crusaders had warned the American middle class for generations. Third, the victims of asbestos-related diseases were laboring men and women, and the assumption was that they should be compensated for harms they sustained as a result of the jobs they worked in order to support their families. Cigarette consumers, in contrast, had made choices to purchase cigarettes for their own pleasure. Notwithstanding the adoption of strict liability based on instrumental policies of loss minimization and loss distribution, the early decisions regarding the liabilities of manufacturers whose products caused diseases were all about morality.

Causation

Neither Clarence Borel nor other asbestos plaintiffs could determine which manufacturer had produced the products that caused their asbestosis and mesothelioma. By interviewing his client at great length and talking with coworkers and union officials, Borel's attorney compiled a list of manufacturers who had produced the asbestos products to which Borel had been exposed at his various work sites. The common law traditionally held that where the tortious acts of two or more defendants are each a proximate cause of an indivisible injury to the plaintiff, the defendants are jointly and severally liable.[79] In *Borel,* Judge Wisdom realized that it was "impossible, as a practical matter, to determine with absolute certainty which particular exposure to asbestos dust resulted in injury to Borel." He found, however, that Borel had been "exposed to the asbestos products of all the manufacturers" and "that the effect of exposure to asbestos dust is cumulative, that is, each exposure may result in an additional and separate injury."[80] On the basis of this circumstantial evidence, Judge Wisdom held that the jury could find that each manufacturer was a cause-in-fact of Borel's asbestos-related diseases. He then stated that under the common law, unless a particular manufacturer could show that it was not responsible for any portion of the harm, all the defendant-manufacturers should be held jointly and severally liable.

Third-Party Misconduct

In the *Borel* litigation, the defendant-manufacturers also asserted that "the insulation contractors, not the manufacturers, had a duty to warn insulation workers of the risk of harm." In his opinion for the court, Judge Wisdom rejected this argument. He noted that the manufacturer's warning "must be reasonably calculated to reach" the ultimate consumer or users and that "the presence of an intermediate party will not by itself relieve the seller of this duty."[81]

The *Borel* court's specific conclusion that a manufacturer could be held liable to the ultimate user of the product on the theory of failure to warn does not necessarily answer the harder question of whether the manufacturer can be held liable to the ultimate victim when the victim is neither a consumer nor a user. The spouse or other family member who was exposed to asbestos dust from the clothes that the insulation worker brought home cannot be characterized as the ultimate user or consumer,[82] in the same way that the victim of secondhand smoke or the child victimized by lead poisoning a century after the application of lead paint is not a user or consumer. If these victims of product-caused diseases, however, were able to prove the other prerequisites of liability under a negligence theory of recovery, no court today would rule that such a foreseeable victim could not recover under a negligence theory of recovery,[83] and most would reach the same result under a cause of action based on strict products liability.[84]

The more difficult remaining question is under what circumstances the original product manufacturer would still be held liable when the wrongdoing of another party, such as the insulation contractor's failure to warn Borel or maintain safe working conditions, has intervened since the time of the manufacture and distribution of the product and is a necessary contributing cause to the plaintiff's harm. The property owner who allowed lead-based paint hazards to exist in a rented dwelling, as described previously, is an obvious example. The issue of whether such neglect should affect the pigment manufacturer's liability will be considered further in chapter 7.

Judge Wisdom narrowly defined the issues associated with the insulation contractor's failure to both warn Borel and his coworkers of the dangers of working with asbestos insulation and provide safe working conditions. By focusing on the manufacturer's obligation to warn the ultimate consumer, he avoided addressing what were perhaps more difficult questions: how culpable had Borel's employers been in failing to both warn Borel and protect him from working with asbestos materials in a hazardous manner, and how had their actions contributed to his death? On one hand, the manufacturers of asbestos products either knew or should have known more about the risks of asbestos than either Borel himself or the contractors who employed him.[85] On the other hand, Borel's employers also had an obligation to provide a safe workplace, and they probably were more aware of the specific conditions—such as the lack of ventilation or Borel's own persistent refusal to wear respirators—that had dramatically increased the risks to Borel's health.

Though Judge Wisdom's opinion did not explicitly address these issues, it nevertheless sent a powerful message at a very early moment in the history of mass products litigation. Manufacturers, not other parties involved in the handling of the products, would often be held liable because they generally were in the best position to minimize product-caused diseases, at least when considered from a societal perspective. Further, product manufacturers were in the best position to distribute losses across society by purchasing insurance to cover liability resulting from product-caused diseases and by increasing the price of their products to account for losses that result from such diseases.

Extending the Time in Which to File

Even prior to *Borel*, courts had recognized the unfairness of rules that barred an action claiming damages for a latent disease before the victim realized that he suffered from such an illness. The most important decision in this respect was the U.S. Supreme Court's 1949 opinion in *Urie v. Thompson*,[86] in which an employee working on steam locomotives and suffering from silicosis, another pneumonia-like, occupational respiratory disease, had sued his employer, a railroad, under the Federal Employers' Liability Act. The Supreme Court rejected the defendant's argument that the plaintiff must have been harmed by each exposure to the harmful substance over the course of his decades-long career, because such a conclusion "would mean that at some past moment in time, unknown and inherently unknowable even in retrospect, Urie was charged with knowledge of the slow and tragic disintegration of his lungs" and that his "failure to diagnose within the applicable statute of limitations a disease whose symptoms had not yet obtruded on his consciousness would constitute waiver of his right to compensation at the ultimate day of discovery and disability."[87] Congress, said the Supreme Court, would not have "intended such consequences to attach to blameless ignorance."[88]

Outside the latent disease context, by the 1960s and 1970s, courts deciding common-law cases of medical malpractice often had adopted the *discovery rule*, providing that the statute of limitations does not begin to run until the plaintiff either knew or, in the exercise of reasonable care, should have known that he had been injured.[89] In a typical case, the surgeon or staff in an operating room left a clamp or sponge in a patient's body during surgery, but that foreign object had not caused discomfort or infection until years later. In *Borel*, relying on both *Urie* and the medical malpractice cases decided under the discovery rule, the court applied the discovery rule and held that Borel's cause of action did not arise until he was aware that he had asbestosis.[90]

Following the adoption of the discovery rule version of the statute of limitations, manufacturers argued that the discovery rule defeated the basic purposes underlying a statute of limitations. At their behest, a number of state legislatures, as a part of tort reform, adopted *statutes of repose* that prohibited a victim from bringing an action more than a specific period of time (usually ten to twenty years) after the date of the manufacture and sale of the product.[91] Obviously, this "reform" posed a huge problem for victims of childhood lead poisoning, because lead-based paint was last manufactured and sold in 1978. Victims of asbestos-related diseases were similarly barred from the courts because the manufacture and distribution of asbestos products also ended, for all intents and purposes, in the 1970s. This blatant unfairness led a number of jurisdictions to explicitly exempt asbestos claims from the operation of the statute of repose, either legislatively[92] or judicially.[93]

CONCLUSION

Borel v. Fibreboard Paper Products Corp., and similar legal opinions that followed on *Borel's* heels, constituted a striking legal development, opening the doors of courthouses for the first time to victims of product-caused diseases.[94] In *Borel*, Judge Wisdom applied Justice Traynor's conception of strict products liability to diseases caused by products decades after the products' manufacture and distribution. This application was justified by the understanding that manufacturers were in a better position to minimize losses and distribute losses than were consumers. Product-caused diseases posed issues far more challenging than those brought up by the typical cases that had once formed the core of American tort law: railroad accidents, workplace injuries, and automobile collisions. The issues posed by product-caused diseases were even more challenging than those that arose in the context of the more routine products liability cases that the restatement drafters originally had envisioned when they framed section 402A of the *Restatement (Second) of Torts*—for example, cases involving malfunctioning automobile steering mechanisms,[95] power saws without adequate guards,[96] or pebbles contained in a can of beans.[97]

Judge Wisdom had fulfilled Justice Traynor's visionary reasoning by establishing the fundamental principle that at least when a manufacturer of products that caused disease failed to warn the user or consumer of the health consequences resulting from product exposure, it was the manufacturer who should bear the financial consequences—not the consumer or an intermediary whose actions arguably also contributed to the harm. Doctrines that previously had precluded liability in latent disease cases (such as those concerning queasiness about whether the plaintiff could really prove which manufacturer's products had caused him harm or the plaintiff's inability to file the legal action in what traditionally had been regarded as a timely manner) were to be regarded as legal niceties, not sacrosanct fundamental principles. At the moment that the *Borel* case was decided and even more so today, when we look backward and compare the law it overturned and what was to come in subsequent decades, Judge Wisdom's opinion can safely be characterized as revolutionary. Within a few years, however, it became clear that *Borel*, at least in the world of litigation arising from product-caused diseases, was a comparatively simple case.

CHAPTER 3

The First Wave of Challenges to the Individual Causation Requirement

By the 1970s, the public understood the strong causal link between lung cancer and smoking, the effects of low doses of lead on children in terms of loss of intelligence, and other links between product exposure and detrimental health consequences. At the same time, changes in the law—particularly the adoption of strict products liability—had overcome many of the legal barriers faced by victims and their lawyers seeking compensation from manufacturers of harmful products. Many asbestos victims, like Borel, were able to recover if they identified specific product brands present at their work sites and brought suit in a jurisdiction applying the principle of "concurrent causation resulting in an indivisible harm" in these circumstances.[1] For many other victims of product-caused diseases, however, their inability to identify the manufacturer of the specific products that caused their harms continued to pose an insurmountable barrier to recovery. For example, the traditional requirement of proof of an individualized causal connection between the injurer and the victim remained a core obstacle to recovery for lead-poisoned children in suits against paint or pigment manufacturers and for many victims of tobacco-related diseases.

During the 1980s, victims' attorneys and the courts invented new legal doctrines and reinterpreted old ones to enable victims, usually those in asbestos litigation, to recover from multiple and indeterminate manufacturers. These efforts were the first attempt to move toward what David Rosenberg has called a "public law model" that recognizes "the intrinsically collective character of the interests of potential victims of the defendants' misconduct."[2] This chapter analyzes these procedural mechanisms and substantive theories of causation that have enabled victims to recover even when they are unable to prove a causal connection between the harms they each experienced and any specific product manufacturer.

I begin by describing how the emergence of law and economics provided the intellectual foundations for allowing liability on a collective basis. Then, I analyze those collective liability theories most familiar to scholars and practitioners of mass products torts: alternative liability, the enticing prospect of market share liability, the now defunct idea of enterprise liability, and concert of action and civil conspiracy. Finally, I turn my attention to actions in which the individual victims of latent diseases are aggregated: namely, class actions and consolidations.[3] These aggregation devices are rarely seen as means of circumventing the requirement of proof that a specific manufacturer's product caused a particular victim's disease, yet in the hands of particularly creative and ambitious practitioners and judges, they have served that role. These novel approaches that circumvented the individualized causation requirement were also the first steps in transforming tort actions involving victims of latent product-caused diseases from contests of "individual against individual" concerning specific, isolated tortious acts and the resulting particular harms, on the one hand, to a very different mechanism for handling product-caused diseases collectively within the legal system, on the other.

Often, tort actions involving mass products combined the collective form of the plaintiff—for example, a class action seeking damages resulting from the latent diseases experienced by hundreds, thousands, or even millions of victims—with an attempt to impose liability on multiple or indeterminate defendants. The combination of the collective plaintiff and the collective defendant was greater than the sum of the parts. This transition toward viewing the liability issue as a matter to be decided between the collective plaintiff and the collective defendant would culminate a decade later in government-sponsored litigation against cigarette and lead pigment manufacturers relying on the substantive claim of public nuisance as a means to collectivize individual harms and to circumvent any requirement of proof of a causal link between a particular victim and a specific manufacturer. Actions by the government against product manufacturers are the principal focus of chapters 6–9 of this book.

THE INTELLECTUAL FOUNDATIONS OF COLLECTIVE LIABILITY

Some of the most influential architects of tort theory during the past generation paved the way for the rejection of the individual causation requirement in torts. In 1975, then Yale professor Guido Calabresi, currently a judge on the Court of Appeals for the Second Circuit, argued that the requirement that a particular plaintiff prove that a particular defendant caused his harm was "far from being the essential, almost categorical imperative it is sometimes described to be."[4] The origins of Calabresi's progressive version of law and economics lay in both the field of welfare economics and the reformist zeal of the Great Society of the 1960s.[5]

For Calabresi, the goal of tort law was to reduce "the costs of accidents," an economic and social problem like any other economic and social problem. To achieve this goal, Calabresi identified several "subgoals" for the tort system. First, he argued that any accident compensation system should "discourage activities that are 'accident prone' and substitute safer activities as well as safer ways of engaging in the same activities."[6] Calabresi's second objective was to distribute the costs of accidents in a manner that inflicts "less pain" than if the accident costs were borne solely by the original victims.[7] The most important way of accomplishing this objective is to distribute the losses resulting from an accident broadly across many people. Calabresi also acknowledged the "deep-pocket" notion that the costs of accidents will cause less pain and disutility if paid for by people who will suffer less "social or economic dislocations as a result of bearing them, usually thought to be the wealthy."[8]

Calabresi thus focused on objectives related to the victim (loss distribution or compensation) and to the injurer (loss avoidance or deterrence) that are not necessarily intrinsically linked. The need to impose liability on the injurer to discourage his harm-producing activity did not require that the financial penalty extracted from the injurer be transferred to the particular injurer's victim. Conversely, any particular victim could receive his compensation from any party capable of distributing his accident losses as broadly as possible; loss distribution did not require that his losses be paid by the injuring party. Calabresi recognized that "for centuries society has seemed to accept the notion that justice required a one-to-one relationship between the party that injures and the party that is injured."[9] He clearly accepted, however, the notion that liability could be assessed on injurers on a collective, as opposed to a particularized, basis, when he concluded that "there is, of course, no logical necessity for linking our treatment of victims, individually or as a group, to our treatment of injurers, individually or as a group." Viewed from the perspective of the goal of loss avoidance, according to Calabresi, the traditional "but for" requirement of causation between a particular injurer's acts and a particular victim's harm "is simply a useful way of toting up some of the costs the [potential injurer] should face in deciding whether avoidance is worthwhile."[10] Thus, Calabresi severed the linkage between deterrence and compensation that had been regarded as inherent in traditional tort law.

Calabresi was not alone, among law-and-economics scholars, in rejecting the traditional requirement that a particular victim be required to prove that his harm was caused by a particular tortfeasor in order to recover. Judge Richard Posner, for example, who subscribed to a much different, more conservative and free-market oriented version of law and economics, agreed.[11] Posner and his coauthor William Landes virtually mocked any requirement of individual causation, arguing that "causation in the law is an inarticulate groping for economically sound solutions."[12] Posner concluded that manufacturers that may have produced the product responsible for the harm of any particular victim, even when the particular manufacturer who produced the particular product causing the victim's harm could not be identified, should be held jointly and severally liable without a right of contribution against other manufacturers, because "joint liability under a negligence standard creates incentives for both potential injurers to take due care."[13] In short, the law-and-economics conception of tort law that emerged during the 1970s and 1980s among tort scholars generally viewed any requirement of particularity in causation as old-fashioned and likely to impede their goals. By 1987, legal philosopher Judith Jarvis Thomson observed, "Fault went first. . . . Now cause is going."[14]

At the same time, the emergence of the law-and-economics conception of tort law and its disavowal of any necessity for a causal connection between a particular injurer and a particular victim provoked a response. In contrast to law-and-economics scholars, corrective justice theorists believe that the essence of tort law is the objective that an injuring party should repair the losses caused by his wrongful conduct, not that tort law should pursue instrumental or utilitarian social goals.[15] Ernest Weinrib, the principal proponent of one of the two versions of corrective justice, the one based on formalism,[16] argues that "*the* basic feature of private law" is that "a particular plaintiff sues a particular defendant."[17] According to Weinrib, tort law is not a matter of promoting welfare or otherwise implementing instrumental objectives; instead, it is a matter of protecting rights. Both the original wrong and "the transfer of resources that undoes it" constitute "a single nexus of activity and passivity where actor and victim are defined in relation to each other."[18] Unlike instrumental conceptions of torts that make it possible to separate the victim's need for compensation from the desire to discourage harm-producing activity by the defendant, Weinrib concludes that the "changes in the value of the parties' holdings" are interdependent and cannot "be restored by two independent operations."[19] The correlative relationship between the right and the duty, he argues, "locks the plaintiff and defendant into a reciprocal normative embrace, in which factors such as deterrence and compensation, whose justificatory force applies solely to one of the parties, play no role."[20] The court's role can only be understood as that of applying liability in the case of a bipolar relationship. As a result, formalist corrective justice principles as articulated by Weinrib appear to leave victims of diseases resulting from exposure to harmful products without a remedy within the tort system when they cannot identify the other half of the reciprocal, bipolar relationship, that is, the specific manufacturer whose products injured them.

Not all theorists typically associated with corrective justice perspectives necessarily share Weinrib's view that the law should require that a particular victim must prove that his harm was caused by a particular injurer in order to recover. Jules Coleman, for example, sanctions liability without proof of individual causation—even when such liability is imposed within the tort system—as something separate and apart from the basic corrective justice principles of tortious liability.[21] Coleman supports it as a form of social insurance or an alternative compensation system (operating within the common-law tort system) in which manufacturers who have imposed the risk are assessed for the damages.

LIABILITY OF MULTIPLE AND INDETERMINATE MANUFACTURERS

Following traditional principles of tort law that foreshadowed Weinrib's version of corrective justice principles, courts during the 1980s usually continued to deny liability in cases in which the victim was unable to identify the particular injurer that caused his harm. Yet during that era, courts influenced by instrumental approaches to tort law creatively applied traditional tort doctrines and invented new ones that enabled—and continue to enable—a victim in some jurisdictions, usually in isolated circumstances, to recover without proof that his harm was caused by a particular injurer. On initial examination, many of these doctrines appear to be only procedural devices that shift the burden of proof to the defendant to prove that the injurer did not cause the plaintiff's harm. The realistic effect of such doctrines, however, generally has been to impose liability without proof of individual causation, because, in actuality, neither the plaintiff nor the defendant can prove which injurer's acts caused a particular plaintiff's harm. Further, some courts have held that a defendant can be found liable on a collective basis even if it can prove that it was not in fact the cause of the harm to the specific plaintiff.

Of the approaches described in this section, concurrent causation resulting in indivisible harm, alternative liability, and market share liability are all applicable to situations in which each of the mass products manufacturers or potential injurers acts independently or in parallel with each other. The remaining two theories of collective liability—industry-wide liability (enterprise liability) and concert of action or civil conspiracy—apply when the defendant-manufacturers have acted jointly or cooperated with each other.

Concurrent Causation Resulting in Indivisible Harm

As discussed in chapter 2, Judge Wisdom's opinion in *Borel v. Fibreboard Paper Products Corp.*[22] broke down many barriers that had prevented victims of latent diseases caused by product exposure from recovering, including the traditional requirement that a particular victim must prove that a specific manufacturer's product had caused his disease. It was not until twenty years after *Borel*, however, that the California Supreme Court articulated the most analytically sound basis for holding all manufacturers of asbestos products present at the plaintiff's workplace jointly and severally liable. In *Rutherford v. Owens-Illinois, Inc.*,[23] the court held, in a consolidated action seeking damages resulting from asbestos-related diseases, that plaintiffs "need not prove with medical exactitude that fibers from a particular defendant's asbestos-containing products were those, or among those, that actually began the cellular process of malignancy."[24] The court went beyond the *Borel* opinion and the often fictitious assumption on which it rested—namely, that the products of each manufacturer had in fact contributed causally to the victim's asbestos-related disease. It acknowledged that it was scientifically unclear whether each exposure to asbestos and the resulting "scarring contributes cumulatively to the formation of a tumor" or whether "only one fiber or group of fibers actually causes the formation of a tumor," in which case "the others would not be legal causes of the plaintiff's injuries."[25] Arguing that "plaintiffs cannot be expected to prove the scientifically unknown details of carcinogenesis, or trace the unknowable path of a given asbestos fiber," the court held that the plaintiff could recover by showing that the defendant contributed "to the aggregate *dose* of asbestos the plaintiff . . . inhaled or ingested, and hence to the *risk* of developing asbestos-related cancer," without proving that "fibers from the defendant's particular product were the ones, or among the ones, that *actually* produced the malignant growth."[26]

Traditionally, tort law generally understood cause-in-fact as something more than an increase in the risk of an injury. In *Rutherford*, however, the effect of the novel application of concurrent causation was to impose liability without identification of the particular defendant that caused a particular plaintiff's injury. The victim was exposed to the asbestos products of many manufacturers. These products may have increased the victim's risk of suffering from cancer, and it was impossible for any of the defendants to show that its products were not an actual cause of the cancer, just as it was impossible for the victim to prove that any particular manufacturer's product did in fact contribute to his cancer. Under the court's holding, these manufacturers were held collectively liable.

Alternative Liability

During the 1980s and in succeeding decades, a number of courts used a tort doctrine known as *alternative liability* to hold defendant-manufacturers liable for latent diseases resulting from exposure to asbestos or other products even when it could not be shown which manufacturer's products harmed a particular plaintiff. The origins of alternative liability lie in the classic case of *Summers v. Tice*,[27] in which a hunter was injured when shot in the eye by one of his two hunting companions. The victim was able to prove that both defendants had fired shotguns negligently, but he could not establish which defendant's shot was in fact the cause of his substantial injuries. The California Supreme Court held that because the two defendants had each acted negligently, although independently, toward the plaintiff, the burden shifted to each defendant to prove that his negligent act was not the cause of the plaintiff's injury. Unless this burden was met, the defendants would be held jointly and severally liable. In short, under alternative liability, the plaintiff need not prove with particularity the identity of the injuring party in order to recover. In the absence of rebuttal, the defendants are held liable collectively.

At least some courts have applied variations of alternative liability in actions brought by victims of asbestos-related and other latent diseases resulting from exposure to mass products.[28] Consider, for example, *Menne v. Celotex Corp*,[29] in which case the plaintiff developed mesothelioma after working for forty years as a pipe fitter and plumbing and heating contractor. Plaintiff proved that he had been exposed to asbestos products manufactured by each of the ten defendants but could not prove that exposure to any particular defendant's product was a substantial cause of his disease, a requirement under Nebraska law. In these circumstances, the court held that the burden shifted to each defendant to prove that "exposure was unlikely to have been frequent or long enough to be a substantial factor in causing Menne's mesothelioma."[30] The court acknowledged that "where a defendant lacks evidence as to the frequency or duration of exposure, the burden shift may well result in a finding that the defendant is a cause of the harm."[31]

Plaintiffs' counsel in asbestos and other mass products tort cases often have attempted to push the use of alternative liability even further, to encompass situations where it cannot be proved that the plaintiff had been exposed to products manufactured by each defendant. Courts identify at least two reasons why these factual scenarios differ from that in *Summers v. Tice*, and they therefore usually decline to expand the applicability of the doctrine of alternative liability beyond a context involving "one plaintiff and two defendants" to asbestos litigation involving, for example, "180 plaintiffs. . . and at least 16 defendants."[32] First, as the Michigan Supreme Court has stated, "in *Summers* each defendant was negligent toward the sole plaintiff,"[33] but in products cases, each defendant may have been negligent toward or distributed defective products to some consumer or user but not necessarily to the particular plaintiff who is suing. Rosenberg contends, however, that alternative liability may be even more appropriate in this context. As he notes, although both hunters committed wrongs in *Summers*, tort law ordinarily imposes no consequences in the absence of harm. The wrong of one of the hunters resulted in no harm in that case, argues Rosenberg, but "in the mass exposure context, . . . harmless wrongs rarely occur."[34]

The second issue that arises in mass tort cases but not in *Summers* is whether all possible manufacturers whose products might have harmed the victim or victims must be named as defendants in the legal action. Courts frequently regard it as unfair to hold defendant-manufacturers jointly and severally liable in a situation in which not all manufacturers have been joined as defendants, because the plaintiff cannot prove that any of the defendants before the court in fact caused the harm.[35] Other courts, however, have been more understanding of the victim's situation and have not imposed on plaintiffs the frequently impossible obligation of suing all manufacturers. Instead, these courts have required plaintiffs to "make a genuine attempt to locate and identify the tortfeasor responsible for the individual injury."[36]

Though some courts have allowed the use of alternative liability to prove causation in a mass products case in which the particular victim cannot identify which specific manufacturers caused his harm, most courts reject the application of alternative liability principles in these cases. This is because the large number of potential injurers would result in too many "false positives"—manufacturers held liable even though only one or a few among many manufacturers were in fact the injurer(s) responsible for the victim's injury—and because the plaintiff typically cannot identify and join all the potential injurers in his legal action.[37] In short, except in cases meeting specific criteria that typically do not occur in the mass products tort context (excluding, perhaps, some asbestos cases in isolated jurisdictions), alternative liability is unlikely to prove effective in overcoming the obstacle of individual causation.

Market Share Liability

No other judicially created mechanism for holding defendant-manufacturers collectively liable during the 1980s was as ambitious, as original, or as tantalizing to academic tort commentators as market share liability,[38] which originated in the California Supreme Court's widely known opinion in *Sindell v. Abbott Laboratories.*[39] The plaintiff had sued on behalf of herself and other similarly situated women suffering from cancerous and precancerous growths that she alleged resulted from their mothers' consumption, at least ten or twelve years earlier, of diethylstilbestrol (DES), a synthetic estrogen compound intended to prevent miscarriages in pregnant women. She lacked the means to identify which pharmaceutical company manufactured the DES consumed by her mother, because eleven drug companies named in the complaint and scores of additional drug companies used an identical chemical formula for the drug. The trial court therefore dismissed the complaint. The California Supreme Court, however, reversed, holding that "each defendant will be held liable for the proportion of the judgment represented by its share of that market unless it demonstrates that it could not have made the product which caused plaintiff's injuries."[40] This holding, claimed the court, resulted in each manufacturer's liability reflecting the injuries caused by its own products, even though the tortious acts of any specific defendant were never causally linked to the harm suffered by any particular victim. The court justified its reasoning on the basis of instrumental goals, including loss minimization: "The manufacturer is in the best position to discover and guard against defects in its products and to warn of harmful effects; thus, holding it liable for defects and failure to warn of harmful effects will provide an incentive to product safety."[41] Further, the opinion reflects the instrumental goal of loss distribution. The court also invoked an argument sounding in corrective justice. It stated that "as between an innocent plaintiff and negligent defendants, the latter should bear the cost of the injury."[42]

Because the defendant could be excused from liability if it proved that it was not responsible for a particular victim's harm, it is possible to interpret *Sindell* as an opinion that merely shifts the burden of proof on the issue of causation to the defendant. In *Hymowitz v. Eli Lilly & Co.*,[43] however, the New York Court of Appeals imposed true collective liability for the creation of risk,[44] when it held that a particular manufacturer of DES that could prove that its product could not have been the one that caused the harm to the particular victim nevertheless would be liable on a market share liability theory. The court forthrightly explained that "because liability here is based on the over-all [*sic*] risk produced, and not causation in a single case, there should be no exculpation of a defendant who, although a member of the market producing DES for pregnancy use, appears not to have caused a particular plaintiff's injury."[45] It conceded "the lack of a logical link between liability and causation in a single case."[46]

Despite considerable scholarly support for the idea of market share liability, the concept met with virtually universal rejection by the courts during the quarter century following the *Sindell* decision, except in cases against DES manufacturers.[47] Until recently, attempts to use market share liability against the manufacturers of lead paint and lead pigment followed this negative trend. Market share liability presupposes that the disputed products pose a uniform risk. According to the court in *Brenner v. American Cyanamid Co.*, unlike DES, which had "an identical chemical composition," lead-based paint "is not a fungible product ; it 'contains varying amounts of lead pigments,'" ranging from 10 to 50 percent.[48] Even if plaintiffs' counsel, as in *Skipworth v. Lead Industries Ass'n*, chose to sue the manufacturer of the lead pigment, an ingredient in paint, courts found that the different compositions of the product ultimately consumed—that is, the lead paint—affected the quantum of risk posed by equivalent amounts of lead pigment.[49]

The other requirement for the principled application of market share liability is the ability of a court to determine the respective market shares of various manufacturers, if not with absolute accuracy, then at least with a meaningful approximation. This requirement alone typically doomed the application of market share liability in cases against manufacturers of lead paint or lead pigment. For a couple reasons, it is impossible, as a practical matter, to determine market shares with a reasonable degree of accuracy in the context of lead paint or pigment. First, as the *Skipworth* court noted, the paint that caused the harm may have been applied at any point during "a more than one hundred year period from the date the house was built until the lead paint ceased being sold for residential purposes."[50] It is one thing to determine the manufacturers' respective market shares during a nine-month period, as in the DES context; it is a far different proposition to determine market shares during a period in excess of a century. Manufacturers of lead paint and lead pigment entered the market, exited the market, and reentered (and perhaps reexited) the market during this period.[51] Second, courts declining to apply market share liability in the lead paint context have found that determining market shares of manufacturers within any given period that occurred as long as 130 years ago may be unrealistic, regardless of the duration of the period during which the products were manufactured. Neither plaintiffs nor defendants possess the necessary records to determine the market shares for lead paint or lead pigment in 1880, 1900, or 1920.

In July 2005, however, in *Thomas v. Mallett*,[52] the Wisconsin Supreme Court allowed the action by a victim of childhood lead poisoning against manufacturers of lead pigment to proceed to trial on a "risk contribution" theory (similar to market share liability), despite plaintiff's inability to identify the specific manufacturers of the product that caused his illness. The court justified its opinion using reasoning similar to that articulated in *Sindell*, stating that on fairness grounds, the cost of the harm should be imposed on the "possibly negligent" manufacturers, not on "an innocent plaintiff who is probably not at fault."[53] On an instrumental basis, reasoned the court, manufacturers are in a better position to distribute losses than is the individual victim.

The court rejected the manufacturers' arguments that their product lacked the fungibility necessary for market share liability, finding that the jury should be allowed to consider the testimony of the plaintiff's expert that the differences in chemical composition between various lead pigments did not affect their bioavailability and hence the consequences of exposure to lead-based paint. The defendant-manufacturers argued that it would not be feasible for the trial court to administer a risk contribution regime and to accurately determine each defendant's market share because the paint containing lead pigment and present in the three houses where the child had lived could have been applied at any point between 1900 and 1978. The Wisconsin Supreme Court, however, responded that the manufacturers "are essentially arguing that their negligent conduct should be excused because they got away with it for too long."[54] In response, one of the dissenting judges noted that "many of the defendants produced white lead carbonate for only a small fraction of the 78-year period during which paint containing white lead carbonate could have been applied to the walls of [the plaintiff's] three residences."[55]

At this time, it is impossible to predict whether the decision in *Thomas v. Mallet* is an isolated opinion driven by the Wisconsin Supreme Court's desire to identify a funding source to address the public health problems posed by childhood lead poisoning[56] or whether it is a harbinger of things to come. As a practical matter, the determination of the risk contribution of each defendant over a seventy-five-year period appears to be an impossible task for any trial court judge and jury. In fact, when the Wisconsin Supreme Court became the first court to accept the risk contribution theory in a DES case in 1984, it recognized the substantial practical problems with market share liability[57] and distinguished risk contribution liability from it. At that time, the court held that in assigning a percentage of liability to each defendant-manufacturer, the jury should consider not only its respective market share but also the relative degree of the egregiousness of its conduct compared to that of other manufacturers. In *Thomas*, in determining the manufacturers' relative market shares, which comprise one of the ingredients in the risk contribution calculation, the jury would be required to consider the following factors: the timing of the various producers' entry, exit, and sometimes reentry into the relevant market; the great differences in the amount of lead pigment contained in various lead-based paints; how much of the plaintiff's exposure occurred at each of three houses where he lived; and the possible effect of bioavailability on the effects of exposure (disputed between the parties). The jury's determination of market share here would be far more challenging than in the DES situation, where the chemical formula of manufacturers' products were identical and where the products causing the harm were consumed by the victim's mother in a specific period lasting less than nine months. The jury then would be required to consider these factors along with its evaluation of the level of egregiousness of each manufacturer's conduct. It is difficult to see how combining "apples and oranges"—the percentage of market share and level of egregiousness of each defendant—in any way makes the jury's calculation more manageable.

When *Thomas v. Mallet* was first decided, many believed that the opinion would set a trend for overcoming the causation obstacles inherent in litigation against manufacturers of lead paint and lead pigment. However, on the heels of the *Thomas* opinion, the Missouri Supreme Court rejected the use of market share liability in litigation against paint manufacturers.[58] Perhaps most ironically, when the *Thomas* case went to trial, the jury found in favor of the defendant-manufacturers, specifically finding that although the child in question had elevated blood lead levels, his disabilities did not result from lead.[59]

Enterprise or Industry-wide Liability

In *Hall v. E. I. Du Pont de Nemours & Co.*,[60] Judge Jack B. Weinstein relied on a theory of "enterprise liability" to hold six manufacturers of blasting caps and their trade association potentially liable, jointly and severally, for damages to children resulting from eighteen separate accidents. Later courts would more often refer to Weinstein's collective liability approach as "industry-wide liability."[61] Judge Weinstein justified his decision on the grounds that defendants had cooperated in a safety program through a trade association and, acting either jointly or in parallel, had adopted common safety features that were inadequate—they did not provide for warnings on dangerous products.

Since Judge Weinstein's decision, courts almost universally have rejected liability based on enterprise or industry-wide liability. For example, in *Ryan v. Eli Lilly & Co.*,[62] the court refused to apply enterprise liability and described it as "repugnant to the most basic tenets of tort law."[63] In cases against asbestos manufacturers, courts have concluded that enterprise liability is inapplicable because "the number of potential defendants is so great that the assumption that all the defendants jointly controlled the risk of injury is weak."[64] When the U.S. Court of Appeals for the Eleventh Circuit similarly rejected enterprise liability claims against asbestos manufacturers, it explained that "elimination of a causation requirement would render every manufacturer an insurer not only of its own products, but also of all generically similar products manufactured by its competitors."[65] Citing essentially these same reasons, courts have refused to allow the use of enterprise liability against manufacturers of lead pigment.[66]

Civil Conspiracy and Concert of Action

Beginning in the early 1980s, courts also began using concepts other than enterprise liability, specifically concert of action and civil conspiracy, to hold manufacturers of products collectively liable to one or more victims harmed by fungible products, even when a causal connection could not be established between a specific manufacturer and a particular victim. Under the well-established doctrine of concert of action, manufacturers of products were held jointly and severally liable if they "acted in concert," that is, if they acted pursuant to a common plan or design. Some courts equated "civil conspiracy" with "concert of action."[67] Other courts distinguished the two concepts, reasoning that civil conspiracy requires the joint tortfeasors to share intent to accomplish an unlawful objective but that tort liability for concert of action merely requires that the tortfeasors engage in tortious conduct while acting in concert.[68]

Courts had little difficulty in holding manufacturers of asbestos and other mass products liable on the basis of concert of action when it could be proved that there was an explicit agreement among manufacturers to engage in tortious conduct.[69] Further, courts hold that manufacturers can be found to have conspired to engage in tortious conduct in the absence of express agreement if they tacitly ratified the conspiracy. For example, in *Nicolet, Inc. v. Nutt*,[70] the Delaware Supreme Court found the evidence sufficient to allow victims of asbestos-related diseases to proceed against manufacturers under a cause of action alleging that the defendants "conspired with other asbestos manufacturers to actively suppress and intentionally misrepresent medical evidence warning of the health hazards of asbestos."[71]

The courts took varying approaches, however, on the issue of whether "consciously parallel conduct" was sufficient to create liability under concert of action by "implied or tacit agreement or understanding."[72] In *Bichler v. Eli Lilly & Co.*, the New York Court of Appeals upheld the jury's finding of concert of action based on the DES manufacturers' "consciously parallel behavior" in marketing DES without adequate testing.[73] Other courts have held that parallel activity by several product manufacturers is insufficient to establish concert of action.[74]

The success of some asbestos and DES plaintiffs in recovering under a claim of concert of action or civil conspiracy has not translated into success for victims of childhood lead poisoning. In *Santiago v. Sherwin-Williams Co.*,[75] a Massachusetts federal district court held that a victim of childhood lead poisoning could not proceed against manufacturers of lead pigment on a cause of action based on concert of action because such a claim was appropriate only when one of the defendants could be found to have been the cause-in-fact of the injury. Obviously this was not possible in the lead pigment context, where there were a number of potential manufacturers and no way of identifying the manufacturer of any of the pigments that caused the plaintiff's harm.

The civil conspiracy theory of liability eventually would emerge as an important claim against manufacturers, including tobacco companies, in the late 1990s and the early twenty-first century. The claim of civil conspiracy or concert of action, however, only allows the victim to overcome the requirement of individual causation in those limited instances in which he can prove an agreement, either explicit or tacit, to engage in tortious conduct.

THE COLLECTIVE PLAINTIFF

Class Actions

The Early Promise

During the 1980s, class actions appeared to be the vehicle that courts would choose to "collectivize" claims of victims resulting from mass products torts and, in the process, perhaps circumvent the traditional individualized causation requirement. A class action permits the judicial resolution of the claims of multiple victims that arise from a common set of facts as long as the legal and factual issues raised by each claim are identical. In an asbestos class action case, the federal Court of Appeals for the Third Circuit described the obvious inefficiencies of litigating claims against mass products manufacturers without the benefit of some form of aggregate litigation.

> Inefficiency results primarily from relitigation of the same basic issues in case after case. Since a different jury is empanelled in each action, it must hear the same evidence that was presented in previous trials. A clearer example of reinventing the wheel thousands of times is hard to imagine. . . . In case after case, the health issues, the question of injury causation, and the knowledge of the defendants are explored, often by the same witnesses.[76]

One of the most ambitious attempts to use the class action vehicle as a means of collectivizing the tort process was the decision by federal district court judge Robert M. Parker in *Cimino v. Raymark Industries, Inc.*[77] He combined the class action vehicle with statistical sampling[78] to effectively eliminate any requirement of individual causation. Judge Parker sat in the same district court that had heard Clarence Borel's case more than twenty years earlier. Since that time, more than five thousand cases had been filed in his district alone by workers claiming that they suffered from asbestos-related disease. In the intervening years, Judge Parker had seen several of the largest manufacturers of asbestos products declare bankruptcy. At least 448 of the victims whose claims were to have been litigated died while waiting for trial. The transaction costs of the litigation were consuming approximately fifty-seven cents of every asbestos litigation dollar, with the plaintiff receiving only 43 percent of the total amount spent on the litigation.[79]

Facing these realities, Judge Parker himself initiated the process of certifying as a class action the claims of 3,031 victims of asbestos disease[80] in their litigation against various asbestos manufacturers. He then permitted the use of statistical evidence to prove the causal connection between the defendants' products and the plaintiffs' injuries and the amount of each claimant's damages. Judge Parker's trial plan included three phases. During phase 1, the jury heard a complete trial of the individual cases of ten class representatives, designed to reach resolution, for all members of the class, on the issues of product defectiveness, the adequacy of warnings, and the appropriateness of punitive damages. The trial took 133 days and yielded 25,348 pages of trial transcript. During the trial, 272 expert witnesses and 292 fact witnesses testified.

Judge Parker then divided the members of the plaintiff class into five disease categories, based on which illnesses the class members allegedly had sustained as a result of exposure to the asbestos products. During phase 2 as originally anticipated, the same jury was to decide the percentage of class members in each category that had been exposed to each defendant's products and the percentage of claims in each disease category barred by statutes of limitations or other specified defenses.[81] During this phase, the jury also would assess a determination of damages in a lump sum for all plaintiffs within each specified disease category. The court skipped phase 2, however, and went directly to phase 3. Following the trial of phase 3, the parties stipulated what the findings of phase 2 would have been, including the relative proportion of financial responsibility of each defendant.

In phase 3, Judge Parker employed a statistical sampling approach: two juries heard a sampling of cases to determine whether each sample plaintiff suffered from an asbestos-related disease and, if so, what damages that sample plaintiff had sustained. The court conducted a hearing to confirm that the randomly drawn sample plaintiffs from each disease category were representative of the population of that group.[82] The court then proposed to award each nonsample member within any given disease category the average damage verdict of the sample plaintiffs within that group whose cases had been heard by the jury.

Though the plaintiffs consented to this approach, the defendants objected. The defendants in *Cimino* argued that even as to the sample plaintiffs, there had been no determination that any particular defendant's product had caused any particular disease sustained by a member of the plaintiff class. This lack of individual causation, according to the defendants, violated both the substantive law of causation that governed the case and each defendant's right to a jury trial under the Seventh Amendment to the U.S. Constitution. Judge Parker rejected this argument, reasoning that the liability of any particular defendant for damage awards, determined by the average of the awards of sample class members within a disease category viewed collectively, would be comparable to what would occur if all the damage awards had been determined individually.

Unfortunately for Judge Parker and the future of attempts to overcome the obstacles posed by the individual causation requirement, the Court of Appeals for the Fifth Circuit disagreed.[83] The court first found that the trial court had not established that a particular defendant's products caused the harm to a particular plaintiff, thereby failing the requirement that causation "be determined as to 'individuals, not groups.'"[84] Further, in the "extrapolation cases," the court held that the determination of damages without either a trial or a jury denied the defendants their due process rights.

Barriers to Class Certification

The *Cimino* litigation marked a key battle between proponents of those who sought to combine the class action with statistical sampling to create a collective mechanism to address mass products torts, on the one hand, and those committed to the traditional model of particular claimant/particular defendant, on the other. Since that time, class action litigation largely has fallen by the wayside as a means of determining collective liability for victims of mass products torts. Today, with rare exceptions[85] or in unusual circumstances,[86] courts almost always deny class certification in mass products torts.[87] A plaintiff seeking class certification must first prove that the proposed class meets the four requirements of Rule 23(a) of the Federal Rules of Civil Procedure:

(1) the class is so numerous that joinder of all members is impracticable;
(2) there are questions of law or fact common to the class;
(3) the claims or defenses of the representative parties are typical of the claims or defenses of the class; and
(4) the representative parties will fairly and adequately protect the interests of the class.[88]

In addition, the party seeking class certification must show that he satisfies one of three alternative requirements of Rule 23(b). At least initially, attorneys who sought class certification for a victim allegedly suffering from a product-caused disease typically relied on subsection (b)(3), which requires that "questions of law or fact common to class members predominate over any questions affecting only individual members, and that a class action is superior to other available methods for fairly and efficiently adjudicating the controversy."[89] Courts usually found that litigants in mass products torts failed to meet this requirement, because these cases turned on individualized proof of causation, reliance, comparative fault, and/or damages.[90] Other courts, in denying class certification to proposed nationwide classes, have noted the difficulty of applying varying substantive principles of state law from different states.[91] Chief Judge Richard Posner of the Seventh Circuit, for example, has described how, in such a case, the trial court might be required to give "a kind of Esperanto instruction, merging the negligence standards of the 50 states and the District of Columbia."[92]

Perhaps the most important and sophisticated attempt to certify a class action of victims of tobacco-related disease occurred in the mid-1990s, in *Castano v. American Tobacco Co.*[93] Wendell Gauthier, the lead attorney in this nationwide class action, appreciated the difficulties of certifying a class action of victims of tobacco-related disease and therefore framed a complaint that did not focus on recovery of traditional personal injury damages resulting from tobacco-related diseases. Instead, the complaint alleged that the tobacco companies fraudulently had failed to inform consumers of the addictive nature of cigarettes and had manipulated the nicotine level in them in order to addict smokers. In other words, the complaint alleged that the harm sustained by class members was addiction, not any of a variety of tobacco-related illnesses. By focusing on addiction shared by all members of the represented class as it was defined, the complaint minimized the importance of issues of fact that were individual and not common.

Acting under what the trial court believed to be the authority of Rule 23(c)(4) of the Federal Rules of Civil Procedure, which allows for certification of a class "with respect to particular issues," the court certified the class only for purposes of trying "core liability issues" and punitive damages.[94] The trial court judge anticipated a multiphased trial similar to that in *Cimino*. Phase 1 of the trial would decide issues of fact and law related to the defendants' liability and other common issues.[95] In phase 2, the jury would determine compensatory issues for sample plaintiffs and establish a ratio of punitive damages to compensatory damages. Phase 3 would consist of a streamlined administrative process to determine the compensatory damages for each of the other estimated fifty million members of the class.

The Court of Appeals for the Fifth Circuit, however, reversed the partial class certification.[96] It held that the trial court had failed to consider how differences in state law affected even those issues that the trial court had found to be common issues.

> The class members were exposed to nicotine through different products, for different amounts of time, and over different time periods. Each class member's knowledge about the effects of smoking differs, and each plaintiff began smoking for different reasons. Each of these factual differences impacts the application of legal rules such as causation, reliance, comparative fault, and other affirmative defenses.[97]

The court of appeals further criticized the trial court for not adequately explaining how it would handle the trial of the causation and damages aspects of each class member's addiction claim. One particular concern was that the trial court's plan would require various phases of a victim's trial to be heard by separate juries, a fact that the court of appeals held to violate the jury trial guarantees of the Seventh Amendment.

Class certification rules also would doom the largest jury verdict ever against tobacco manufacturers. When a Florida jury awarded punitive damages of $145 billion in a class action against tobacco manufacturers in July 2000, Matt Myers, president of the Campaign for Tobacco-Free Kids, hailed it as "a true sea change in the public perception of who is most responsible for the death toll of tobacco." Dan Webb, an attorney representing Philip Morris, one of the defendants, indicated that neither his company nor any other "company in the world" could withstand such a verdict.[98] Three years later, however, a Florida appellate court reversed the certification of the class and the resulting judgment,[99] holding that because of the need for causation and damages issues to be determined individually and the need to apply the laws of numerous other states, the requirements for class certification had not been met.

Global Settlements and the Supreme Court

In two celebrated opinions issued during the 1990s, *Amchem Products, Inc. v. Windsor*[100] and *Ortiz v. Fibreboard Corp.*,[101] the U.S. Supreme Court struck down attempts to use the class action vehicle to accomplish an even more ambitious goal: to establish administrative compensation systems through agreements among the litigants that would compensate those who had been exposed to asbestos products but who had not yet displayed symptoms of asbestosis, cancer, mesothelioma, or other disabling diseases. In 1991, the Judicial Panel of Multidistrict Litigation, created by statute and consisting of federal judges, had consolidated 26,639 asbestos lawsuits in the federal courts and transferred them to the court of Judge Charles Weiner of the federal District Court for the Eastern District of Pennsylvania for pretrial proceedings.[102] The asbestos manufacturers hoped that Judge Weiner would be able to achieve a "global settlement" of all federal cases already filed against them, as well as those that might be filed against them in the future. After settling the claims of the "inventory" claimants (those with pending claims), counsel for the plaintiffs and twenty asbestos manufacturers reached agreement to settle the future claims of those who had not yet filed claims and who, in many cases, were asymptomatic even though they had been exposed to asbestos. The agreements between the parties were estimated to provide $1.3 billion to one hundred thousand claimants during the first ten years.[103]

Within a single day, lawyers purporting to represent all those who had been exposed to asbestos products but who had not yet filed a claim against any of the multiple defendants, together with opposing counsel representing the defendant-manufacturers, filed a complaint initiating litigation on behalf of this class, the defendant's answer to the complaint, a joint motion seeking class certification, and a proposed settlement agreement.[104] The settlement agreement, running more than one hundred pages, outlined in detail a schedule of payments to be made to those who had not yet filed lawsuits but who were at risk of developing any of the specified asbestos-related illnesses, as well as an administrative mechanism for considering the claims and disbursing the payments. In short, through settlement, counsel for both the plaintiffs and the defendants tried to accomplish for future victims what Judge Parker had attempted to do for current victims in *Cimino.*

In *Amchem Products, Inc.,* the U.S. Supreme Court held that those individuals who had not yet filed an asbestos-related lawsuit against one of the defendant-manufacturers but who had either been exposed to asbestos in their work or through contact with family members could not be properly certified as a class under Rule 23 of the Federal Rules of Civil Procedure.[105] The Supreme Court found that common issues did not predominate over individual issues. Some of the members of the proposed class already suffered from any of a number of very different asbestos-related diseases; others did not. The Court further noted, "Class members were exposed to different asbestos-containing products, for different amounts of time, in different ways, and over different periods. . . . Each has a different history of cigarette smoking, a factor that complicates the causation injury."[106]

Perhaps more important, the Supreme Court also held that certification was not proper because the named class representatives could not "fairly and adequately protect the interests of the class" as required by Rule 23. The Court noted that "named parties with diverse medical conditions sought to act on behalf of a single giant class rather than on behalf of discrete subclasses." According to the Court, "the critical goal" of "generous immediate payments" for the currently injured "tugs against the interest of the exposure-only plaintiffs in ensuring an ample, inflation-protected fund for the future."[107] These issues are present in any global settlement seeking to settle the claims of both current and future victims of product-caused diseases. The Court also rejected the parties' argument that these requirements should be applied less rigorously in the case of settlement-only class actions.

Two years later, in *Ortiz v. Fibreboard Corp.*,[108] the Supreme Court rejected another attempt to use class action certification to achieve a global settlement, albeit one limited to a single manufacturer. The case arose in the federal District Court for the Eastern District of Texas, the same court where Clarence Borel's widow had prevailed many years earlier. The Fibreboard Corporation found itself squeezed between the thousands of new asbestos-related claims filed against it each year and coverage disputes with its insurers. In 1993, Fibreboard and its principal insurer, Continental, agreed to settle the forty-five thousand pending claims against Fibreboard in a global settlement that also settled all future claims for $1.535 billion, with Continental and one other insurer providing almost all the funds. Under the proposed agreement, claimants would seek compensation from a trust funded with these proceeds, and their rights to sue in court would be extremely limited.

As agreed to by the parties, a class action was filed on behalf of future claimants, those who had been exposed to Fibreboard's products containing asbestos but had not yet brought suit or settled claims. *Amchem* appeared to have made it impossible to certify such a class under Rule 23(b)(3) of the Federal Rules of Civil Procedure, requiring that common issues predominate over individual issues, so counsel sought to certify the exposure-only plaintiffs as a class under a different subsection of the rule, Rule 23(b)(1)(B), which allowed class action treatment if plaintiffs' separate actions would risk "adjudications with respect to individual class members that, as a practical matter, would be dispositive of the interests" of other victims or "would substantially impair or impede their ability to protect their interests."[109] Fibreboard argued that the coverage available from its insurers and its own very modest contribution of five hundred thousand dollars to the settlement together constituted a "limited fund."[110] It asserted that the full payment of earlier claims of asbestos victims would mean that funds would not be available to pay victims who filed later, and therefore the situation fit within the requirements of the Rule 23 subsection.

The Supreme Court rejected this argument, noting that the advisory committee that originally recommended the text of Rule 23 had not contemplated that a mandatory class action could "be used to aggregate unliquidated tort claims on a limited fund rationale."[111] In Fibreboard's situation, it was not clear that all the funds that could have been made available would have been inadequate "to pay all the claims."[112] Further, all available funds must "be devoted to the overwhelming claims." The members of the class must be given "the best deal"; the defendant should not get "a better deal than *seriatim* litigation would have produced."[113] As it had stated earlier in *Amchem*, the Court reiterated that the trial court must scrupulously follow the class certification rules even if it also was to assess the fairness of the eventual settlement.

Despite its rejection of a "limited fund" class certification under the facts of *Ortiz*, the Supreme Court left open the possibility that it might approve a limited fund plan in another mass products tort action.[114] A decade later, however, there is little evidence that class actions brought by victims of product-related diseases against manufacturers have reemerged as a major factor in compensating such victims or in punishing culpable manufacturers. Notably, in *In re Simon II Litigation*, federal district court judge Jack B. Weinstein had certified a nationwide class of smokers against tobacco companies alleging fraudulent denial and concealment of the health risks posed by smoking and seeking only punitive damages, therefore arguably avoiding many of the individualized issues of causation and damages inherently present in actions seeking compensatory damages.[115] Judge Weinstein found that the Supreme Court had imposed due process limitations on the total amount of punitive damages assessable against any defendant by multiple courts for the same conduct and that these limits therefore established a limited fund under Rule 23(b)(1)(B) of the Federal Rules of Civil Procedure. However, the Court of Appeals for the Second Circuit reversed the class certification, holding that the asserted "upper limit" of constitutionally allowed punitive damages was too "difficult to ascertain" and too speculative to fit within the far more restrictive understanding of a "limited fund" as understood by the drafters of Rule 23.[116]

The Inability of Class Actions to Remedy Product-Caused Diseases

In short, the prospect of class actions as the vehicle that would enable victims of latent diseases resulting from exposure to mass products to recover either compensatory or punitive damages from manufacturers appears to be dead. Perhaps the final blow to that prospect occurred in 2005, with the enactment by Congress and the signing into law by President George W. Bush of the Class Action Fairness Act.[117] That act made it more difficult for victims of mass torts to file actions in state courts where the class certification requirements sometimes are more flexible than in the federal courts.

Richard A. Nagareda recently convincingly analogized the ultimately unsuccessful attempts of counsel to achieve global settlements in class action litigation with the enactment of workers' compensation systems during the initial decades of the twentieth century.[118] Nagareda observes, "In broad-brush terms, the basic compromise reached in workers' compensation and in the asbestos class settlements remains strikingly similar. Both traded the delay and uncertainty of tort litigation for a more streamlined administrative approach."[119] Similarly, in *Amchem* and *Ortiz*, the justices repeatedly emphasized that the "elephantine mass of asbestos cases . . . defies customary judicial administration and calls for national legislation."[120] Suggesting that Congress should consider the "sensibly made" argument that "a nationwide administrative claims processing regime" should be implemented,[121] the Court expressed frustration that Congress had failed to act.[122] In other words, the Supreme Court went on record suggesting that at least in some contexts, legislatively adopted administrative systems, not courts, should provide compensation for victims of product-caused diseases.

Consolidation

In a few instances, plaintiffs' attorneys and courts in mass products tort cases have sought to use the procedural device of consolidation as a vehicle for satisfying individual causation requirements that they otherwise could not overcome. As used by most courts, consolidation is most accurately viewed as a procedural device for joining many individual actions for determination of one or more issues that otherwise would need to be tried repetitively in individual trials of particular plaintiffs, but not as a means of truly imposing collective liability on a group of defendants to benefit multiple victims or to eliminate a requirement of individual causation.[123] Yet a few trial courts have employed consolidation in conjunction with statistical evidence to establish true collective liability to benefit a group of undifferentiated plaintiffs and, in the process, eliminate the individual causation requirement.

In *In re Brooklyn Navy Yard Asbestos Litigation*,[124] for example, Judge Jack B. Weinstein consolidated for trial on all issues sixty-four actions brought by victims of asbestos-related disease. The defendants contended that the plaintiffs had failed to identify the particular manufacturer whose product injured each particular plaintiff, thus failing to satisfy the causation requirement.[125] The Court of Appeals for the Second Circuit, however, upheld Judge Weinstein's finding of causation, noting that the plaintiffs had established that they or their decedents had spent time at a common work site, the Brooklyn Navy Yard, where they were exposed to asbestos; that the products of each defendant had been used at the shipyard and contributed to the asbestos fibers in the air; and that each plaintiff or plaintiff's decedent had developed diseases linked to the defendants' products. The Court of Appeals concluded, "Because the events happened years ago, and many of those exposed to the asbestos are deceased, to require precision of proof would impose an insurmountable burden."[126] By loosening the standard of evidentiary sufficiency on the issue of whether any particular defendant's product contributed to any particular plaintiff's injury, Judge Weinstein and the Court of Appeals for the Second Circuit used consolidation to implicitly impose a form of collective liability, with little or no proof of individual causation linking a particular defendant and a particular plaintiff.[127]

More often in consolidated cases, however, collective liability is not anticipated, and the trial court judge goes to great lengths to stress to the jury the requirement of a causal link between each plaintiff and a specific defendant.[128] For example, in another consolidated asbestos case, the court outlined the steps to be taken to assure that the jury separately considered the causal connection between each specific plaintiff and each specific defendant.[129] After noting that common issues of law and fact favored consolidation, the court noted that other issues, including those related to causation, required "effort to prevent confusion and prejudice," including "separate notebooks for the jurors, tabbed for each plaintiff and each defendant, careful attention to the presentation of evidence, and cautionary instructions reminding the jurors that, during their deliberations, they would have to consider each of the plaintiff's claims separately."[130]

As illustrated by these examples of consolidated cases, most instances of consolidation impose collective liability only in the sense that multiple plaintiffs are able to prove certain elements of liability related to a defendant's conduct in a single proceeding. In and of itself, consolidation does not eliminate the need to prove a causal link between the acts of a specific defendant and the harm sustained by a particular plaintiff. A very small number of trial court judges, however, including Judge Weinstein, appear to have used consolidation as a means of accomplishing collective liability when proof of the causal connection between a particular victim and a particular plaintiff appears insufficient to meet the ordinary civil burden of proof.

EPILOGUE ON THE FIRST WAVE OF CHALLENGES TO THE INDIVIDUALIZED CAUSATION REQUIREMENT

By the 1990s, victims of latent diseases resulting from product exposure had attempted to overcome the historical requirement of proof of individualized causation between a particular victim and a specific manufacturer and had met with mixed success. Often, the factual contexts present in asbestos litigation enabled victims of asbestos-related diseases to meet the requirement. A worker exposed to asbestos was usually able to identify which manufacturers' products had been used at the various work sites where he had been employed. Beginning with these facts, counsel often was able to prove causation through the "substantial factor" test of causation, concurrent causation contributing to an indivisible harm, or alternative liability. The parameters of each of these traditional doctrines clearly were stretched by the compelling facts of asbestos litigation. Further, the egregious conduct of asbestos manufacturers in conspiring to hide the health effects of asbestos products sometimes enabled plaintiffs to proceed with claims of civil conspiracy or concert of action. Finally, both the sheer volume of asbestos litigation and the difficulty in satisfying traditional causation standards led innovative judges to experiment with various forms of the "collective plaintiff" in asbestos litigation, including class actions and consolidation. While Judge Weinstein's efforts to consolidate cases in the *Brooklyn Navy Yard Litigation* met with the approval of the federal appellate courts, Judge Parker's attempt to use a class action for this purpose was rejected on grounds that it violated both the substantive law of causation and the manufacturers' due process rights.

During roughly the same period of time that asbestos plaintiffs achieved their successes in overcoming traditional causation barriers, victims of tobacco-related illnesses and childhood lead poisoning were stalemated. The tobacco litigation cycle had yet to reach the point in the development of the law where causation was the critical issue in most cases. Most early tobacco plaintiffs were chosen partly because of their ability to readily identify the manufacturer of the cigarettes that they smoked exclusively. As described previously, the failure of these plaintiffs to prove requisite elements of liability other than causation prevented their recoveries. Ultimately, courts rejected the use of class actions consisting of victims of tobacco-related diseases, concluding that individual characteristics of each victim's case, including such factors as causation and affirmative defenses, overwhelmed common issues.

The obstacles facing victims of childhood lead poisoning were even more daunting. They had far greater difficulty than even victims of asbestos-related diseases in proving the traditional individualized connection between a particular victim and a specific manufacturer. Counsel simply had no practical way to prove which manufacturers' products were those to which the client had been exposed. The doctrines that asbestos attorneys used to overcome the individualized causation requirement—substantial factor, concurrent causation, and alternative causation—did the lead-poisoned child no good.

PART II

PUBLIC PRODUCTS LITIGATION AS A RESPONSE TO REGULATORY FAILURE

CHAPTER 4

The Seeds of Government-
Sponsored Litigation

The inability of victims of tobacco-related illnesses and childhood lead poisoning to recover against manufacturers by using the usual product liability claims might have spelled the end of such litigation. Instead, four developments occurring between the mid-1960s and the mid-1980s set the stage for a new form of litigation, actions brought by states and municipalities against product manufacturers, which began in 1995. During this era, the emergence of both modern environmental law and automobile crashworthiness cases afforded attorneys and public health advocates new perspectives on how to pursue cases involving product-caused diseases. Further, beginning with *Borel v. Fibreboard Paper Products Corp.* in 1974, a discrete subset of plaintiffs' personal injury attorneys honed their abilities to litigate mass products tort claims while handling hundreds of thousands of claims against asbestos manufacturers. In the process, they also garnered huge war chests for future litigation against well-financed industries. Both byproducts of the asbestos litigation cycle were necessary for the eventual assaults on the tobacco and lead pigment industries. Finally, public health advocates, frustrated by the failure of legislatures and regulatory authorities to address public health problems caused by such products as cigarettes, handguns, and lead pigment, increasingly saw litigation against product manufacturers as the last, best hope.

PRODUCTS AS CAUSES OF ENVIRONMENTAL HARM

Perhaps the most important precursor of the newly invigorated attempts to hold manufacturers of tobacco and lead pigment/paint liable during the

1990s and the twenty-first century was the recognition that a product that caused public health problems could be viewed as an environmental threat as well as the object of a series of commercial transactions that caused a clustering of individual illnesses through product exposure. To a large extent, whether the health problems of multiple victims are seen as a matter of public health and therefore environmental law or merely as a series of individual harms, each of which was inflicted by a product, determines if it is legitimate to ignore the obstacles to recovery that exist under traditional products liability law. For example, should the millions suffering from tobacco-related illnesses, particularly those whose illnesses were caused by secondhand smoke, be relegated to individual products liability actions, where it is clear that they most likely will not be able to recover? Or is there another lens through which to view such cases, one with its origins in environmental law, which categorizes carcinogenic cigarette smoke, at least for the nonsmoker, as just one more type of air pollution? Is the victim of childhood lead poisoning limited to a products liability action, or is his claim really one that categorizes the presence of lead-based paint hazards in his home as an environmental hazard? It is unlikely that environmental activists in the 1960s, 1970s, or 1980s ever envisioned either the application of principles of environmental law to product-caused diseases or the dissonance in the law that would result when such principles clashed with the more traditional law governing products liability. Nevertheless, mass products attorneys and state attorneys general saw the opening created by the emergence of environmental law when they began filing actions on behalf of states and municipalities against product manufacturers in the mid-1990s.

The Emergence of Modern Environmental Law

Federal legislation enacted during the 1960s and the 1970s established the modern federal regulatory framework for protecting the environment, which, with modifications, continues to exist today. In quick order, Congress passed a series of acts that, for the first time, mandated comprehensive national regulation of the environment, including the National Environmental Policy Act,[1] the Clean Air Act,[2] the Clean Water Act,[3] the Safe Drinking Water Act,[4] the Toxic Substances Control Act (TSCA) of 1976,[5] the Resource Conservation and Recovery Act of 1976,[6] and the Comprehensive Environmental Response, Compensation, and Liability Act (CERCLA),[7] as well a number of other others. In addition, the Environmental Protection Agency was established by executive order.[8]

Several of these new federal regulatory approaches directly regulated toxic products, such as pesticides, asbestos products, and lead pigment. For example, in 1992, Congress enacted the Residential Lead-Based Paint Hazard Reduction Act,[9] amending TSCA and adding a federal legislative framework for reducing the incidence of childhood lead poisoning resulting from exposure to lead-based paint hazards. Similarly, amendments to the Safe Drinking Water Act authorized regulations governing the amount of lead in drinking water.[10] Of course, the statutes requiring health warnings to be included on cigarettes were enacted during the 1960s and reflected parallel concerns about the impact of hazardous substances on public health.[11]

In terms of its significant, albeit indirect, impact on the development of the law governing litigation against manufacturers of products that cause disease, probably the most important of the federal environmental legislative enactments was CERCLA, or, as the statute often was referred to at the time, "Superfund." Unlike most environmental statutes, CERCLA does not regulate conduct creating pollution. Instead, it governs the enforcement authority and funding mechanisms for cleaning up hazardous waste sites and for responding to spills of hazardous substances. Congress enacted CERCLA in the face of widespread publicity about the Love Canal hazardous waste site, which I discuss later in this chapter. In many ways, because the liability provisions of CERCLA enable the federal government to seek the costs of remediating the hazardous waste site from those parties that contributed to causing it, these provisions more closely resemble a common-law tort claim than they do the regulatory provisions of most federal environmental statutes. Most important for our purposes, under CERCLA, the government is able to seek reimbursement of cleanup expenses not only from the current owner and operator of the property where the hazardous waste site lies but also from those whose actions causally contributed to the existence of the hazardous waste site, including past owners and operators of the site, those who generated the hazardous wastes and arranged for them to be deposited there, and those who transported the wastes to the site.[12] The analogy to the contributions of manufacturers of lead pigment and lead paint in creating lead-based paint hazards, even if they are not currently in possession or control of lead-based paint hazards, is apparent.

Speaking more generally, the development of the federal environmental regulatory system has impacted litigation against manufacturers of tobacco products and manufacturers of lead pigment in a number of ways. First, the development of the modern federal environmental regulatory system clearly placed the government—specifically the federal government—in the position of protecting citizens from the health risks resulting from exposure to environmental harms. That statement sounds almost axiomatic in the twenty-first century—after all, is that not what governments do? Yet, until the 1970s, protection from environmental threats lay largely in the hands of states and localities or, often more realistically, in the hands of the victims or potential victims themselves, who were required to turn to the courts to seek redress through common-law tort claims, such as those for private nuisance. It is difficult to overstate the extent to which the federal government's assumption of the role of protector from environmental threats was a change from previous American practice. Even more surprising was the government's new role under CERCLA, in which it pursued financial recompense from polluters for cleanup costs. Unbeknownst to lawmakers at the time, this new role of the federal government would become a precedent decades later for state governments pursuing litigation against product manufacturers, notably the manufacturers of cigarettes and lead pigment, for the harms caused by their products.

Second, federal environmental regulation obviously created the field of environmental law and lawyers specializing in it, a bar prone to pursue litigation in order to eliminate the public health risks posed by toxic substances. As David Sive, one of the original leaders of the environmental law movement, has remarked, "litigation has been more important in the development of environmental law than in any other body of public law."[13] Environmental litigation forced the government to enact regulations when it was reluctant to do so.[14] In short, environmental litigation has exhibited a remarkably powerful and creative impact on public policy, an impact that has inspired and given hope to attorneys from a different professional realm, plaintiffs' personal injury attorneys, who heretofore had been frustrated in their unsuccessful efforts to hold product manufacturers accountable for public health problems.

The third significant way that the emergence of environmental law affected subsequent actions against manufacturers whose products caused latent diseases was the manner in which courts interpreted causation in environmental cases resting on either private nuisance or CERCLA. As shown in chapter 2, many victims of product-caused diseases were traditionally unable to recover because they could not identify the manufacturer(s) of the fungible products that caused their respective harms. Judge Wisdom's opinion in *Borel v. Fibreboard Paper Products Corp.* was a breakthrough in allowing individual victims of product-caused disease to recover, but it usually was confined to those very limited circumstances where a small number of manufacturers each exposed the victim to a risk and where all the manufacturers were before the court.

Causation doctrines that emerged during environmental litigation brought under private nuisance law in the 1970s and 1980s were even more flexible than *Borel* in allowing victims of environmental harms to recover without proving the identity of the specific tortfeasor that caused any particular victim's harm. Historically, under the law of private nuisance, governing claims for damages to another landowner's property, courts had held that a plaintiff was obligated to show how much harm each tortfeasor contributed in order to recover from any of them.[15] If the plaintiff could not meet this burden, as was often the case with claims alleging air or water pollution, recovery was denied.[16] However, in a series of environmental cases during the 1970s, courts reached the opposite conclusion and imposed joint and several liability in private nuisance cases.[17] For example, in 1974, in *Michie v. Great Lakes Steel Division*, where three corporations operated seven plants that emitted noxious gases that combined and caused the plaintiffs' indivisible harms, the federal Court of Appeals for the Sixth Circuit held that under Michigan law, the plaintiffs should be able to recover against all three corporations on a joint and several liability basis, even if plaintiffs could not show what portions of their harms resulted from the emissions of each specific plant operator. To hold otherwise and deny recovery to the plaintiffs because they could not prove the specific amounts of damages caused by each polluter would result, said the court, in "manifest unfairness."[18] Similarly, even though the statutory language of CERCLA did not explicitly state that liability under the act was to be joint and several where the harm was indivisible and could not be reasonably allocated to the various responsible polluters, courts reached that result when interpreting the act.[19]

The consequence of holding polluters jointly and severally liable for creating an environmental nuisance, whether under private nuisance law or CERCLA, has important implications for actions against product manufacturers in which plaintiffs allege that cigarettes and lead paint cause cancer and childhood lead poisoning, respectively. If the accumulation of multiple instances of product-caused latent diseases experienced by many individual members of the public is redefined as a collective public health or environmental problem, some courts would view the resulting public health problem as an indivisible harm. Viewing the victims' harms through the lens of environmental law instead of through a more traditional lens of products liability enables a collective plaintiff, such as state government, to hold multiple and indeterminate product manufacturers jointly and severally liable for creating the indivisible, collective harm. As a result, the state recovers without proof that any specific manufacturer contributed to the harm experienced by any particular victim.

BACK TO THE FUTURE: PUBLIC NUISANCE AS AN
ENVIRONMENTAL TORT

The litigation brought by states and municipalities against product manufacturers during the past twenty years did not rely directly on any of the environmental statutory enactments of the 1960s, 1970s, or 1980s. The tort most commonly relied on by states and municipalities suing lead pigment manufacturers—and probably those suing tobacco manufacturers as well—was public nuisance. This tort has origins more than eight hundred years old but its more recent history is inextricably intertwined with the statutory expansion of environmental law.

In the field of law known as torts, where vague definitions, rules, and doctrines abound, perhaps no other tort is as vaguely defined or poorly understood as public nuisance. The Florida Supreme Court recently proclaimed, for example, that "a public nuisance may be classified as something that causes any annoyance to the community or harm to public health."[20] In 1982, the Illinois Supreme Court acknowledged, "The concept of common law public nuisance does elude precise definition."[21] The more than eight-hundred-year history of public nuisance law suggests that public nuisance's use as a means to end environmental hazards, such as air pollution and water pollution, is supported by precedent, but holding manufacturers liable for the production and distribution of their products under such a theory of liability is more questionable.

Historically, conduct found to be a public nuisance was treated, more often than not, as a crime and not a tort.[22] If something was declared to be a public nuisance, government authorities not only treated it as a crime but had the power to terminate the defendant's conduct by obtaining an injunction. As a result, state legislatures during the nineteenth century declared a wide range of conduct to be public or "common" nuisances, including participating in lotteries and other forms of gambling and wagering, keeping a disorderly house or tavern, enabling prostitution, and using obscene language. These statutes allowed public authorities to terminate the defendant's ongoing conduct by obtaining an injunction from a court, authority they would not ordinarily have possessed under an ordinary criminal statute.

Aside from such offenses against public morals and from other matters declared by the legislature to be public nuisances, the other principal category of public nuisance offenses was one in which the defendant interfered with what might be regarded as "public property." These cases were principally ones involving the obstruction of highways[23] or diversion of navigable waterways.[24] Beginning as early as the 1840s, the onset of industrialization prompted public nuisance claims alleging new types of injuries—namely, water pollution[25] and air pollution.[26] Most often, courts failed to explicitly explain the rationale for extending the public or common nuisance cause of action to these new fact patterns. However, in *People v. Gold Run Ditch & Mining Co.*, a case involving water pollution, the California Supreme Court traced the lineage of the use of the public nuisance cause of action for water pollution to earlier cases involving obstructions to public waterways. The court reasoned that "to make use of the banks of a river for dumping places . . . is . . . an unauthorized invasion of the rights of the public to its navigation," constituting "a public nuisance."[27] The precedents supporting use of the public nuisance action against those engaged in polluting the air are somewhat more difficult to trace. Presumably, the justification for extending public nuisance to these fact patterns lay either in the long tradition of finding noxious trades to be public or common nuisances[28] or in the fact that courts often found the "unpleasant and unwholesome vapors"[29] resulting from the fouling of water to be a public nuisance.[30]

Early Twentieth-Century Parens Patriae *Environmental Litigation*

In the early twentieth century, the U.S. Supreme Court repeatedly sanctioned the use of public nuisance claims in cases brought by states against their sister states alleging environmental harms. In the first of these, *Missouri v. Illinois*,[31] the State of Missouri sought to prevent officials of Illinois and the City of Chicago from discharging sewage through a newly constructed canal that would reverse the flow of the Chicago River and cause the sewage to be carried into the Mississippi River. Justice Holmes, writing for the Court, ultimately denied the requested injunctive relief because of both a failure of proof and Missouri's own practice of discharging untreated sewage into the Mississippi River. More pertinently, however, the Court at least implicitly held that nuisance was a proper basis for seeking relief under federal common law.

During the very next year, 1907, Justice Holmes more explicitly held that a state had standing to sue as *parens patriae* to vindicate claims arising from environmental harms on a public nuisance theory. In *Georgia v. Tennessee Copper Co.*,[32] Georgia had sought an injunction against a Tennessee manufacturing company for the discharge of noxious gases over its territory. In his classic opinion for the Court, Justice Holmes described the state's quasi-sovereign interests as follows: "In that capacity the State has an interest independent of . . . its citizens, in all the earth and air within its domain. It has the last word as to whether its mountains shall be stripped of their forests and its inhabitants shall breathe pure air."[33] Two decades later, the Supreme Court entertained two common-law public nuisance actions brought by the states of New York and New Jersey against each other, with each state suing the other for dumping sewage and garbage into New York Bay.[34]

To a large extent, the enactment of federal environmental legislation during the 1960s, 1970s, and 1980s created federal regulatory regimes that supplanted the federal common law of public nuisance as the basis of actions to prevent environmental degradation. In a series of decisions subsequent to the legislative enactments, the U.S. Supreme Court held that provisions of various environmental acts preempted federal and state public nuisance actions seeking to abate environmental hazards that also were regulated under the environmental legislative enactments. In these opinions, the Supreme Court expressed its view that Congress and its authorized administrative agencies are in a better position to regulate environmental harms than are courts through the use of public nuisance law. The legal trend began with the Supreme Court's decision in *City of Milwaukee v. Illinois*,[35] in which the Court held that the 1972 amendments to the Federal Water Pollution Control Act preempted an action brought by the State of Illinois against Milwaukee city officials whose efforts to abate overflows of sewage resulted in their discharge into Lake Michigan and its tributaries, which Illinois asserted constituted a public nuisance. The Court found that Congress had "occupied the field through the establishment of a comprehensive regulatory program supervised by an expert administrative agency" and that it had "not left the formulation of appropriate federal standards to the courts through application of often vague and indeterminate nuisance concepts."[36] Further, the Court reasoned that if the judicial system attempted to regulate water pollution, it would find "the technical problems difficult," and it noted that Congress had characterized such efforts in the past as "'sporadic' and 'ad hoc.'"[37]

In other instances, courts have declined to allow common-law public nuisance claims concerning the environment to proceed in view of the existence of a comprehensive federal regulatory system, on the grounds that such claims intrude on matters properly decided by the political branches of government and therefore are nonjusticiable. Most relevantly for our purposes, in *People v. General Motors Corp.*,[38] the State of California sued various automobile manufacturers for creating or contributing to global warming, which it identified as a public nuisance under federal common law. The trial court held that these public nuisance claims were nonjusticiable because they would involve the court "injecting itself into the global warming thicket" and would "require an initial policy determination of the type reserved for the political branches of government."[39] Further, stated the trial court, judicial entry into the global warming controversy "would have an inextricable effect on interstate commerce and foreign policy—issues constitutionally committed to the political branches of government."[40]

In short, during the early twentieth century, the U.S. Supreme Court had pioneered a federal common law of public nuisance in the environmental context. The onset of federal regulatory regimes addressing key aspects of environmental degradation relegated the federal common law of public nuisance to background status in these specific arenas. However, for many categories of environmental harms not comprehensively regulated, public nuisance remained a viable remedy in addressing environmental harms.

Public Nuisance, the Restatement (Second) of Torts, and the Environmental Movement

William Prosser, one of the most influential tort scholars of the mid-twentieth century and the reporter for the *Restatement (Second) of Torts*, viewed the tort of public nuisance less expansively than one might have expected from the Supreme Court's early twentieth-century opinions. When he first submitted the proposed draft of the public nuisance provisions of the *Restatement (Second) of Torts*, he sought to limit recovery to those situations in which there had been a violation of a criminal statute. The available appellate decisions during the preceding decades often, but not always, supported this proposition.[41] Section 821B of his proposed draft provided, "A public nuisance is a criminal interference with a right common to all members of the public."[42] Under section 821C, liability in tort was limited "only to those who have suffered harm of a kind different from that suffered by other members of the public exercising the public right." When first brought before the American Law Institute (ALI), these provisions, after some discussion, were easily approved.

At the very next meeting of the ALI a year later, however, members of the institute who saw Prosser's proffered language as a way to restrict the use of public nuisance in recent environmental cases sought to have it reconsidered. John P. Frank, a practitioner from Arizona, eloquently led the charge for reconsideration.

> What is happening at the moment all over America is that the people are seeking to deal with pollution of air and of water and of land, that in this connection a developing body of law is beginning to formulate which is breaking the bounds of traditional public nuisance. . . .

Pollution may be a crime against God and its [sic] nature, but it is not usually a crime against the laws of the state, so that by putting in that definition we make it impossible to reach the problem of the black cloud of filth which hangs in the air over my community and, I suspect, yours.[43]

In an article published after the debate, two staff attorneys for the Natural Resources Defense Council argued that it was important to have public nuisance available as grounds for a remedy, even if there were no criminal violation, in order to protect "public" interests other than the "private" interests protected by the tort of private nuisance.

Modern courts should not permit the right to challenge pollution to . . . focus only on the effect on a plaintiff's enjoyment of his land. . . .
. . . [P]rivate nuisance actions may be reserved for situations in which the principal problem is invasion of property interests, and environmental threats to the health, comfort, and beauty of the community will be treated—as they logically should—as public nuisances.[44]

The two public nuisance sections were sent back to the reporter for further study and reconsideration.

In May 1971, the ALI adopted a version of section 821B of the *Restatement (Second) of Torts* that rejected Prosser's argument that liability in public nuisance required a violation of criminal law. Section 821B defined public nuisance as "an unreasonable interference with a right common to the general public." The restatement provision then listed three factors for courts to consider in deciding whether "an interference with a public right is unreasonable":

(a) Whether the conduct involves a significant interference with the public health, the public safety, the public peace, the public comfort or the public convenience, or
(b) whether the conduct is proscribed by a statute, ordinance or administrative regulation, or
(c) whether the conduct is of a continuing nature or has produced a permanent or long-lasting effect, and, as the actor knows or has reason to know, has a significant effect upon the public right.[45]

The rejection of Prosser's substantially more restrictive language following the strong reaction from environmentalists and their supporters was intended to make the tort available against various forms of environmental pollution. However, most of the cases the environmentalists intended to cover, such as water pollution and air pollution, fell within the historical understanding of the tort, even if they did not fall within Prosser's restrictive interpretation of these precedents. The language of the *Restatement* and its reliance on a multifactorial approach to determine when a court should impose liability for a public nuisance went beyond what was necessary to combat the more traditional forms of pollution and was destined to open the courthouse doors decades later to those who sought to use public nuisance as a principal cause of action against manufacturers of products that caused public health problems.

Love Canal

Beginning in the 1970s and 1980s, both the federal government and several states successfully pursued public nuisance actions in the environmental context, against defendants who had created or contributed to the creation of an injurious condition at some point in the past, even if they were no longer conducting such activity. These cases broke the heretofore prerequisite link between the defendant's ability to abate the injurious condition, which required its control of the instrumentality causing the harm, and the court's ability to find the defendant liable for public nuisance.

This expansion of liability for public nuisance perhaps can best be explained by examining *United States v. Hooker Chemicals & Plastics Corp.*, or *Hooker II*,[46] one of the many opinions in the Love Canal litigation of the 1980s. Beginning in 1942, the Hooker Electrochemical Company disposed of chemical wastes by dumping them in an abandoned canal, approximately three-quarters of a mile long, located in Niagara Falls, New York. The City of Niagara Falls also dumped "municipal wastes" into the canal. In 1953, Hooker sold the site to the city's Board of Education, and a written agreement between the parties informed the board that chemicals had been disposed of in the old canal. The city agreed to assume all liability for claims brought by any party resulting from the chemical wastes. An elementary school and a number of private homes were built on or near the site during the 1950s and 1960s. Subsequently, "hazardous substances were . . . detected in the surface water, groundwater, soil, the basements of homes, sewers, creeks, and other locations in the area surrounding the Love Canal landfill."[47] By the late 1970s, the residents of Love Canal had become the focus of nationwide attention.[48] Both the federal and state governments declared the site a public health emergency, and nearly a thousand families were evacuated from the area. To confine the twenty-one thousand tons of chemical waste that remained at the site, the Environmental Protection

Agency capped the canal with a thick covering of clay, installed a drainage system that prevented the runoff of chemically contaminated water, replaced the sewage system that previously had drained the area, and surrounded the site with a fence. Federal and state authorities spent four hundred million dollars to clean up the site. They then sued in federal district court to hold Occidental Chemical Corporation liable, under both CERCLA and the common law of public nuisance, for their costs in cleaning up the toxic chemical wastes stored in the Love Canal more than thirty-five years earlier by Occidental's corporate predecessor, Hooker.

In an earlier opinion in the same case, *Hooker I*,[49] the trial court had found Occidental liable under CERCLA for the costs incurred by the federal and state governments in cleaning up the Love Canal site. Section 107(a) of CERCLA outlines four groups of potentially liable parties, including "past owners and operators of hazardous waste facilities."[50] The court rejected Occidental's argument that it could not be held liable under the statute for past acts and explicitly found that "CERCLA's legislative history suggests that the statute was enacted as a means of compelling the waste disposal industry to correct its past mistakes and to provide a solution for the dangers posed by inactive, abandoned waste sites."[51]

In *Hooker II*, the court borrowed the CERCLA-based analysis from *Hooker I* and applied it to the common-law public nuisance claims, finding that Occidental was liable for "creation of the 'public health nuisance.'"[52] It did so despite the absence under the public nuisance claim of any statutory provision, similar to the one that the court had legitimately relied on in *Hooker I*, governing the liability of those no longer in control of the property giving rise to the public nuisance. Occidental was not in control of the toxic waste site at the time of the litigation, as had typically been required under the common law. The court was able to rely on the reasoning of an earlier opinion in a New York state trial court decision, *State v. Schenectady Chemicals, Inc.*, which atypically had held that "while ordinarily nuisance is an action pursued against the owner of land for some wrongful activity conducted thereon, 'everyone who creates a nuisance or participates in the creation or maintenance . . . of a nuisance are [sic] liable jointly and severally for the wrong and injury done thereby.'"[53] The New York trial court had been quite explicit in identifying both the sympathetic fact patterns and the policy analysis that drove its conclusion.

> The Common Law is not static. . . . The modern chemical industry, and the
> problems engendered through the disposal of its byproducts, is, to a large

extent, a creature of the twentieth century. Since the Second World War hundreds of previously unknown chemicals have been created. The wastes produced have been dumped, sometimes openly and sometimes surreptitiously, at thousands of sites across the country. Belatedly it has been discovered that the waste products are polluting the air and water and pose a consequent threat to all life forms. Someone must pay to correct the problem.[54]

The court in *Hooker II* also acknowledged that the boundaries of the term *nuisance* were very open-ended: "the term nuisance . . . means no more than harm, injury, inconvenience, or annoyance."[55]

Hooker II and *Schenectady Chemicals* are good examples of cases that blur the distinction between public nuisance and private nuisance. Both courts borrowed, from the law of private nuisance, the liability-expanding principle that a defendant can be held liable for creating a nuisance even if the defendant does not continue to own or possess the land or otherwise carry on or maintain the nuisance at the time of the litigation.[56] Other courts have approached public nuisance cases in the same way, also guided by private nuisance precedents that do not require the defendant to own or occupy the property where the injurious condition exists or otherwise to continue to maintain or carry on the activities constituting the nuisance.[57]

In this manner, a number of courts abandoned the traditional limits on public nuisance actions that arose from the historical core purpose of the tort, the termination of the harmful conduct. As a result, the courts in *Hooker II* and *Schenectady Chemicals* expanded the possible range of defendants in public nuisance actions in environmental cases, far beyond those in the past. Previously, the remedy in a public nuisance action had been an injunction ordering the defendant currently in control of the property or instrumentality causing the public nuisance to terminate its conduct that was injuring the public health or welfare. Now, courts transformed public nuisance claims into a means of forcing financial contributions to the government from any party that had causally contributed to any of the wide variety of conditions that fall within the amorphous definition of the tort. In so doing, these courts planted the seeds that eventually would create what some courts would see as an environmentally based body of products liability law that served as an alternative to more traditional products liability claims, such as strict products liability, negligence, or misrepresentation, at least in those instances when state and local governments sued manufacturers and alleged that their products had caused public health problems.

PRODUCTS LITIGATION AND AUTOMOBILE SAFETY AS A PUBLIC HEALTH ISSUE

During the same era, public health experts began to define death and disablement resulting from highway traffic accidents as a public health problem. In the process, they shifted the focus of efforts to prevent injuries resulting from traffic accidents from one resting solely with the driver to a broader perspective that included such factors as highway and automobile design. Accordingly, a few years later, when plaintiffs' attorneys began to sue automobile manufacturers for automobile design defects that increased the injuries resulting from traffic accidents, they became the first attorneys seeking to hold manufacturers liable for the costs of product-caused public health problems.

With the decline of mortality and morbidity caused by infectious illness during the first half of the twentieth century, previously described in chapter 1, accidental death and disablement were increasingly defined as a public health issue in their own right.[58] Attention focused on motor vehicle accidents. In 1960, John F. Kennedy, then a senator, characterized traffic accident prevention as "one of the greatest, perhaps *the* greatest of the nation's public health problems."[59] The traditional assumption had always been that accidents resulted from bad drivers or at least from bad driving. From this perspective, the way to minimize the harm to society resulting from traffic accidents was to force drivers to be more careful, through vigorous enforcement of traffic laws and drivers' education. As late as 1960, the recommendations of the President's Committee for Traffic Safety emphasized compliance with traffic laws.[60] This focus was quite consistent with the American legal system's approach to preventing accident injury at that time, one that typically required negligence or fault as a prerequisite for liability. Yet studies showed that there was no significant correlation between the level of driver skill and the likelihood that a driver would be involved in an accident.[61]

Looking at automobile traffic injuries from a public health perspective opened up new avenues for reducing the costs of accidents. In 1967, William Haddon, a public health expert then serving as director of the National Highway Safety Bureau, outlined three approaches for reducing the impact of traffic accidents.[62] The first of these involved avoiding the initial accidental collision itself, whether by improving driving behavior or through altering the driving environment by designing safer highways or improving signage. The second approach, of greatest concern here, involved eliminating or reducing the impact between the automobile occupant's body and the interior of his own automobile or whatever other objects the occupant might collide with as a result of the accident—perhaps another car or a bridge abutment. Haddon's third approach sought to mitigate the consequences of the harm from the injuries sustained from the impact—for example, through provision of better emergency response services.

Haddon's second factor pointed those interested in lessening the consequences of automobile accidents to the design of the automobile itself. In 1960, Daniel P. Moynihan, later to serve as a U.S. senator, argued that "enormous improvements could be made in the interior design of automobiles so as to minimize the injuries which result from accidents to make accidents safe."[63] He suggested that 75 percent of all fatalities could be prevented by such changes. As Ralph Nader pointed out in his devastating critique of the tendency of Chevrolet Corvairs manufactured during the early 1960s to "go out of control and flip over," automobile design often contributed to the occurrence of the traffic accident in the first place.[64]

The new perception of automobile accidents as a public health problem led to congressional enactment of the National Traffic and Motor Vehicle Safety Act in 1966. The act authorized the Department of Transportation to require "minimum" safety standards for all automobiles sold in the United States.[65] In 1970, because fewer than 20 percent of all drivers and passengers at the time were found to fasten their seat belts, the department's National Highway Traffic Safety Administration (NHTSA) began promulgating requirements for passive restraint devices, such as automatic seat belts or air bags.[66] For example, one of NHTSA's early attempts to lessen the severity of injuries resulting from auto accidents was an interlock device designed to prevent a driver from starting a car unless the seat belts were fastened. These devices met with enormous consumer resistance, and drivers frequently discovered means of bypassing them. In the short term, at least, strong notions of individual freedom trumped sound public health policy. Conservative Republican senator James L. Buckley proclaimed his opposition to the mandated interlock systems by arguing, "I view such coercive measures as the interlock as an intolerable usurpation by Government of an individual's rights under the guise of self-protection."[67] Despite estimates that interlock systems would prevent 340,000 injuries and save seven thousand lives annually,[68] opponents regaled Congress with tales in which drivers were unable to start their stalled cars in the face of an onrushing locomotive and in which women were unable to start their cars in time to flee rapists.[69] As a result, for decades, both Congress and the courts blocked the efforts of NHTSA to require air bags or other passive restraint systems.[70]

During this same period of time, victims of automobile accidents increasingly turned to the courts to sue auto manufacturers for design decisions that they alleged caused their injuries. These victims not only sought compensation for their injuries; perhaps frustrated by the inability of Congress and NHTSA to regulate automobile safety through the legislative and regulatory process, they also hoped that litigation would create incentives for auto manufacturers to design safer cars. As previously described, Calabresi, Posner, and other law and economics scholars view loss minimization as an important goal of tort liability. In this context, holding automobile manufacturers liable for injuries exacerbated by unsafe designs is seen as providing financial incentives for manufacturers to design automobiles that minimize both the frequency of accidents and the severity of injuries resulting from those accidents that do occur.

In the first subset of cases seeking to hold manufacturers liable for the design of their products, the design flaw is a cause of or causally contributes to the initial collision itself.[71] For example, in the early 1960s, design defects present in the Chevrolet Corvair caused the car to be difficult to steer and to sometimes flip over.[72] In other well-known examples, certain sport utility vehicles were unstable and "rolled over."[73] The automatic transmission in some Ford vehicles in the 1970s and 1980s tended to slip into "reverse" when the driver had placed it in "park."[74] Both numerous lawsuits involving these products and, more important, the accompanying publicity caused manufacturers to change their designs and hence ended their contributions to the public health threat of vehicular accidents.

Cases concerning so-called crashworthiness or second-collision automobile design trace their origins even more directly to the conception of automobile accident injuries as a public health problem. In these cases, the injured victim argued that the design of the automobile unreasonably failed to protect him from injury or exacerbated the extent of his injuries when it was involved in an accident. David G. Owen, a leading products liability scholar, graphically describes the possible consequences as follows:

> A vehicle's crashworthiness is improved or diminished by the extent that its structure can absorb the forces of a crash without collapsing into the passenger compartment against the occupants; that its dashboard and head restraints are appropriately padded, rather than being made of solid steel; that the glass in its windows crumbles harmlessly, rather than fracturing into lethal slivers; that the steering wheel telescopes to absorb the force of a collision with the driver's chest, rather than remaining rigid as a wall; . . .

that the fuel tank is located in a safe position and securely protected against the varying insults it may encounter in different types of collisions; that the doors and windows have sufficient latches to hold them closed in accident situations, to keep the occupants contained, rather than popping open and allowing the occupants to be flung outside; that airbags protect the occupants from injury, rather than activating spontaneously or with explosive force; and that safety belts, harnesses, and head rests effectively restrain and protect the occupants rather than serving as instruments of death.[75]

In its 1966 opinion *Evans v. General Motors Corporation*,[76] the federal Court of Appeals for the Seventh Circuit rejected the "crashworthiness doctrine," holding that a "manufacturer is not under a duty to make his automobile accident-proof or fool-proof,"[77] because the intended purpose of an automobile does not include its participation in collisions. However, within a few years, American courts rejected the position of the *Evans* court and accepted the crashworthiness doctrine. In the leading opinion of *Larsen v. General Motors Corp.*,[78] the Court of Appeals for the Eighth Circuit found that automobile accidents and the resulting injuries were "readily foreseeable."[79] Hence, the court concluded that manufacturers had a duty of reasonable care to minimize the effect of accidents.[80]

In one important example of litigation seeking to make automobiles more crashworthy, the U.S. Supreme Court held that the action was blocked by a NHTSA regulation. The plaintiffs in *Geier v. American Honda Motor Co.*[81] alleged that the serious injuries that Alexis Geier sustained when the Honda Accord she was driving struck a tree would have been significantly lessened if her car had been equipped with air bags. The Supreme Court held that this state common-law tort action was impliedly preempted by the federal regulatory framework that envisioned a mix of various passive restraint options during a transitional period, including not only air bags but also other devices. According to the court, the NHTSA regulation that allowed a variety of approaches to minimizing damage to car occupants after auto collisions was designed to allow car manufacturers to experiment with varying approaches and also was designed to give the public a longer period of time to adjust to the idea of mandatory air bags. Of course, only a few years later, in 1997, air bags became mandatory in all automobiles.[82] The *Geier* opinion represents an early conflict between product litigation intended in part to minimize the extent of a product-caused public health problem and the regulatory regime established by Congress and regulatory agencies. The efforts of the political branches prevailed.

There is little consensus about whether product litigation actually improves the safety of automobiles or other products or, for that matter, reduces the public health risks posed by gun and cigarette manufacturers.[83] On one hand, public health experts claim that product liability is "an important tool for the prevention of injuries."[84] They point to a telephone survey conducted as a part of a 1977 study by the Federal Interagency Task Force on Product Liability, showing that a majority of large business firms had responded to the risk of product liability by implementing efforts at loss prevention.[85] On the other hand, when considering overall effect, Robert A. Kagan concludes that "the spotty existing evidence suggests that American tort law has an erratic effect on safety."[86]

In their classic study of federal regulation of automobile safety, Jerry L. Mashaw and David L. Harfst conclude that the liability system is unlikely to effectively regulate automobile manufacturers' decisions regarding the safety features of the automobiles they design.[87] They find that legal decisions rarely yield consistent results of a quantifiably sufficient magnitude to justify design changes, and they characterize liability judgments as mere background noise. According to Mashaw and Harfst, the financial impact of liability judgments rarely, if ever, exceeds a very modest percentage of manufacturers' total revenues, and the impact generally is widely dispersed among manufacturers and market segments. Further, engineering design teams seldom take such feedback into account. Exceptions certainly do occur. For example, a very large and well-publicized punitive damage award against Ford Motor Company, resulting from a design flaw in the gas tank in the Ford Pinto that caused the tank to explode when struck even by another car driven at slow speed,[88] clearly tipped the balance. When combined with similar ongoing lawsuits, the award resulted in a product recall, retrofitting of the existing tanks, and redesign of the problematic gas tanks in Pintos manufactured subsequently.

Regardless of whether products liability in fact had a major impact on the safety of automobile design during the past generation, plaintiffs' lawyers and many members of the public certainly perceived that it did. In all probability, market pressure generated by lawsuits alleging that design features of automobiles were unsafe probably played a more important role in encouraging safety improvements than did the litigation itself. In any event, this cycle of litigation concerned with automobile safety design was the first major campaign to use product liability law to ameliorate what the reformers of earlier decades had identified quite accurately as a public health problem. It would not be the last.

THE EMERGENCE OF THE MASS PRODUCTS LIABILITY BAR

In the three decades following *Borel v. Fibreboard Paper Products Corp.*, the first success in holding asbestos manufacturers liable, more than six hundred thousand claimants alleged either that they suffered from asbestos-related diseases or that they had been exposed to such diseases and experienced changes in their lungs,[89] more than fifty-four billion dollars was paid in compensation and litigation expenses, and at least sixty defendant companies filed for bankruptcy as a result. Perhaps the single most significant change in the legal landscape arising from this initial breakthrough by asbestos plaintiffs in holding manufacturers liable was the arrival on the scene of a relatively small group of mass products liability firms that fundamentally changed the handling of claims alleging that manufacturers' products resulted in public health problems. From the perspective of plaintiffs' counsel, litigation against asbestos manufacturers was more complex and resource-intensive than the more typical fare of automobile accidents or even medical malpractice claims. The handling of such claims therefore required a highly specialized and sophisticated plaintiffs' bar.[90] The filing of asbestos claims exploded during the late 1970s and early 1980s, and by 1985, ten law firms accounted for one-quarter of all such claims. The trend toward concentration continued, and by 1992, the top ten firms (a group that substantially overlapped with, but was not identical to, the leading firms in 1985) accounted for three-quarters of all new claims.

These specialized firms handled the repetitive claims of asbestos-caused diseases much differently than the manner in which a series of automobile accidents or other tortious harms had been addressed in the past. By handling these mass claims in bulk, plaintiffs' asbestos lawyers focused on the aggregate picture when planning litigation and settlement strategies. The emergence of these firms was a necessary precondition for later litigation filed by states and municipalities against manufacturers of cigarettes, handguns, and lead pigment. After the eruption of massive asbestos claims and their eventual settlement, these firms not only offered the litigation expertise and sophistication to tackle tobacco manufacturers; they also possessed the financial resources necessary to bankroll upfront the enormous costs of such a daunting enterprise. In short, plaintiffs' counsel handling asbestos litigation begat those who filed tobacco litigation,[91] and in turn tobacco litigation counsel begat those who pursued litigation against the manufacturers of handguns and lead pigment.[92] Indeed, in many instances, the same counsel handled the litigation in the three successive litigation cycles.

PRODUCTS LITIGATION AS A PUBLIC HEALTH MEASURE

Inherent in the conception of the science of public health is governmental action that improves health through society-wide measures. Government statutes and regulations assure sanitation in the food preparation of restaurants and the disposal of wastes, as well as mandating childhood immunizations and the wearing of seat belts. Litigation has long been a tool of public health officials. For centuries, state and local governments sued to prevent the community-wide harmful health consequences of malarial ponds and the discharge of toxic pollutants into the water and air. It is not surprising, therefore, that public health activists played a key role in promoting and facilitating litigation against manufacturers of products that contributed to perceived public health crises.

As early as 1986, public health advocate Stephen P. Teret outlined the case for using litigation against manufacturers of products—particularly cigarettes, guns, and automobiles—as a public health measure.

> Unlike rodents and mosquitoes, the modern day vehicles of injury and disease have vested interests, lobbyists and political action committees that sometimes thwart effective legislative and regulatory attempts to enhance the public's health. When this happens, public health advocates have turned to the third branch of government, the judiciary, to seek relief from juries.[93]

Public health advocates pushed states and municipalities to sue product manufacturers and then worked alongside the attorneys general and privately retained counsel in planning litigation strategies and testifying as expert witnesses. By 1995, the American Medical Association advocated pursuing "all avenues of individual and collective redress . . . through the judicial system."[94] In November 1997, during the pendency of the states' actions against the tobacco manufacturers, the American Public Health Association, the world's oldest public health organization, with more than fifty thousand members, issued a resolution in favor of litigation against "manufacturers of products that contain lead." Noting that "the cost of abating the nation's residential lead paint hazards and treating the short- and long-term effects of lead poisoning is many billions of dollars," the resolution recalled "that the principle of 'polluter pays' is well-established in cases of environmental damage and public health problems."[95]

In short, the identity of public health advocates as professionals routinely involved in government regulation of conditions posing widespread health risks led them to endorse and participate in government litigation against product manufacturers. Public health advocates traditionally had turned first to legislative and regulatory solutions. However, when they perceived that such initiatives had failed in addressing the health problems caused by tobacco products and lead-based paint, they became the partners of state attorneys general and mass plaintiffs' attorneys in endorsing litigation against product manufacturers. Like their close cousins, the environmentalists, public health professionals imparted a collective, aggregate, and social perspective on viewing the harms resulting from tobacco products, handguns, and lead pigment, in contrast to the more individually focused, one-on-one lens of traditional tort lawyers. The combined effort of the environmental law community and public health community in reframing mass products torts stands as a major intellectual contribution to the law.

CONCLUSION

The evolution of the common law during the 1960s, 1970s, and 1980s in two disparate litigation cycles not directly related to tobacco and lead pigment litigation paved the way for government-sponsored litigation during later decades. In the environmental context, courts departed from the traditional prerequisite of liability requiring a direct link between a particular victim and a specific tortfeasor and reexamined the law governing the historical boundaries of the tort of public nuisance. In automobile crashworthiness litigation, public interest advocates viewed the design of the automobile as being every bit as much a cause of traffic accidents as were drivers. When attorneys suing automobile manufacturers alleged that the manufacturers' design decisions contributed to traffic accidents, they planted the seed, for the first time, of using products liability litigation to reduce the impact of a public health problem. During roughly the same period of time, the two professional groups that would later serve as architects of government actions against product manufacturers emerged: highly sophisticated and specialized mass products torts attorneys and public health officials who were committed to suing product manufacturers to remedy public health problems. These seeds would germinate for another decade or so before they would sprout as government-sponsored litigation against tobacco and lead pigment manufacturers at the turn of the twenty-first century.

CHAPTER 5

A Failure of Democratic Processes? Legislative Responses to the Public Health Problems Caused by Tobacco and Lead Pigment

The responsibility for assuring that products are safe for consumers is most often handled in the American political structure by legislatures and the administrative agencies they create. To be sure, any time a court orders a manufacturer to pay damages to consumers or others injured by its products, such judgment produces incentives for manufacturers to avoid creating harmful products, particularly if a pattern of damage awards emerges. In the first instance, however, it is the legislature and administrative agencies that provide the *ex ante* macroregulatory framework for products. Similarly, legislatures and their administrative agencies have the primary governmental responsibility for addressing society-wide public health problems.

When state attorneys general and the plaintiffs' mass products tort attorneys retained by the states filed *parens patriae* actions against the manufacturers of cigarettes, handguns, and lead pigment in the waning years of the twentieth century and the first decade of the twenty-first, they often justified their actions by explaining that the ordinary regulatory processes of the political branches had failed. Manufacturers' lobbying and disproportionate campaign contributions, they asserted, left tobacco interests with a stranglehold on legislatures and administrative agencies. It was their intent to bring manufacturers to the table to negotiate a new framework for regulating their products.[1] In this chapter, I explore the record of the legislature and administrative agencies both in regulating tobacco products and in formulating effective governmental solutions to eliminate childhood lead poisoning. Further, I evaluate the extent to which plaintiffs' attorneys and public interest advocates are correct when they assert that *parens patriae* product litigation is warranted because the nonjudicial, political processes have failed.

REGULATION OF TOBACCO PRODUCTS

Federal Regulation

Whether tobacco manufacturers' lobbying and campaign contributions rendered the ordinary regulatory processes ineffective during the three decades preceding the filing of the *parens patriae* litigation is a closer question than state attorneys general and plaintiffs' attorneys would later contend. Despite the pronouncements of those who urged states to sue tobacco manufacturers, Congress, state legislatures, and federal administrative agencies had not ignored issues of tobacco regulation during the generation following the 1964 surgeon general's report.[2] To the contrary, scholars in 1996 concluded, "Over the course of the past three decades, Congress has micromanaged the cigarette labeling and advertising issue."[3] What frustrated public health and antismoking activists and some attorneys general was that the regulatory schemes adopted by the federal and state legislative branches did not go as far as they would have liked, particularly in the early years following the 1964 surgeon general's report.

When the report was issued, the Federal Trade Commission (FTC) responded by indicating its intention to require the now very familiar health warnings on packages of cigarettes. The proposed FTC regulation would have required the following language on all packages and in all advertisements: "cigarette smoking is dangerous to health and may cause death from cancer and other diseases."[4] Tobacco manufacturers lobbied heavily for Congress to prevent the FTC's proposed regulation from becoming effective and were successful when Congress enacted the Federal Cigarette Labeling and Advertising Act of 1965.[5] The act required a weaker warning label on cigarette packages, one that read, "Caution: Cigarette Smoking May Be Hazardous to Your Health." No warnings were required in advertisements, and at least for a period of three and a half years, the FTC was prohibited to mandate warning labels. Medical historian Allan M. Brandt has characterized this congressional action as "an unprecedented attack on the federal regulatory structure of consumer protection."[6] The act also preempted any state or local law or regulation that would have required stronger warning labels.[7] Finally, as previously described in chapter 2, it set the stage for tobacco companies to claim that because smokers were constantly exposed to these warning labels, they assumed the health risks of smoking and could not sue manufacturers for damages. In short, lobbyists for the tobacco industry had transformed a purported public health measure into legislation that protected the industry.

Reacting to a petition from antitobacco activist John F. Banzhaf III, a separate federal regulatory agency, the Federal Communications Commission (FCC), ruled in 1967 that under the provisions of its fairness doctrine, if a television or radio station broadcast cigarette manufacturers' advertisements promoting cigarettes, it had "the duty of informing its audience of the other side of this controversial issue of public importance—that, however enjoyable, such smoking may be a hazard to the smoker's health."[8] The FCC did not literally mandate equal time for antismoking advertisements, but it did require that stations run approximately one antismoking advertisement for every three cigarette commercials.[9] The American Cancer Society and other public health organizations took advantage of the FCC ruling by sponsoring an antismoking campaign. In 1969, the tobacco industry responded by agreeing to end all television advertising, a concession subsequently codified in the Public Health Cigarette Smoking Act of 1969,[10] which banned television and radio advertising of all cigarettes and mandated that cigarette packages contain more explicit warning labels, such as ones that specifically mentioned lung cancer, heart disease, and the potential harm to pregnant women and their fetuses. Of course, the end of broadcast advertising also meant the end of the stations' obligations under the fairness doctrine to run the ads informing listeners of the health dangers of smoking. In 1984, Congress required even more explicit package warnings that identified various health hazards.[11] These rotating warnings remain in effect today.

During this entire period of two decades, Congress kept control of cigarette package warnings and prevented the FTC from engaging in stronger action, as it was inclined to do at several points. During this same time, Congress also explicitly exempted tobacco products from the regulatory provisions of numerous other congressional acts.[12] Few, if any, congressional enactments significantly impacted smoking until 1989, when smoking was banned on domestic airline flights.[13] In short, the federal regulatory approach to smoking between the release of the surgeon general's report and 1990 can best be characterized as tepid.

Environmental Tobacco Smoke, Antitobacco Activism, and State and Local Regulatory Legislation

During the late 1970s and early 1980s, the public began to appreciate the unhealthful consequences of environmental tobacco smoke (ETS), also known as secondhand smoking or passive smoking.[14] Family members of smokers experienced higher cancer rates than those who lived in a smoke-free residence. Employees exposed to smoke in the workplace or passengers in an enclosed airplane similarly may have been at risk. These discoveries dramatically altered the public opinion and political equations concerning smoking. The most powerful argument of tobacco companies—that smoking was a matter of individual choice and that each citizen should be entitled to make his own choice, balancing the risks of smoking with the pleasure of smoking—no longer sufficed.

The discoveries of the dangers posed by ETS energized the antitobacco health movement, which consisted of a few traditional public health organizations, such as the American Cancer Society, and groups organized more recently, such as Action on Smoking and Health (ASH), founded by John F. Banzhaf III. In the 1980s, community groups organized throughout the nation and pursued legislation limiting smoking in public places. The tobacco industry found it more difficult to combat antismoking efforts at the local level than it had at the federal level. By 1995, before the state tobacco litigation, forty-seven states required smoke-free environments at least to some degree or in some indoor places, such as restaurants or places of employment.[15] In addition, hundreds or perhaps thousands of municipalities had enacted similar ordinances. In short, public health officials and antitobacco lobbyists, arguing that they had been blocked by tobacco lobbyists in Washington, achieved considerable success in geographically dispersed city councils and county commissions.

Meanwhile, the rising tide of public opinion against smoking and the political appeal of protecting nonsmokers from secondhand smoke ultimately led several federal agencies and even Congress to adopt "smoke-free" policies, at least in some instances. As early as 1973, the Civil Aeronautics Board adopted restrictions on smoking during commercial air flights, and beginning in 1988, Congress passed a series of measures prohibiting smoking on such flights.[16] In 1994, the Department of Defense prohibited smoking in all workplace settings under its control, but it exempted living and recreational facilities.[17] The Synar Amendment, named after its sponsor, Representative Mike Synar, was enacted during the early 1990s. It provided in part that federal agencies should withhold funds from block grants to states for mental health and substance abuse programs if states did not enact laws prohibiting the sale of cigarettes to minors. It also required states to conduct random, unannounced inspections to assure that such prohibitions were being enforced.[18] Unfortunately, it appears that the amendment was rarely, if ever, enforced.[19]

Federal and State Excise Taxes

Public interest activists also note that before the state litigation against tobacco manufacturers, federal and state excise taxes on cigarettes were significantly less in the United States than in many other countries.[20] They attribute these lower taxes to the lobbying influence of the tobacco industry, conveniently failing to note that taxes in the United States are also lower on most other products, such as gasoline. This comparatively low level of cigarette taxation also appears to have reflected the preferences of the electorate before the state tobacco litigation. When voters in a number of states were asked to decide the issue through public referenda during the 1990s, voters in approximately half the states defeated proposals for higher cigarette taxes, and when similar proposals did pass, it often was by a very narrow margin.

Dramatic New Revelations of Nicotine Manipulation

During the early 1990s, the rising tide of antitobacco public sentiment was fueled by disclosures from several former tobacco company employees that tobacco companies had known for decades that nicotine was addictive and, in fact, that the companies had manipulated the nicotine content to assure better delivery of the addictive substance. In March 1994, Merrell Williams, a former paralegal at a law firm that represented Brown & Williamson Tobacco Corporation, one of the major manufacturers, delivered more than four thousand pages of documents (arguably stolen from his former employer) to attorneys preparing to sue cigarette manufacturers.[21] One memorandum written by Brown & Williamson's general counsel in 1963 stated, "We are in the business, then, of selling nicotine, an addictive drug."[22] A 1985 memorandum sent by a company executive specified scores of documents that were to be shipped out of the country, presumably to escape the reach of the legal process.

Jeffrey Wigand, a disgruntled former research director for Brown & Williamson, spoke both with television journalists and, later, with the federal Food and Drug Administration (FDA) beginning in April 1994.[23] Wigand explained how manufacturers mixed high-nicotine varieties of tobacco into cigarettes to enhance their addictive qualities. Similarly, he revealed, ammonia extracts were added to cigarettes to increase their potency. The broadcast media revealed the disclosures of Williams, Wigand, and other insiders shortly after they were shared with the FDA and with private attorneys involved in the soon-to-be filed Mississippi *parens patriae* litigation. In stark contrast to this new evidence, at about the same time, on

April 14, 1994, the heads of six of the largest cigarette manufacturers testified before Congress under oath that they did not believe that nicotine was addictive.[24]

Proposed FDA Regulation

In this new political environment, the FDA initiated the process for assuming regulatory jurisdiction over tobacco products before the states' filing of their litigation against the tobacco companies. Short of prohibition, FDA regulation of tobacco products was probably the strongest possible response that the political branches of government (Congress and its authorized administrative agencies) could have offered to the public health problems caused by tobacco products. For several years, both public health proponents and staff members within the FDA had been advocating that the agency assume jurisdiction over cigarettes and regulate them.[25] In 1980, in *Action on Smoking and Health v. Harris*,[26] the federal Court of Appeals for the District of Columbia Circuit upheld the agency's rejection of a petition filed by Action on Smoking and Health (ASH), an antismoking public interest organization, requesting that the agency assume jurisdiction over cigarettes containing nicotine. The court held that ASH presented no evidence that manufacturers had any intent "to affect the structure or any function of the body of man," as required by the legislation authorizing the FDA.[27] But in the 1990s, the newly discovered evidence of the manipulation of nicotine levels by tobacco companies provided persuasive evidence that these companies had in fact intended to affect bodily functions and therefore that the FDA had jurisdiction over cigarettes.

In February 1994, FDA commissioner David Kessler publicly released a letter indicating that the agency was considering such regulation.[28] In August 1995, President Bill Clinton endorsed Kessler's approach.[29] The proposed FDA rule would have

(1) prohibited the sale of cigarettes to individuals under the age of 18;
(2) required retailers to verify a purchaser's age;
(3) prohibited cigarettes in vending machines except where such machines were inaccessible to minors; and
(4) regulated the advertisement and promotion of cigarettes in ways likely to affect minors.[30]

The proposed regulations provoked a firestorm. Thirty-two senators signed a letter declaring their opposition, and the FDA received a record number of comments, more than 710,000 during the comment period, most of them opposing the regulations.[31] More important, within hours, tobacco manufacturers filed a lawsuit challenging the FDA's authority to regulate cigarettes.

Not surprisingly, the manufacturers filed their legal action in the federal court in Greensboro, North Carolina, in the heart of tobacco country. Yet, to their surprise, the trial court upheld most of the proposed regulations.[32] The federal Court of Appeals for the Fourth Circuit, however, reversed the trial court's decision,[33] and in a five-to-four opinion in 2000, the U.S. Supreme Court affirmed the court of appeals' reversal. Justice O'Connor, writing for the Court, concluded, "Reading the FDCA [Food, Drug, and Cosmetic Act] as a whole, as well as in conjunction with Congress' subsequent tobacco-specific legislation, it is plain that Congress has not given the FDA the authority that it seeks to exercise here."[34]

The core of the Court's reasoning must have seemed ironic to the FDA. The FDCA required the FDA to assure that the products it regulated were safe for the public. Justice O'Connor reasoned that because the FDA had extensively documented that cigarettes could not be made safe, giving the agency jurisdiction would require it "to remove them from the market entirely."[35] Such prohibition would be inconsistent with "Congress' decisions to regulate labeling and advertising" and its other legislation in support of trade in tobacco.[36] The Court's decision, of course, left the issue of whether the FDA could regulate tobacco in the hands of Congress. The proposal died there. Even if the Court had ruled that the FDA had jurisdiction over tobacco products, it is quite possible that Congress would have amended the FDCA to exclude such jurisdiction.

TOBACCO INDUSTRY INFLUENCE AND PUBLIC HEALTH REFORM IN THE POLITICAL BRANCHES

In 1993, Public Citizen, a public interest organization focused on consumer advocacy, published a report entitled *Contributing to Death: The Influence of Tobacco Money on the U.S. Congress*, which concluded, "The fact that tobacco money buys pro-tobacco results is clear, consistent, and irrefutable."[37] Similarly, Representative Richard Durbin (later a senator) remarked that the tobacco companies "have friends in many high places and have managed to make certain those friends were in key decision-making positions in virtually every administration and Congress."[38] In 1992, for example, the tobacco industry contributed over five million dollars to members of Congress, including individual campaign contributions, soft money contributions, and PAC (political action committee) contributions, ranking it thirty-second, among more than eighty industries, in campaign contributions.[39] While the amount of these campaign funds is significant, it pales in comparison to the amounts contributed by other industries: lawyers ($57 million, ranking first), health professionals (almost $28 million, ranking fourth), commercial banks (almost $15 million, ranking eighth), and pharmaceutical manufacturers (almost $8 million, ranking twenty-second).

Public interest advocates argue that the failure of Congress and federal agencies to more stringently regulate tobacco products was something more egregious than bad policy. They allege that campaign contributions and lobbying from the tobacco industry corrupted the legislative and administrative processes. This argument appeared to be confirmed by an empirical study conducted by researchers at the Public Citizen's Health Research Group, which concluded, "Tobacco industry contributions to members of the US Congress strongly influence the federal tobacco policy process."[40] This conclusion is less convincing, however, than observers would be led to believe by the public interest advocates and plaintiffs' lawyers who ultimately advocated state litigation against tobacco manufacturers.

In an empirical study of congressional voting behavior on issues of tobacco control between 1980 and 2000, John Wright, a political scientist specializing in interest group politics, concludes that the receipt of campaign contributions from the tobacco industry was a less important factor in voting patterns than were the legislators' generally probusiness and antiregulatory philosophies. Surprisingly, Wright found that "the effect of campaign contributions on voting is statistically indistinguishable from no effect."[41] Wright's analysis, specifically addressing the impact of contributions from the tobacco industry, is consistent with political scientists' long-held view that campaign contributions generally do not affect legislative votes.[42] An influential, more recent meta-analysis of earlier empirical studies concludes, however, that "a noteworthy minority" of these studies suggested that "money did have a statistically significant impact on how legislators voted."[43] In any event, it is likely that the centrality of smoking to the American consumer society, as well as the argument that smoking was a personal decision, played a larger role than did lobbying activities and campaign contributions. Not until the dangers of environmental tobacco smoke became well publicized and generally accepted (i.e., the mid-1980s) was the American voter ready for Congress, state legislatures, and city councils to take strong action to prevent tobacco-related illness.

Obviously, a few well-placed congressional committee chairs—usually from regions of the South, where economic well-being was thought to depend on tobacco—did block or at least delay many federal antitobacco legislative proposals. Further, the political climate during the presidency of Ronald Reagan and in the years to follow was not generally conducive to strong legislation favoring product regulation. Thomas W. Merrill has described this era as one "characterized by widespread pessimism about the capacity of any governmental institution to achieve results that will promote the public interest."[44] Despite this climate, antismoking legislation moved forward during this period, particularly at the state and local level. It did not happen as quickly as public interest advocates and antismoking activists would have liked, but the regulation of smoking by the nonjudicial branches of government by the mid-1990s was more stringent in the United States than in most other nations. In addition, at the time states filed litigation against the manufacturers, the FDA already was moving forward on a proposal to include tobacco within its regulatory regimes.

Looking backward from the early twenty-first century, it is easy to be critical of Congress and federal administrative agencies for failing to do more to regulate tobacco products during the previous decades. Today, the link between smoking and cancer and other risks is widely accepted among members of the electorate, as are the risks posed by environmental tobacco smoke. From the current vantage point, it is easy to say that the political processes failed when it came to preventing the public health problems caused by cigarette smoking. When viewed from the perspectives of earlier decades, however, the argument that legislative bodies failed because they did not represent the will of the electorate, used so often to justify the resort to state *parens patriae* litigation against tobacco manufacturers as an alternative to the usual processes of regulation in a constitutional democracy, rests on more questionable grounds.

REGULATORY RESPONSES TO CHILDHOOD LEAD POISONING

When Congress, state legislatures, or federal and state administrative agencies seek to prevent childhood lead poisoning, they confront a situation much different than that posed by tobacco-related illnesses. The manufacture and distribution of residential paint containing lead pigment, the product that causally contributes to childhood lead poisoning, has been banned by federal law since 1978.[45] However, a 2002 Department of Housing and Urban Development (HUD) study indicated that lead-based paint remained in place in 38 million housing units in the United States and that 24 million of those units had significant lead-based paint hazards that likely posed a risk to children.[46]

Eliminating the risk caused by lead-based paint already covering surfaces of older residential premises has proved to be a daunting task. In addition to sponsoring government programs that educate homeowners and parents of young children about the risks of lead-based paint hazards, government action to eliminate existing lead-based paint hazards addresses two separate issues. First, standards adopted either legislatively or by administrative agencies may impose requirements on the property owner to eliminate or at least substantially reduce the risk posed to a child by the presence of lead within the residential environment in which he lives. Second, legislation may determine which party bears the costs of implementing these measures. Should the costs be paid by the property owner, the former manufacturers of lead pigment or lead-based paint, the current manufacturers of paint (which, of course, does not contain lead), or the government?

For the most part, the role of regulating housing containing lead-based paint and, thus, of preventing the harm caused in part by the presence of such paint has rested with states and municipalities.[47] Federal statutes and regulations do require federally owned residential properties and housing paid for with federal housing assistance funds to comply with so-called interim control standards, as described in the following paragraphs.[48] In addition, statutes and regulations enforced by the Environmental Protection Agency require safe work practices for contractors removing lead or otherwise working in residences containing lead-based paint.[49]

From a public health perspective, one of the critical decisions that must be made in deciding how best to eliminate or at least substantially reduce the incidence of children with elevated blood lead levels (EBLs) is the choice between requiring "total abatement" or mandating only interim controls. The federal government defines "abatement" as the permanent elimination of lead-based paint or lead-based paint hazards.[50] Abatement includes such measures as the removal of lead-based paint, the permanent enclosure or encapsulation of such paint, the replacement of windows or other fixtures painted with lead-based paint, and the removal or permanent covering of soil that has been contaminated with lead paint or leaded gasoline.

In contrast, interim controls are measures designed to temporarily reduce the exposure of children to lead-based paint hazards.[51] Examples include the removal of all chipping, peeling, or flaking paint and repainting the surface and repairing problems, such as leaks, that may be causing paint to deteriorate. Several interim control standards address the creation of dust containing lead when "friction surfaces" rub against each other, such as when doors or windows that are improperly hung rub against their frames as they are opened. Interim control standards require that caps of vinyl, aluminum, or a similar material be installed in window wells to make them smooth and cleanable and that windowsills be stripped and repainted or covered with vinyl. In addition, all floors must be smooth so that dust containing lead can be removed by routine cleaning.

Relying on interim controls to eliminate the risks of childhood lead poisoning requires an ongoing maintenance program to assure that housing conditions do not deteriorate and once again pose health hazards. For this reason, from a public health perspective, total abatement would be preferred in a hypothetical world in which resources to address childhood lead poisoning were unlimited. In a world of limited resources, however, the cost differential between the two approaches probably tips the balance in favor of interim controls. A presidential task force in 2000 estimated that the mean cost of interim controls was one thousand dollars per unit and that the mean cost of abatement was nine thousand dollars per unit.[52] On an aggregate nationwide basis, the Task Force reported that it would take $1.84 billion to implement interim controls and $16.6 billion for total abatement. Both sets of estimates are probably unrealistically conservative. Based on my own work with public health and housing officials in several states between 2000 and 2008, I would estimate that while the costs of interim controls vary considerably depending on how poorly the residential property had been maintained, the average expenditure would probably considerably exceed one thousand dollars per unit (perhaps even coming to several times that amount), while total abatement costs would average somewhere between fifteen and twenty-five thousand dollars per unit. Consistent with earlier studies, a recent study by the Battelle Institute and the National Center for Healthy Housing shows that interim controls are effective at controlling the risk of childhood lead poisoning and that there were no statistically significant differences in the lead content of floor dust over time between units treated with interim controls and those that had undergone total abatement.[53]

At least by the mid-1990s, a handful of states and municipalities had enacted laws requiring rental property owners to protect the children of tenants from the risks posed by lead-based paint hazards by forcing most landlords (typically all except small landlords renting one to three units) to implement interim controls. Among the stronger statutory frameworks for preventing childhood lead poisoning were those in the states of Maryland[54] and Massachusetts.[55] Both require, at a minimum, that property owners implement interim control measures similar to those contained in the HUD requirements. Both statutes require inspection by state officials or independent inspectors. Landlords who fail to comply with statutory requirements are subject to criminal sanctions. The Massachusetts statute provides that a landlord is strictly liable for damages sustained by a lead-poisoned child,[56] while Maryland grants a limited immunity to a landlord in full compliance with the act.[57] Rhode Island enacted a similar statutory framework in 2002.[58] With rare exceptions, other states do not require remediation of lead-based paint hazards until one or more children in a residential unit already have been tested and found to have EBLs that pose a threat to their health.[59] In short, children in these states are treated like the proverbial canary lowered into the mine to check for the presence of dangerous gases.

Not surprisingly, the Massachusetts and Maryland statutes appear to have had a significant effect in reducing the incidence of children with EBLs. A multiple regression analysis completed in 1999 compared the incidence of children with EBLs in Worcester County, Massachusetts, with those in Providence County, Rhode Island, where, at that time, effective state policies were not in place. The areas were selected because they had approximately the same populations and similarities in demographics, housing age, climate, industrial base, and lead concentrations in the air.[60] The authors found that the incidence of childhood lead poisoning was two-thirds lower in Massachusetts because of the state's housing policy as outlined in its statutes. Similarly, from 1996, when the Maryland act was enacted, through 2004, the percentage of children who were tested for lead that showed a blood lead level above 10 µg/dL decreased 89.7 percent,[61] a rate of decrease substantially greater than that found in most other states.

Who Pays for Remediation of Lead Hazards?

The cost of remediating lead hazards through implementing interim controls is considerable, though only a fraction of full lead abatement. Who pays these costs at the current time, and who should pay? When state statutes or regulations mandate interim controls, the responsibility for paying such costs ordinarily lies with landlords.

From one perspective, this assessment appears reasonable. Rental property ownership and management is, after all, a business. Landlords should not be able, any more than should any other business, to engage in their business in a manner that poses considerable risk to families, who are their consumers. The risk posed to the toddler who crawls around on the floor of an apartment that is coated with dust contaminated by lead parallels the risks posed to a consumer of food processed in a plant that does not comply with sanitary standards; in fact, it may be far worse.

At the same time, popular intuitions often are at odds with this reality.

On one hand, large companies that manage rental properties fit comfortably within the categories of businesses that generally are expected to conduct their operations in a manner safe to their consumers. On the other hand, many residential rental properties in the United States are owned and managed by small property owners who control only one, two, or three units. The detailed regulatory standards contained in the HUD standards or Maryland act appear to many of them to be foreign and heavy-handed. In addition, landlords and their advocates claim that profit margins are small, particularly for properties rented to low-income tenants, where most childhood lead poisoning occurs.

Property owners frequently point to the manufacturers of lead pigment or paint as the parties that should be paying for the costs of remediation of lead hazards. Their argument is a simple one: manufacturers of paint or lead pigment originally placed their toxic products into the stream of commerce, and their products now pose risks to children and therefore must be remediated. The property owners who used lead-based paint decades ago rarely, if ever, understood the risks to children exposed to lead-based paint.

One of the difficulties with imposing the costs of remediation on paint manufacturers, as contrasted with lead pigment manufacturers, is that residential paint containing lead has not been distributed since 1977. Many manufacturers who once produced paint containing lead are no longer in business. Conversely, many corporations that sell paint today sold little or no paint containing lead before 1978.

Two different avenues exist for forcing either paint manufacturers or pigment manufacturers to contribute to the cost of remediation. The first is the approach that Rhode Island and a number of municipalities and counties attempted to take when they sued pigment manufacturers. This litigation cycle, the obstacles it faces, and the considerable disadvantages of this approach are fully described in later chapters of this book, particularly chapters 7 and 8. The second approach is for either Congress or the state legislature to impose an excise tax or "fee" on the sale of paint and to use the proceeds to assist property owners in remediating lead-based paint hazards. Currently two states, California and Maine, impose fees or excise taxes on the sale of paint.[62] Not surprisingly, the paint industry opposed the enactment of legislation mandating these fees. Both states use the proceeds of the fees for other aspects of the effort to prevent childhood lead poisoning (e.g., public education or screening) and not for remediation of lead-paint hazards. In the conclusion to this book, I briefly evaluate a fee on paint as a possible component of a plan to tackle childhood lead poisoning.

In addition to the alternatives of imposing the costs of remediation on either property owners or manufacturers of paint or of lead pigment, the third source of funds for remediation is the state itself. The federal and state governments routinely spend funds to prevent or ameliorate public health problems, ranging from malaria and Lyme disease to HIV, breast cancer, and diabetes. A few states do provide low-interest or even no-interest loans to property owners to assist them in implementing interim standards, and some even provide grants for such purposes, but the funding for such programs is meager.[63]

LOBBYING, CAMPAIGN CONTRIBUTIONS, AND STATE LEGISLATIVE EFFORTS TO PREVENT CHILDHOOD LEAD POISONING

Legislatures in such states as Massachusetts and Maryland have enacted strong legislation that has substantially reduced the incidence of childhood lead poisoning. Why has similar legislation not been adopted in other jurisdictions? An important aspect of the answer to this question is the pervasive influence of rental property owners and other real estate interests in state and local legislative politics.

In 2006, real estate interests contributed more than any other industry to the campaigns of legislators and other statewide offices in many states, including New York, Florida, Maryland, and Missouri.[64] Common Cause, a citizens' advocacy group, issued a special report in 2003 outlining how campaign contributions from landlords and other real estate interests appeared to be preventing the enactment of stronger childhood lead poisoning prevention legislation in New York City,[65] though such legislation, with amendments, was later adopted. The member of the New York City Council who represented the district with the highest rate of childhood lead poisoning had spoken out against a bill advocated by public health advocates and physicians and cosponsored by a majority of the members of the council.[66] Fifteen of his largest twenty campaign contributions had come from real estate interests.

Legislative hearings considering legislation for prevention of lead poisoning often take place in the housing committee of the state legislature or city council. Many members of such committees receive sizable campaign contributions on a regular basis from groups representing landlords. Testimony before the committee from representatives of landlords unrealistically suggests that if they are forced to comply with even interim control measures, the costs of rental housing for low-income tenants will increase to a point that many such tenants will not be able to pay their rents. These landlords claim that they will be forced to abandon such properties and that other investors will not be interested in purchasing them.

The battle as to who should pay for preventing childhood lead poisoning resulting from exposure to lead-based paint hazards has become largely a struggle between two large industries—the residential rental industry and the paint industry. In comparison with rental property owners and other real estate interests, paint companies and lead pigment manufacturers typically have a surprisingly limited presence in most state legislative political arenas. When state legislation requiring landlords to implement interim controls is introduced, state lobbyists representing real estate interests claim that the paint industry is trying to shift the blame from itself, while representatives of paint companies point out that only property owners can eliminate lead-based paint hazards and that their own contributions to the problem ended no later than 1978.

CONCLUSION

Public interest advocates and plaintiffs' attorneys assert that state *parens patriae* litigation against tobacco and lead pigment manufacturers is justified because of a failure of the legislative process. But these assertions are not particularly convincing. It is certainly the case that tobacco manufacturers concealed the risks of their products and lied for decades. However, during the same year that Mississippi filed the first state litigation against the tobacco companies, the FDA initiated the process to begin the comprehensive regulation of cigarettes. Further, Congress, administrative agencies, and state and local legislators had already taken a number of critical steps to regulate cigarettes. A recent empirical analysis suggests that the link between a member of Congress's receipt of "tobacco money" and his vote on tobacco-related issues during this time period was less axiomatic than might have been expected.

When it comes to litigation against lead pigment manufacturers, their product became subject to the regulatory death sentence in 1978: its use in residential paint is prohibited. The legislation that has proven to be substantially, if not totally, effective in reducing the incidence of childhood lead poisoning is state legislation regulating the conduct of residential landlords. Unfortunately, real estate interests may be one of the most potent industries in influencing state legislative politics, and only a handful of states have enacted effective legislation for prevention of childhood lead poisoning.

Parens patriae litigation against lead pigment manufacturers is best understood as a process to decide whether they should be the parties to bear the enormous costs of eliminating childhood lead poisoning or whether such costs should instead be paid by property owners, perhaps subsidized by federal and state governments. Even assuming, for purposes of argument, that the paint or lead pigment industry should, as a policy matter, pay a portion of the costs to remediate dangerous conditions in older residential real estate, the question remains as to whether such a wealth transfer from one industry to another, albeit with considerable public health benefits, is best accomplished through the common-law judicial process.

CHAPTER 6

The Government as Plaintiff: *Parens Patriae* Actions against Tobacco and Gun Manufacturers

Until the mid-1990s, the manufacturers of tobacco and lead products still had not been forced to contribute financially to resolving the public health problems resulting from the use of their products. To recapitulate, courts most often rejected claims against tobacco manufacturers because of the cigarette consumer's own conduct in deciding to smoke. Meanwhile, the victim of childhood lead poisoning was never able to overcome the fact that she was inherently unable to identify the specific manufacturers who produced the paint or pigment that caused her disease.

Within the short period of four and a half years that elapsed between May 1994 and November 1998, the law governing the liability of manufacturers whose products cause latent diseases shifted cataclysmically. Tobacco manufacturers agreed to pay more than $250 billion in damages, by far the largest civil settlement ever. Dramatic disclosures of the manipulation of the nicotine content of cigarettes by manufacturers and of the outright lying of corporate executives (described in chapter 5) prompted this development. At the same time, the legal vehicle that enabled it to happen was *parens patriae* litigation against manufacturers, in which states asserted a number of substantive claims that were new to the product liability context, notably public nuisance. In short, viewing tobacco-related diseases as an environmental harm and borrowing the intertwined concepts of public nuisance and *parens patriae* from environmental law transformed the law governing liability for product-caused latent diseases.

Of course, settlements themselves do not create formal legal precedents. Nevertheless, a new genre of litigation had been unleashed. By the end of the decade, the combination of *parens patriae* standing and public nuisance claims had become staples in state and municipal litigation against the manufacturers of handguns and lead pigment. In this chapter, I consider this new form of litigation against the manufacturers of tobacco products and handguns. I reserve until chapter 8 consideration of both the settlement ending the tobacco litigation and the lawmaking process inherent in it.

"THE STATE OF MISSISSIPPI HAS NEVER SMOKED A CIGARETTE"

Despite congressional hearings, presidential pronouncements, and aborted attempts at FDA regulation, the judicial process became the focal point of those seeking to regulate the tobacco industry during the mid-1990s. The centerpiece of the new wave of litigation strategies was the use of *parens patriae* actions by state governments against tobacco manufacturers. The fifty states had spent hundreds of millions of dollars each year paying the medical expenses for victims of tobacco-related illnesses. Beginning as early as 1985, visionary Northeastern University law professor Richard Daynard began thinking about litigation that states might bring against tobacco manufacturers.[1]

By 1989, Raymond E. Gangarosa, a physician who contacted Daynard, was promoting the idea that states or public hospitals could sue to recover their unreimbursed costs in treating victims of tobacco-related illnesses. As may be recalled, smokers' own conduct had blocked their lawsuits against tobacco companies for decades. But Frank J. Vandall, an Emory law professor who coauthored an article with Gangarosa, realized that in a direct action brought by a state against tobacco companies, any alleged misconduct of the victims themselves would not bar the lawsuit. Mississippi plaintiffs' attorney Don Barrett would later say, "The state of Mississippi has never smoked a cigarette."[2] A suit on behalf of the state, seeking reimbursement for the expenses resulting from tobacco-related illnesses, would force manufacturers to pay these expenses and would punish them. It could not be blocked by the doctrines of common knowledge, assumption of risk, or comparative fault.[3]

On May 23, 1994, Mississippi attorney general Michael Moore filed a complaint on behalf of the state of Mississippi against thirteen tobacco companies.[4] This complaint would revolutionize legal actions seeking to hold product manufacturers liable for pervasive product-caused public health problems. Moore's ingenious approach combined two elements. First, it asserted that the state had standing to sue as a collective plaintiff on behalf of its citizens suffering from product-related diseases. Second, it employed three causes of action that previously had not been used against product manufacturers: public nuisance, unjust enrichment, and indemnity. The combination of these novel torts with *parens patriae* standing enabled states to avoid defenses based on the conduct of the victims of smoking-related diseases, as well as dismissals resulting from the inability of any particular victim to prove that any specific manufacturer caused her harm. In addition, the aggregation of the medical expenses of millions or even tens of millions of victims made the potential contingent fees at stake for private attorneys assisting the states large enough to attract the highly competent plaintiffs' attorneys necessary to match up with counsel retained by manufacturers.

In short, beginning with the filing of the Mississippi complaint, the face of legal assaults on industries that caused public health problems was transformed. Within three years of the filing of the Mississippi complaint, at least forty states had filed suits against tobacco manufacturers.[5] Municipalities,[6] health care insurers,[7] and labor union insurers[8] soon followed, filing similar complaints seeking reimbursement for the costs they claimed to have sustained as a result of tobacco-related illnesses.

In their *parens patriae* reimbursement actions, different states pursued a wide variety of substantive claims, including common-law misrepresentation,[9] deceptive advertising,[10] antitrust violations,[11] and violations of the federal Racketeer Influenced and Corrupt Organizations (RICO) Act.[12] Some states totally bypassed the causes of action advanced by Mississippi—public nuisance, unjust enrichment, and indemnity. Florida, for example, relied instead on a state statute specifically passed by the legislature to abrogate all affirmative defenses that would have blocked the state's recovery in *parens patriae* litigation.[13] The statute also allowed the state to pursue its actions without identifying the specific individuals who received medical assistance benefits as a result of smoking-related illness, but the Florida Supreme Court subsequently ruled that this provision violated the due process clause of the state constitution.[14] Five years later, in 1999, the Clinton administration filed a somewhat similar action on behalf of the federal government under the civil remedy provisions of RICO, seeking "disgorgement" of $280 billion in profits that allegedly resulted from tobacco companies' "intentional, coordinated campaign of fraud and deceit."[15] The Court of Appeals for the District of Columbia Circuit held, however, that disgorgement was not a remedy available under the RICO statutory provisions.[16]

State *parens patriae* actions against product manufacturers arguably became the most important form of litigation against manufacturers whose products cause disease. I analyze this form of state standing next. I then consider several of the substantive causes of action employed by states in these *parens patriae* cases that were not typically available to individuals suing product manufacturers.

EXPANSIVE INTERPRETATIONS OF THE SCOPE OF
PARENS PATRIAE STANDING

In order for the state to sue product manufacturers in a *parens patriae* role, that concept must be understood more expansively than it has been in the past. For example, the tobacco litigation marked the first time states had brought *parens patriae* actions to recover for harm to the common good that served as a proxy for the amalgamation of individual injuries caused by product exposure. These actions sought to protect the state's "quasi-sovereign" interests in the health, safety, and welfare of its citizens.[17] For the most part, the state's damages consisted of the increased health costs it sustained through its medical assistance program.

In *parens patriae* litigation against product manufacturers, the state nominally asserts that the damages result from damages to the state itself. Some damages claimed by the state clearly do arise, in the first instance, in this manner. In its action against lead pigment manufacturers, for example, Rhode Island sought to recover its costs in educating its residents about the risks of childhood lead poisoning. More often, however, as in the tobacco litigation, the damages claimed by the state result largely from injuries initially sustained by its residents, for which the state later assumed financial responsibility through its medical assistance (Medicaid) program.

In both tobacco and lead pigment litigation, the plaintiff-state serves as a conduit through which money paid by product manufacturers is to be delivered to those directly harmed by the manufacturers' products or to medical providers that have treated these victims. The principal difference between the tobacco and lead pigment situations is that in the tobacco litigation, the state is reimbursed for money it already has paid to health care providers of victims of tobacco-related illness. In contrast, in the action brought by the State of Rhode Island against the manufacturers of lead pigment, the state had not yet expended the funds to abate the nuisance prior to the litigation. Instead, any expenditure of funds to eliminate or reduce lead-based paint hazards in Rhode Island residences would not have occurred until the state (or perhaps a special master appointed by the court) received the funds from the defendant-manufacturers or their insurers.

In either instance, *parens patriae* litigation enables the private individual directly harmed to be compensated. The state is reimbursed for the medical bills it has paid for the victim of tobacco-related disease. The Rhode Island property owner improves his property by abating lead-paint hazards at the manufacturer's expense. *Parens patriae* litigation thus accomplishes what the victim herself could not have accomplished as a litigant. As stated previously, the individual victim of tobacco-related disease probably could not have recovered her medical expenses or any other damages, because defendants would argue that the dangers of cigarette smoking were "common knowledge" and that she either knew or should have known of these risks.[18] Similarly, the Rhode Island property owner almost certainly would not have won in a direct action against the pigment manufacturers, because of a whole host of impediments to recovery, including his inability to identify the specific producer of the lead pigment that now covers the walls of his house, the long-ago expiration of the statutes of limitations, and his own failure to use reasonable care to avoid lead-based paint hazards.

Of course, when appropriate, as an alternative to filing *parens patriae* litigation, the state always can sue as a subrogee of a resident's claims against a manufacturer. For example, the federal statutes governing the Medicaid program require states to seek reimbursement from any tortfeasor whose actions caused a Medicaid recipient harm resulting in medical expenses paid by Medicaid.[19] As would be expected, the states have enacted subrogation statutes enabling them to sue the tortfeasor in these circumstances.[20] From the state's perspective, however, the subrogation cause of action poses a problem. As a subrogee, the state "steps into the shoes" of the Medicaid recipient and takes her cause of action subject to any defenses that would be available if the recipient herself had sued the manufacturer.[21] In short, under the subrogation cause of action, the state's claims in tobacco litigation often would be denied because of the smoker's knowledge of the risks of smoking, and for all intents and purposes, they always would be denied in actions against lead pigment manufacturers, because of the inability to identify the specific manufacturer that produced the product causing an individual victim's harm. Because these defenses would not preclude recovery if the state is allowed to sue as *parens patriae*, the availability of this form of standing is critical to the state in litigation against product manufacturers.

Litigation by states against the manufacturers of most types of products extends the *parens patriae* doctrine far beyond its historical boundaries. The concept of *parens patriae* is fundamentally a means of granting standing to the state.[22] Most often, determining whether the state has *parens patriae* standing to sue a product manufacturer is a question of state law, but state courts frequently are guided by Supreme Court decisions on *parens patriae* standing under federal law.[23] In *Alfred L. Snapp & Son v. Puerto Rico*, perhaps the U.S. Supreme Court's most important *parens patriae* opinion, the Court considered whether the Commonwealth of Puerto Rico had *parens patriae* standing in federal courts to represent its "quasi-sovereign" interest "in the well-being of its populace."[24] The Court recognized the right of the commonwealth (or a state) to sue to protect such quasi-sovereign interests, as long as it is more than a nominal party and articulates "an interest apart from the interests of particular private parties."[25]

In *parens patriae* litigation against product manufacturers, it is generally easy for the state to satisfy an injury-in-fact test of standing,[26] such as when it has sustained financial injury by paying medical assistance payments to victims of tobacco-related illnesses that result from exposure to cigarettes. The state's ability to satisfy the injury-in-fact requirement, however, does not necessarily mean that it should be granted *parens patriae* standing, enabling it to recover damages initially inflicted on other private parties and then "passed through" to the state. Chief Justice John Roberts recently stated, "Far from being a substitute for Article III injury, *parens patriae* actions raise an additional hurdle for a state litigant: the articulation of a 'quasi-sovereign interest' '*apart* from the interests of particular private parties.'"[27] Even though virtually all states filed *parens patriae* actions against tobacco manufacturers in the late 1990s, many state appellate courts have since concluded that a state or municipality does not have standing to sue as *parens patriae* against a product manufacturer, because the damages sustained by the government are "derivative" or "too remote."[28] For example, in *Ganim v. Smith & Wesson Corp.*,[29] the Connecticut Supreme Court held that the City of Bridgeport lacked *parens patriae* standing to bring an action against gun manufacturers, because the harms allegedly sustained by the city—predominately increased costs for police and other municipal services resulting from the illegal use of guns—were simply too remote from the manufacturers' conduct.[30]

In addition to the state and federal opinions holding that a state does not have *parens patriae* standing against a product manufacturer because the state's harms are too remote or derivative, a careful reading of Supreme Court opinions provides a second reason for denying such standing. The early history of *parens patriae* cases based on quasi-sovereign interests almost entirely consists of disputes between the interests of separate states with regard to natural resources and territory.[31] For example, in *Georgia v. Tennessee Copper Co.*,[32] Georgia sought an injunction against a Tennessee manufacturing company for the discharge of noxious gases over its territory. Other Supreme Court decisions recognize a state's ability to sue as *parens patriae* to protect its residents against economic discrimination when such discrimination results from the victims' identities as citizens of the state.[33] These precedents set the stage for the decision in *Snapp*, in which the Supreme Court held that Puerto Rico has a quasi-sovereign interest in protecting its residents from employment discrimination, because "the State has an interest in securing observance of the terms under which it participates in the federal system."[34]

In every one of these precedents, the harms suffered by the original (individual) victims were causally connected to their residency within a particular state. In each case of economic discrimination, the economically harmed individuals were harmed because they were citizens of a particular state and not another jurisdiction. Similarly, in the pollution cases, the harms to private property owners occurred because their properties, polluted by the defendant's activities, were physically located at a particular location that was within the plaintiff-state's territorial boundaries. In other words, in either instance, the victim's harm was directly related and causally connected to her identity as a resident of the state that sought to vindicate her interests through *parens patriae* litigation.

In contrast, the state of residence and the harm sustained are independent variables in *parens patriae* actions against product manufacturers. Assuming that the marketplace for tobacco products is nationwide (a seemingly safe assumption), the victim's residence in Iowa neither increases nor decreases her risks of contracting a tobacco-related illness. Her residence in Iowa is unrelated to her harm. The risk of childhood lead poisoning to the child occupying a dwelling with deteriorated lead-based paint is no greater because that house is located in New York than it would be if the dwelling were located in Illinois.

Dicta in the Supreme Court's 2007 opinion in *Massachusetts v. EPA*[35] at least suggest, as persuasive authority, that if harm caused by mass products is viewed as environmental harm, a state's *parens patriae* standing might be interpreted broadly enough to encompass an action against the manufacturers. In that case, the Court held that states had standing, under a federal statute, to challenge the refusal of the federal Environmental Protection Agency to exercise its rule-making authority to regulate greenhouse gas emissions under the Clean Air Act. The Court quoted the *Snapp* opinion in stating, "One helpful indication in determining whether an alleged injury to the health and welfare of its citizens suffices to give the State standing to sue *parens patriae* is whether the injury is one that the State, if it could, would likely attempt to address through its sovereign lawmaking powers."[36]

Certainly, viewed in isolation, this statement suggests that states have standing to sue as *parens patriae* to prevent either tobacco-related diseases or childhood lead poisoning. Further, on the surface, the facts in *Massachusetts v. EPA* are remarkably similar to those in *parens patriae* litigation against product manufacturers. The alleged injury to Massachusetts is not at all limited to air within the commonwealth's territorial boundaries but, indeed, includes the entire earth's atmosphere, making it even more ubiquitous than product-caused mass torts. Allowing the state to recover for injury to the physical environment is also, of course, consistent with earlier Supreme Court precedents, such as *Georgia v. Tennessee Copper Co.*

The key distinguishing factor between *Massachusetts v. EPA* and products litigation driven by the state, however, is unrelated to the issue of whether diseases caused by mass products can be legitimately viewed as an environmental harm. Instead, the decision rests on the state's ability, as a sovereign state in a federal system, to sue the federal government, at least under some circumstances, to enforce federal regulatory powers. This important aspect of the Supreme Court's opinion is not relevant to *parens patriae* litigation against private actors, the manufacturers of mass products. In his opinion for the majority, Justice Stevens stressed that when Massachusetts entered the Union, it gave up its sovereign rights to protect its environment by, for example, invading "Rhode Island to force reductions in greenhouse gas emissions" or negotiating "an emissions treaty with China or India."[37] The Constitution transferred these powers to the federal government, and the state "is entitled to special solicitude in our standing analysis" in order to protect "its quasi-sovereign interests." The state's role in public products litigation obviously is not the unique one present in *Massachusetts v. EPA,* where only the state, as a key partner in the constitutional structure, can protect its interests as a sovereign by enforcing the implementation of powers it has yielded to the federal government.

PARENS PATRIAE AND PUBLIC PRODUCTS LITIGATION

Allowing the state attorney general to act as "superplaintiff" and to sue on behalf of all its residents in actions against products manufacturers turns the usual structure of compensation in our legal system on its head. This approach envisions state social welfare programs, such as medical assistance (Medicaid), as the initial sources of compensation for victims of mass harms caused by product manufacturers. In order to be reimbursed, the state then sues the manufacturers that allegedly caused the mass harms. In doing so, the state avoids the obstacles caused by the inability of most individual victims to recover directly from manufacturers, because they either would not be able to prove the necessary individualized causal connection between a specific manufacturer and a particular plaintiff or because their own conduct would otherwise preclude or at least reduce recovery. This disruption of fundamental common-law compensation doctrines ultimately requires an expansive interpretation of the state's standing to sue as *parens patriae*, one that goes far beyond an understanding grounded in Supreme Court precedents.

PUBLIC NUISANCE AND OTHER SUBSTANTIVE CAUSES OF ACTION

In the *parens patriae* tobacco litigation, state attorneys general utilized substantive claims (e.g., public nuisance) for which the boundaries of the torts were extremely indeterminate, and their use against product manufacturers was not well grounded historically. When the states sued the tobacco manufacturers, their legal claims often did not include strict products liability, negligence, implied warranties, or other well-understood products theories. For the state to act as the collective plaintiff in the tobacco litigation, the attorney general required innovative substantive tort claims that treated the harm as harm to society as a whole, instead of as discrete harms to a series of individual victims. These substantive theories of recovery, along with the state's *parens patriae* role as a collective plaintiff, enabled the states as plaintiffs in litigation against manufacturers of lead pigment to bypass the traditional requirement of an individualized causal connection between the manufacturer and the victim. Similarly, states suing cigarette manufacturers were able to avoid defenses to their claims based on the smokers' conduct, which often had proved to be insurmountable for victims of tobacco-related disease in the past.[38]

At the time of the filing of the tobacco litigation, it was not clear which of the claims contained in the states' *parens patriae* actions represented the strongest chance of recovery. The answer probably varied from one state to another, depending on the law of a particular jurisdiction and the causes of action that the state elected to include in its complaint. However, by the time of the subsequent waves of *parens patriae* litigation against the manufacturers of other products (e.g., handguns, lead pigment, and automobiles that contributed to global warming), public nuisance had emerged as the most important substantive claim in actions brought by states and municipalities seeking damages for collective harm. Courts in these public nuisance actions brought by states and municipalities sometimes have described a "public right" (the interest protected by the public nuisance tort) more expansively—and less accurately[39]—than its traditional understanding as "an indivisible resource shared by the public at large, like air, water, or public rights of way."[40] Instead, some courts have characterized a statewide or citywide accumulation of private harms as a violation of the entitlement protected by the public nuisance tort.[41] Because of the central role played by the tort of public nuisance in *parens patriae* litigation against manufacturers of products that cause diseases, I focus specifically on the legitimacy and wisdom of the expansion of the tort of public nuisance in chapter 7.

The other two substantive claims included in Mississippi's complaint against the tobacco companies were unjust enrichment and indemnity, closely related claims based on a notion of restitutionary justice. Like public nuisance, these claims are also extremely vague and novel in their use against product manufacturers. Again, the eventual tobacco settlement short-circuited any judicial testing of these claims during the tobacco litigation, but in the subsequent litigation by the State of Rhode Island against lead pigment manufacturers, the trial court initially denied the defendants' motion to dismiss the unjust enrichment claim,[42] before eventually dismissing it.[43] In its first opinion allowing the claim to move forward, the court quoted from a Rhode Island Supreme Court precedent that stated that the doctrine of unjust enrichment "permits the recovery in certain instances where a person has received from another a benefit, the retention of which, would be unjust under some legal principle, a situation which equity has established or recognized."[44] In order to recover, said the trial court, the state or other plaintiff must show that it conferred a benefit on the defendant that the defendant both appreciated and accepted "in such circumstances that it would be inequitable for a defendant to retain the benefit without paying the value thereof."[45] The Rhode Island trial court found that the state's payment of expenses caused by childhood lead poisoning at a time when defendants continued to profit from the sale of lead added to the defendants' benefit and therefore was sufficient to avoid a motion to dismiss.[46] The attorneys who drafted the complaints of Mississippi and many other states in their litigation against tobacco manufacturers evidently subscribed to similar reasoning. Most courts, however, have rejected unjust enrichment claims for injuries generally regarded as sounding in the more traditional theories of strict products liability or negligence.[47] In *Perry v. American Tobacco Co.*, for example, the court rejected the plaintiffs' claims on the grounds that the tobacco companies had not been enriched, "because Defendants [had] no legal duty to smokers to pay their medical costs."[48] In the process, the court implicitly rejected the argument that liability might exist if the defendants had violated an ethical, if not a legal, duty.[49]

The trial court in the Rhode Island paint litigation also initially denied the defendants' motion to dismiss a separate claim based on indemnity, before ultimately dismissing it.[50] In allowing the claim to move forward, the court stated, "The concept of indemnity is 'based upon the theory that a party who has been exposed to liability solely as a result of the wrongdoing of another should be able to recover from the wrongdoer.'"[51] In order to recover on an indemnity theory, according to the court, the plaintiff must prove three elements: "First, the party seeking indemnity must be liable to a third party. Second, the prospective indemnitor must also be liable to the third party. Third, as between the prospective indemnitee and indemnitor, the obligation ought to be discharged by the indemnitor."[52] The court initially upheld Rhode Island's allegations that "as between the State and the defendants, . . . the defendants ought to bear the burden of the lead-related expenditures resulting from the damages due to the lead."[53]

Again, most other courts have rejected the use of the indemnity theory of recovery in what essentially is a tort claim brought by a state, municipality, health insurer, or similar party.[54] In rejecting such a claim, the court in *Allegheny General Hospital v. Philip Morris, Inc.*[55] correctly noted that the right of indemnity exists only when one party without active fault on its part is legally obligated to pay damages caused by the actions of another party.[56] For example, under the doctrine of vicarious liability, an employer, faultless in its own right, generally is obligated to pay for the torts committed by its employees acting within the scope of their employment. If the employer pays a claim on an employee's behalf, it has legal grounds to pursue (but probably will not) an indemnity claim against the employee whose conduct was negligent or otherwise tortious. According to the court in *Allegheny General Hospital*, the liability of the party who pays—in the case of *parens patriae* litigation, the plaintiff-state—must rest on fault that is imputed or constructively imputed to the state because of either some legal relationship between the parties or some "positive rule" of statutory or common law.[57]

In the case of *parens patriae* actions against product manufacturers, the state is not legally obligated to make medical assistance payments to recipients as a result of the manufacturers' tortious conduct being imputed to the state. In short, the indemnity claim adds nothing to a *parens patriae* action unless the state can show both (1) another basis for the legal obligation of the manufacturer to reimburse victims of product-related harms and (2) a legal obligation that the state pay for the product-related harms because of some relationship between the state and the manufacturer. Since the initiation of the tobacco litigation, complaints filed by attorneys on behalf of states or municipalities in mass products tort actions have commonly included indemnity claims against product manufacturers for damages resulting from product-caused public health or safety problems, but courts almost universally reject such claims in this context.[58]

These three vague torts applied against product manufacturers for the first time—public nuisance, unjust enrichment, and indemnity—comprised all of the causes of action in Mississippi's complaint against tobacco manufacturers, the first state complaint filed in the tobacco litigation. As previously indicated, other states included a variety of causes of actions in their complaints against the tobacco manufacturers. Even though similar litanies of causes of actions continue to be included in *parens patriae* complaints against manufacturers of products causing public health crises, none has proved viable when tested in the courtroom.

NEGOTIATIONS LEADING TO A NEW REGULATORY REGIME

By early 1996, executives of tobacco companies were ready to explore settlement of the states' *parens patriae* litigation. The public was outraged by the newly disclosed revelations of nicotine manipulation. The FDA and its aggressive commissioner were moving toward regulation and perhaps prohibition of tobacco products. Lawyer fees and other legal expenses exceeded six hundred million dollars per year for the six largest tobacco companies.[59] A Florida jury had returned a $750,000 verdict in favor of Grady Carter against Brown & Williamson Tobacco Corp. in the first action brought by an individual cancer victim after the disclosure of the tobacco documents showing nicotine manipulation.[60] With that verdict as background, the potential liability of tobacco companies in future litigation appeared virtually limitless. Within six months after the Carter verdict, one of the more modestly sized tobacco companies, Liggett, reached separate settlements of the *Castano* litigation (later nullified by the class decertification) and, shortly thereafter, with twenty-two states.[61] As part of the agreement, Liggett agreed to turn over documents proving that the tobacco industry had covered up the risks of smoking and to admit that cigarettes both caused cancer and were addictive.[62]

Perhaps most important, Trent Lott, the Republican U.S. senator from Mississippi who was soon to serve as Senate majority leader, had approached his brother-in-law Dickie Scruggs, one of the plaintiffs' attorneys involved in the states' tobacco litigation, and asked him to explore settlement. Senator Lott perceived that unless the litigation was settled, strong voter sentiment against the tobacco industry might harm the political prospects of the Republican Party, which generally was perceived to favor the interests of the tobacco industry.[63] With the encouragement of Senator Lott and representatives of the Clinton White House, the two sides began negotiations through intermediaries in 1996, later followed by many rounds of direct negotiations between two teams representing the tobacco industry and the state attorneys general and *Castano* attorneys.[64] The state attorneys general and their partners from the mass plaintiffs' bar announced the Global Settlement Agreement on June 20, 1997.

The Global Settlement Agreement never took effect, because it required congressional action to implement at least two aspects of its provisions. First, the settlement recognized the Food and Drug Administration's regulatory authority over tobacco products. Second, legislation was necessary because the agreement would have eliminated or reduced the legal entitlements of groups that had not been parties to the agreement, including individual victims of tobacco-related illnesses, individuals who might have been represented in class actions in the future, and those tobacco companies that had neither negotiated nor signed the agreement. Such legislation was never passed. Instead, on November 16, 1998, the states involved in the *parens patriae* litigation reached agreement on the so-called Master Settlement Agreement (MSA), a separate, weaker settlement that did not require the approval of Congress. The negotiations between the state attorneys general and the tobacco company representatives that eventually led to the Global Settlement Agreement and the MSA represented a lawmaking process that served as an alternative to legislation or administrative rule making.

I evaluate the Master Settlement Agreement in chapter 8. Although the MSA was a weaker version of the original agreement, it nevertheless imposed a new regulatory regime governing how the tobacco industry conducted its business. The MSA meant that *parens patriae* litigation had enabled the judicial system for the first time to hold tobacco manufacturers financially accountable (albeit through settlement) for the public health problems that the consumption of their products created. Further, whatever the shortcomings of the settlement agreement, the negotiated resolution was—and remains—the largest settlement of a civil action ever. Tobacco manufacturers paid for at least a portion of the harm they had been a factor in creating. What the tobacco litigation did not yield, because it settled before significant rulings on the merits of its claims, were judicial tests of either the novel, more broadly interpreted conceptions of the states' *parens patriae* standing or the use of vague causes of action, such as public nuisance, against product manufacturers.

A NEW TARGET: GUNS

The perceived success of the tobacco litigation led government officials, the tobacco lawyers, and public health advocates to consider *parens patriae* litigation as a means of solving other public health problems. A handful of firms specializing in mass products torts had both compiled huge war chests as a result of the settlement and developed highly specialized expertise that would prove useful in actions against the manufacturers of other products. Attention initially focused on the epidemic of inner-city violence that arguably resulted from the ready availability of handguns. Since the early 1980s, individual victims of gun violence had attempted to recover from gun manufacturers.[65] Now, public health advocates reasonably viewed the harms resulting from the proliferation of handguns as a public health problem, opening the door to state *parens patriae* and municipal actions.[66] Within a few years, Congress effectively ended such litigation against handgun manufacturers by passing the Protection of Lawful Commerce in Arms Act,[67] but in the meantime, such litigation yielded a considerable portion of the common law that continues to govern *parens patriae* litigation against product manufacturers.

The link between the *parens patriae* litigation against tobacco companies and the subsequent assertion of similar claims against gun manufacturers is not difficult to trace. Professor David Kairys, a key figure in at least two major legal actions brought by municipalities against gun manufacturers, reveals that in 1996, the early tobacco litigation "did play some role in my thinking about a possible governmental suit."[68]

In October 1998, New Orleans filed the first municipal litigation against the gun industry, alleging that firearm manufacturers, retailers, and distributors had "caused the city to pay out large sums of money to provide services including . . . necessary police, medical, and emergency services, health care, police pension benefits and related expenditures, as well as to have lost substantial tax revenues due to lost productivity."[69] The city's petition asserted liability under Louisiana's products liability statute,[70] in other words, a traditional products liability cause of action. Its proponents were the Castano Safe Gun Litigation Group, one of the groups of tobacco lawyers who had garnered both expertise and resources from the tobacco litigation.[71] Within two weeks of the filing of the New

Orleans litigation, the City of Chicago filed an action against gun manufacturers, but one with a different genesis and legal theory.[72] Chicago's suit originated in Mayor Daley's request to corporation counsel to find some way to hold the gun industry liable, eschewed the use of private plaintiffs' attorneys, and focused on a public nuisance cause of action.[73] Within a couple of years, dozens of cities and counties and two states had filed actions against gun manufacturers.

These *parens patriae* actions against gun manufacturers, like those against tobacco and lead pigment manufacturers, resulted from a perception that the regulatory processes established by the political branches of government were failing. In theory, the federal Bureau of Alcohol, Tobacco, Firearms and Explosives regulated the distribution of firearms by gun dealers.[74] Yet in 1998, the bureau conducted onsite inspections of less than 5 percent of all federally licensed dealers, and over a fifteen-year period, it revoked the licenses of only 373 dealers. Before the filing of the New Orleans gun litigation, Mayor Morial had failed in his attempts to have the state legislature pass stronger gun legislation.[75] The National Rifle Association (NRA) had been and continues to be a powerful lobbying organization in opposition to such legislation at both the federal and state levels.[76] As one plaintiffs' lawyer noted, however, "You don't need a legislative majority to file a lawsuit."[77] The leading plaintiffs' attorney who promoted and led the litigation efforts in New Orleans and in many other cities boasted that the plaintiffs' bar was "a *de facto* fourth branch of government."[78]

Many of the state and municipal lawsuits against gun manufacturers, as with those against tobacco manufacturers, rested on novel legal theories (e.g., public nuisance) that, when combined with the nature of *parens patriae* litigation, enabled the plaintiff to circumvent problems of proof and other legal defenses. The assertions in the complaint in *Camden County Board of Chosen Freeholders v. Beretta, U.S.A. Corp.*,[79] for example, alleged that "defendants. . . have knowingly created, facilitated and maintained an oversaturated handgun market that makes handguns easily available to anyone intent on crime, including legally prohibited purchasers. This constitutes a public nuisance by unreasonably interfering with public safety and health."[80]

Other states and cities employed negligence law in an innovative manner to create new claims for negligent distribution or negligent marketing against gun manufacturers.[81] These claims alleged that manufacturers distributed or marketed their products in an unreasonable manner that foreseeably led to harm to members of the public.[82] Again, recovery under such claims did not require proof of individualized causal connections between residents directly experiencing the harm and specific gun manufacturers.

State and municipal actions against gun manufacturers, though clearly the progeny of the tobacco litigation, pushed the judicial process beyond what had happened in that earlier litigation cycle. At least in part, tobacco litigation seemed driven by a genuine desire to compensate the states for the huge medical bills they paid as a result of tobacco-related diseases. The regulatory intent of *parens patriae* handgun litigation was far more evident: such proponents as Kairys and such cities as Chicago blatantly sought to regulate gun manufacturers in a manner that state legislatures and federal regulatory authorities were unwilling to do.

In addition, in comparison with tobacco-related diseases, handgun violence is a more complex public health issue. A number of additional groups of actors, other than manufacturers, have a role in producing handgun violence. Most obvious, of course, are criminal assailants who use guns. The role of the manufacturer is comparatively tenuous. In dismissing actions against gun manufacturers, courts often have justified their decisions on the grounds that any harm suffered by the government plaintiff was too indirect or remote from the manufacturer's conduct.[83] The requirement of directness can be analyzed as one of either proximate cause[84] or the standing of the governmental unit to sue,[85] but it is essentially "based on policy considerations, of setting some reasonable limits on the legal consequences of wrongful conduct."[86] In *Ganim v. Smith & Wesson Corp.*, the Connecticut Supreme Court described the attenuated chain of causation between the gun manufacturer's actions and the injuries for which states and municipalities claim compensation.

Necessarily implicit in this factual scenario are the following links in the chain connecting the defendants' conduct to the plaintiffs. The manufacturers sell the handguns to distributors or wholesalers. . . . The distributors then sell the handguns to the retailers. . . . The next set of links is that the retailer then sells the guns either to authorized buyers, namely, legitimate consumers, or, through the "straw man" method or other illegitimate means, to unauthorized buyers, sales that likely would be criminal under federal law. Next, the illegally acquired guns enter an "illegal market." From that market, those guns end up in the hands of unauthorized users. Next, . . . unauthorized buyers misuse the guns to commit crimes or other harmful acts. . . . [T]he plaintiffs then incur expenses for such municipal necessities as investigation of crime, emergency and medical services for the injured, or similar expenses. Finally, as a result of this chain of events, the plaintiffs ultimately suffer the harms delineated.[87]

The causal connection between the manufacturer and the plaintiff's harm is far more tenuous in handgun litigation than in litigation against tobacco manufacturers, where the state most often seeks reimbursement of the medical expenses of the victim of tobacco-related disease who consumed the manufacturer's products. This factor, coupled with the other parties who contribute to the societal problem of handgun violence, helps explain the dismissal of most *parens patriae* cases filed against firearm manufacturers.[88] There were some notable exceptions.[89] Even after surviving initial judicial review, however, two of the major municipal lawsuits, those brought by the cities of Boston and Cincinnati, were voluntarily dismissed after city officials evaluated both mounting litigation costs and the small probability of success.[90]

State and municipal litigation against handgun manufacturers provoked a legislative reaction—at both the state and federal levels—that prohibited such suits. Legislation barring the industry's liability was passed in at least thirty-two states,[91] and in 2005, Congress passed the Protection of Lawful Commerce in Arms Act, prohibiting such litigation throughout the nation.[92] This legislative response might be viewed as proof that the constitutional structure of government works and that coordinate branches can "correct" an assertion of power by state attorneys general that is out of line with the sentiments of the popularly elected legislature. It is important to remember, however, that no other industry targeted by the state attorneys general is likely able to draw on the support of a lobbying organization as effective and well financed as the NRA, and probably no other industry is in a position to evoke the penumbra of freedoms protected by the Bill of Rights.[93] These unique circumstances cut short the attempts of states and municipalities to employ *parens patriae* litigation to address yet another public health problem, the devastating effects of the proliferation of guns, particularly in inner cities.

The explanation for why the tobacco litigation resulted in a settlement while the handgun litigation did not lies more in the domain of economics and politics than in the law. As mentioned previously, at the time the Global Settlement Agreement was reached between the states and the tobacco industry, many congressional leaders believed that the agreement was not sufficiently harsh in dealing with the industry. In contrast, the litigation against gun manufacturers led Congress to enact legislation protecting the industry from liability. In addition, while virtually all states eventually joined the litigation against the tobacco manufacturers, only a modest number of municipalities and counties sued the gun industry. In short, the forces arrayed against the tobacco industry were more substantial than those who fought the firearms industry.

Given the firearms industry's comparatively stronger political position, it is not surprising that handgun manufacturers decided to fight rather than settle. Any settlement by manufacturers would have encouraged other cities to sue. The industry is much smaller than the tobacco industry, and settlements or judgments from proliferating lawsuits might have bankrupted it. Though less important, there also were differences in the liability exposures facing the two industries. Most important, as described earlier, it was easier for courts to view the damages caused by the handgun violence as "remote" from the manufacture and distribution of the product and therefore not recoverable.

CONCLUSION

During the years straddling the beginning of the twenty-first century, state and municipal *parens patriae* actions afforded the first means of holding manufacturers financially responsible for the public health problems that resulted from their products. *Parens patriae* litigation avoided the need for the victim to identify the specific manufacturer that produced the cigarettes or lead pigment that caused her harm. It also rendered defenses based on the smoker's own conduct irrelevant. The tobacco litigation resulted in few judicial precedents, but its lessons were not lost on either public health officials or mass products tort attorneys. Litigation against handgun manufacturers yielded a handful of legal precedents helpful to these groups, but Congress ultimately intervened and put an end to such litigation.

CHAPTER 7

Judicial Rejection of Recovery for Collective Harm: Public Nuisance and the Rhode Island Paint Litigation

The viability of *parens patriae* litigation seeking to hold manufacturers accountable for product-caused health problems would ultimately be decided not in the tobacco or handgun litigation but in state actions against lead pigment manufacturers. Shortly after the parties in the tobacco litigation finalized the Master Settlement Agreement, Ron Motley, one of the lead attorneys representing the states, turned his attention to lead pigment. He prophesied that the actions of these manufacturers in concealing the dangers of lead-based paint "would make the tobacco companies look like choirboys."[1] Motley noted, "In tens of thousands of public housing units across this country, . . . lead-based paint is still on the walls and still causing serious health problems for children." He suggested, "There is no legal reason why the manufacturers should not be required to pay for the cleanup."[2] Motley's firm established an office in Providence, Rhode Island, headed by partner Jack McConnell.

On October 12, 1999, the State of Rhode Island, assisted by Motley's firm, filed the first state *parens patriae* action against the manufacturers of lead pigment, *State v. Lead Industries Ass'n.*[3] Later, one other state, Ohio, and dozens of municipalities and other local governments filed *parens patriae* actions against lead pigment or lead paint manufacturers. The eyes of the nation's public health and legal communities, however, were focused on the small state of Rhode Island. It was here, both mass plaintiffs' attorneys and manufacturers seemed to acknowledge, where the viability of *parens patriae* actions against manufacturers whose products caused public health problems would be decided.

The focus of the attention was on the once lowly tort of public nuisance. Although states employ a variety of legal theories in claims against product manufacturers, public nuisance claims have increasingly become the most important. As described previously, until the tobacco litigation, public nuisance law and products liability had never been intertwined. Chapter 4 described how the movement to use public nuisance law to remedy public health problems required an abrupt transformation that followed from seeing product-caused public health problems through the lens of environmental law instead of as a clustering of individual product liability claims.

EXPLICIT ACKNOWLEDGMENT OF *PARENS PATRIAE* LITIGATION
AS A REMEDY TO PUBLIC HEALTH PROBLEMS

It was no accident that Ron Motley and his partners chose the state of Rhode Island in which to file their first suit against the lead pigment industry. Rhode Island's vague definition of public nuisance law appeared to afford the state a better opportunity to prevail than was afforded in most other jurisdictions. The Rhode Island Supreme Court previously had stated, "The essential element of an actionable nuisance is that persons have suffered harm or are threatened with injuries they ought not have to bear." The court had further stated that liability for nuisance might exist "despite the otherwise nontortious nature of the conduct which creates the injury."[4]

More important, the incidence of lead poisoning among Rhode Island children was twice the national rate.[5] The state's attempts to regulate rental housing to avoid lead-based paint hazards were woefully inadequate. A leading state housing official admitted, "We have the oldest housing stock and one of the weakest laws in the country (to protect people from lead poisoning)."[6] The state failed to inspect hundreds of houses where children already had been poisoned, and even following inspections, the lead-based paint hazards in more than half of the houses (385 houses) remained uncorrected. In 1997, the state initiated enforcement proceedings in only five cases. The state's single environmental risk assessor in 1997 concluded, "We were sinking like the Titanic."[7] Four years after the Rhode Island litigation against pigment manufacturers began, the attorney general's office closed only six cases against landlords for lead paint violations and devoted only one-third of a single attorney's time to such enforcement actions, even though it committed substantial resources to the litigation against pigment manufacturers.[8]

The incidence of childhood lead poisoning was highly concentrated in poorer neighborhoods in Providence and in other cities. Within one thousand feet of an apartment where a brother and sister became lead poisoned, another fifty-five children had been poisoned within the previous seven years. The director of a Providence nonprofit center focused on prevention of lead poisoning blamed "structural racism." She noted that if there were boarded up houses in more affluent neighborhoods, "neighbors would be up in arms and have the political power to do something."[9]

THE COMBINATION OF *PARENS PATRIAE* STANDING AND THE PUBLIC NUISANCE CLAIM

Earlier in the litigation, on April 2, 2001, the presiding trial court judge, Michael A. Silverstein, ruled on the defendants' motion to dismiss the claims against the pigment manufacturers.[10] Issuing one of the few written opinions addressing a state government's standing to sue a product manufacturer in these circumstances, Judge Silverstein upheld the state's *parens patriae* standing to sue to collect damages resulting from a product-caused disease and to collect the costs to abate the alleged nuisance causing the disease. He relied on the broad language of the Supreme Court's opinion in *Alfred L. Snapp & Son* regarding "the special role that a State plays in pursuing its quasi-sovereign interests in 'the well-being of its populace.'"[11] He ignored the historical limitations on *parens patriae* actions described in chapter 6.

Judge Silverstein quickly dismissed the traditional causes of action based on claims of strict liability, negligence, negligent misrepresentations, and fraudulent misrepresentations. He found that damages under these claims were "entirely derived from alleged damages to others" and therefore were "too remote to be recoverable by the State."[12] The court ruled, however, that the State of Rhode Island had *parens patriae* standing "to vindicate certain interests of . . . its citizens,"[13] including standing to seek the abatement of a public nuisance.

Judge Silverstein also ruled that the claims of strict liability, negligence, negligent misrepresentations, and fraudulent misrepresentations were time-barred by the statute of limitations, because they could have been filed by individual victims, property owners, and victims of childhood lead poisoning at the time the injuries occurred. In contrast, one of the advantages of viewing the presence of lead-based paint through the lens of public nuisance law is that the public nuisance is a continuing one. Lead pigment remains on the walls of residences and arguably continues to represent a public nuisance. Because of both the continuing nature of the nuisance caused by lead pigment and the fact that the remedy sought, abatement, is prospective in nature, the statute of limitations arguably never runs—that is, the time during which the plaintiff must file its action never expires—under traditional public nuisance law.[14] Accordingly, actions to abate a public nuisance are never time-barred.

In addition to dismissing the traditional products liability claims, the trial court also limited liability under the state's Unfair Trade Practices Act to damages flowing from the conduct of defendants after 1970, when the statute was enacted.[15] Eventually, the claim was dismissed.[16] Four claims survived—those resting on the law of public nuisance, unjust enrichment, indemnity, and civil conspiracy. I previously detailed the court's reasoning in allowing the claims for unjust enrichment and indemnity to proceed, but the court subsequently dismissed these two claims,[17] as well as the civil conspiracy claim.[18] What remained was the public nuisance cause of action.

THE COMMON-LAW PROCESS

Any trial court judge hearing claims asserting that a product manufacturer should be held liable for the costs resulting from widespread product-caused health problems is in a tough position. He listens to testimony from lead-poisoned children or survivors of lung cancer victims. He is aware that legislative efforts to end such public health problems have been half-hearted, underfunded, or inept. He senses that because manufacturers have a disproportionate influence with legislators and administrators, he may be the last hope to remedy these problems. Sometimes, he knows that plaintiff's counsel selected his state in which to file bellwether litigation because of what she perceived as favorable law and that she perhaps even chose him as a judge inclined to be receptive to her socially conscious litigation effort.

The judge also knows that counsel for the plaintiff—whether a state, a municipality, a class action, or an individual victim—has presented him with arguments for liability premised on unusual, vague, and seemingly malleable causes of action. As a trial court judge, the authority of his decisions flows from the idea that, at least presumptively, he follows the past rulings of appellate courts in his jurisdiction. Yet relatively few courts have issued written opinions governing cases that seek to hold product manufacturers liable for latent diseases by using novel legal doctrines, such as public nuisance law.

The legitimacy of the common law differs from that which gives authority to statutes enacted by legislatures and to regulations and rulings of administrative agencies. Judges' rulings are not meant to be a reflection of the will of the electorate. Instead, the legitimacy of judicial lawmaking rests on the idea that judges begin their reasoning process with a presumption that they will follow precedents in earlier cases on similar facts. On one hand, few American judges, lawyers, and legal scholars today subscribe to the views of legal formalism or "mechanical jurisprudence" of the late nineteenth and early twentieth centuries, where the common law was conceived of as nothing more than rules deduced from precedents applied syllogistically to the facts of the present case. By the 1920s and 1930s, American legal realists argued that legal rules contained in precedents often did not and should not dictate the results of subsequent cases through deductive reasoning. Instead, legal reasoning, in its actual operation, reflected a wide variety of political, economic, and psychological judicial motivations that often were unacknowledged. Further, the legal realists argued that these factors played an appropriate and desirable role in the common law process. Yet even Karl Llewellyn, a leading spokesperson for legal realism, acknowledged that "it is prior decisions of judges that constitute the primary materials for all law-related activity."[19] Precedents, wrote Llewellyn, "indicate the experiential basis and the approved direction for developing norms, and thus the foundations of existing law."[20] The legal profession offers an "operating technique," according to Llewellyn, which enables the lawyer "to derive basic *guidelines* from legal rules, guidelines which do not enable him to derive the solution of the new case from old law, but which will bring the solution of the new case into harmony with the essence and spirit of existing law."[21] This understanding of the development of the common law is seldom articulated but appears to be deeply embedded in the professional philosophies of most lawyers and judges.[22]

During the past half century, legal scholars have found legitimacy in adjudication by acknowledging that although the law is indeterminate, other constraints should operate on judicial decision making. For example, during the 1950s, Henry Hart and Albert Sacks established a new model of legal philosophy, the legal process school, which sought to restore the legitimacy of neutral judicial decision making by establishing process standards of "reasoned elaboration."[23] According to Hart and Sacks, certain types of decisions should be made by the legislature, the political branch that expresses the will of the people. Other decisions would be made by the unelected judicial branch, but such judicial decision making should be checked by procedures designed to assure self-restraint and objectivity of process. Thus, the subjectivity often criticized by the legal realists would be avoided. In their classic text, Hart and Sacks identify justifications for the principle of *stare decisis*, the idea that courts should follow past legal precedents, including

1. enabling people to plan their affairs [so as to know] they will not become entangled in litigation;
2. [t]he desirability, from the point of view of fairness to the litigants, of securing a reasonable uniformity of decision; and
3. maximizing the acceptability of decisions [by basing them on a] reasoned foundation.[24]

THE TIDE TURNS AGAINST THE INDUSTRY

The liability phase of the Rhode Island litigation lasted over seven years in the trial court alone. Judge Silverstein issued eighteen written opinions and 174 separate orders.[25] The parties took over 412 depositions. In March 2002, Judge Silverstein ruled that the trial should proceed in three phases. In phase 1, the jury would be asked, "Does the presence of lead pigment in paint and coatings in homes and public buildings constitute a public nuisance?"[26] In phase 2, the jury would consider the liability of the various defendants regarding any such public nuisance, as well as any other claims remaining against the defendants. Finally, in phase 3, the court would consider damages and other remedial orders.

In the first trial, lasting seven weeks, the jury addressed only the first of these questions.[27] During the second day of its deliberations, the jury quite astutely posed the following question to Justice Silverstein: "Does the presence of lead paint mean: totally intact, totally flaking, peeling, chipping, etc.—and/or a combination of both[?]" The judge answered that the jury should consider it to mean a combination of both.[28] Rhode Island law requires a unanimous verdict from the six-person jury. After deliberating for four days, the jury was hopelessly deadlocked, and Silverstein declared a mistrial on the basis of the hung jury. According to the jury foreman, four jurors favored a verdict in favor of the defendants, and two sided with the state.[29] Shortly thereafter, Sheldon Whitehouse, then Rhode Island's attorney general, expressed his intention to retry the case.

The case came to trial for a second time on November 1, 2005.[30] The court and the parties had earlier agreed to abandon the originally planned multiphase trial structure and to consolidate all issues related to the liability of the defendants into a single trial. After three months of testimony from the state's witnesses, the four remaining lead pigment manufacturers rested their case without presenting any testimony.[31] Before submitting the case to the jury, Judge Silverstein dismissed the state's claims for compensatory damages for funds that already had been spent by the state in response to childhood lead poisoning.[32] This jury, like that in the earlier trial, again appeared to be deadlocked, and one of the jurors later reported that on the fourth day of deliberations, when another mistrial seemed imminent as a result of the jurors' stalemate, four jurors favored a defense verdict, and only two favored a verdict for the state.[33] On February 22, 2006, however, after eight days of deliberation, the jury returned with a verdict finding three of the four manufacturers liable for the costs of abating the public nuisance caused by lead in buildings throughout the state.[34] Jack McConnell, lead trial attorney for the plaintiffs, later remarked, "That's where the tide turned against the industry."[35] Shortly after the jury verdict, Judge Silverstein denied the state's claims for punitive damages, ruling that the level of egregiousness of the manufacturers' conduct did not warrant such retributive damages.[36] On July 1, 2008, the Rhode Island Supreme Court reversed the trial court's judgment, holding that the defendants were not liable for a public nuisance.[37]

PUBLIC HEALTH PROBLEMS = COLLECTIVE HARM = PUBLIC NUISANCE

After Judge Silverstein had dismissed the other causes of action in the Rhode Island litigation, what remained was the state's first cause of action, alleging that "the Defendants created an environmental hazard that continues and will continue to unreasonably interfere with the health, safety, peace, comfort or convenience of the residents of the State, thereby constituting a public nuisance."[38] In the defendants' motion to dismiss, they had asserted that the state had not alleged a violation of "a public right," the interest protected by the public nuisance claim.[39] Without substantial discussion, the trial court defined that requirement as "interfere[nce] with the health, safety, peace, comfort or convenience of the general community." It concluded that the manufacturers' conduct "unreasonably interfered with the public health, including the public right to be free from the hazards of unabated lead."[40]

On appeal, however, the Rhode Island Supreme Court reached the contrary conclusion. In *Lead Industries Ass'n*, it stated that "interference with a public right" is "the *sine qua non* of a cause of action for public nuisance."[41] In defining a public right, the court turned to a comment of the *Restatement (Second) of Torts* defining such a right: "'A public right is *one common to all members of the general public.* It is collective in nature and not like the individual right that everyone has not to be assaulted or defamed or defrauded or negligently injured.'"[42] The court stated that "a public right is the right to a public good, such as 'an indivisible resource shared by the public at large, like air, water, or public rights of way.'"[43] Perhaps most important, the court clarified that "the sheer number of violations does not transform the harm from individual injury to communal injury."[44] Applying this understanding of the concept of "public right" to the facts in the case considered, the Rhode Island Supreme Court concluded that the production and sale of lead pigment used in lead-based paint does not constitute a violation of a public right.

The court's ruling on this issue conformed to both precedent and the historical origins of public nuisance law. In interpreting the phrase "a right common to the general public" in 2001, the Connecticut Supreme Court had stated,

> Nuisances are public where they violate public rights, and produce a common injury, and where they constitute an obstruction to public rights, that is, the rights enjoyed by citizens as part of the public. . . . [I]f the annoyance is one that is common to the public generally, then it is a public nuisance. . . . The test is not the number of persons annoyed, but the possibility of annoyance to the public by the invasion of its rights.[45]

Understanding the traditional interpretation of the term *public right* requires examination of the historical origins of public nuisance, as described in chapter 4.

By way of review, the earliest cases involving public nuisances encompassed obstructions of public highways or navigable waterways and little else. At least by the sixteenth century, noxious and offensive trades that interfered with the health and comfort of those in surrounding areas sometimes were regarded as public nuisances and implicitly as violations of public rights, though that exact terminology was not yet used. By the mid nineteenth century, those who polluted navigable waterways and those who polluted air by conducting noxious trades or businesses were viewed as potentially liable under the tort. During this same time, state legislatures increasingly passed statutes declaring specific activities, such as disorderly taverns, gambling dens, or similar establishments, to be public nuisances and, implicitly at least, to be violations of public rights, on the grounds that they endangered public health, comfort, and morals. These legislative enactments greatly expanded the range of activities regarded as public nuisances, and on rare occasions, they led courts to expand the boundaries of public nuisance liability on their own initiative.

The manufacture and distribution of products rarely, if ever, causes a violation of a public right as such a right has been understood in the law of public nuisance. Products generally are purchased and used by individual consumers. Any harm they cause—even if the use of the product is widespread and if the manufacturer's or distributor's conduct is unreasonable—is not an actionable violation of a public right. If the owners of a fast-food chain were to sell millions of defectively produced hamburgers causing harm to millions of people who ate them, the violation of rights is a series of separate violations of private rights—typical tort or contract rights that the consumers might have—not a violation of the rights of the general public, or of the public *as the public*. The sheer number of violations does not transform the harm from individual injury to communal injury. Nor does the existence of widespread public health problems necessarily satisfy the requirements of a violation of a public right as such a right was understood both historically and at the time of the adoption of the *Restatement (Second) of Torts*.

What does a proper understanding of what constitutes a public right suggest about the viability of the claims brought by states against the manufacturers of cigarettes, handguns, and lead pigment? Regardless of the vast numbers of residents of the United States and other nations of the world who smoke cigarettes or use other tobacco products, any tobacco-related illnesses from which they suffer do not result from a violation of their public rights, unless the term *public right* is given a much different and more expansive reading than it has had during the previous centuries of public nuisance law. The decision to smoke—even one caused by youthful addiction at a time that the tobacco industry concealed or failed to disclose the dangers of cigarette smoking—is a private choice and results from a smoker's consumption of a product individually. While the user of tobacco products who is now suffering from cancer or other smoking-related disease may have potentially viable causes of action based on misrepresentation or tort claims traditionally used in product cases, he does not have a viable claim based on public nuisance. Conceivably, however, a victim of tobacco-related illness who could prove that his disease resulted from environmental tobacco smoke (secondhand smoke) inhaled while he was in a public building or other public space might be able to satisfy this first requirement of public nuisance.

In much the same way, the claims of those injured by criminals using handguns may satisfy this first requirement of public nuisance, interference with a public right.[46] Public authorities traditionally have criminally prosecuted or used injunctive relief to abate public nuisances threatening public safety and public health. Protecting citizens from crime and violence traditionally has been viewed as one of the core functions of public authorities. Further, specific state statutes occasionally declare that unlawfully manufactured or distributed firearms are public nuisances,[47] and in these instances, the statutory provision trumps the absence of the violation of the public right or any of the other common-law elements of public nuisance.

As recognized by the Rhode Island Supreme Court in *Lead Industries Ass'n*, the concept of public right does not appear to include the right of state and municipal governments to seek abatement of the presence of lead-based paint in private residences. In *City of Chicago v. American Cyanamid Co.*, the Illinois Appellate Court, quoting my earlier scholarship, stated,

> The concept of public right as that term has been understood in the law of public nuisance does not appear to be broad enough to encompass the right of a child who is lead-poisoned as a result of exposure to deteriorated lead-based paint in private residences or child-care facilities operated by private owners. Despite the tragic nature of the child's illness, the exposure to lead-based paint usually occurs within the most private and intimate of surroundings, his or her own home. Injuries occurring in this context do not resemble the rights traditionally understood as public rights for public nuisances purposes—obstruction of highways and waterways, or pollution of air or navigable streams.[48]

Even though the law regulates rental housing, in contemporary American law—for better or for worse—the provision of safe and adequate housing is not understood as an enforceable public right. Any rights granted tenants and their families under legal regulation applicable to rental housing are private rights, enjoyed by many renters in the community in their capacity as renters, not as members of the public.

In short, it is an unusual situation in which the manufacture or distribution of a product can be said to violate a public right as such a right has been understood in public nuisance law. Injuries and harm caused by products typically result from private consumption. However widespread the use of the product and however repetitive the harms it causes, such harms rarely violate a public right. I defer until later in this chapter the question of whether the Rhode Island Supreme Court would have been justified in departing from precedent on this critical "public right" element of public nuisance, as well as on the other traditional requirements for establishing liability premised on public nuisance.

SHOULD PIGMENT MANUFACTURERS OR PROPERTY OWNERS PAY?

A pivotal battle during the Rhode Island litigation involved which group of actors—the manufacturers of lead pigment or property owners—bore the legal responsibility for abating the alleged public nuisance. Pointing to the fact that most cases of childhood lead poisoning occurred in a handful of residential properties that the owners had allowed to deteriorate, defendants argued that intact paint (i.e., paint that has not been allowed to deteriorate) is safe. But the state contended, as the trial court eventually found, that even "intact paint is likely to deteriorate and cause harm in the future"[49] and that the presence of lead pigment therefore necessarily constitutes a public nuisance. One of the state's expert witnesses described lead paint as "an accident waiting to happen."[50]

Control of the Offending Instrumentality When the Harm Occurred

The war between the property owners and the lead pigment manufacturers was fought on several significant "fronts"—that is, legal issues—during the Rhode Island pigment litigation. In their appeal to the Rhode Island Supreme Court in *Lead Industries Ass'n*, for example, defendants argued that "a particular defendant will not be liable in nuisance unless s/he controls the offending instrumentality at the time the harm occurs."[51] Judge Silverstein rejected this argument and instructed the jury that a defendant could be held liable if it "participates to a substantial extent in the activity that causes a public nuisance . . . even after it has withdrawn from or stopped the activity and even if it is not in a position to stop the harm or abate the condition."[52] The supreme court again disagreed, holding that "a defendant must have *control* over the instrumentality causing the alleged nuisance *at the time the damage occurs*" in order for there to be liability in public nuisance.[53] If this element of the public nuisance tort applies, few, if any, product manufacturers can be held liable under a public nuisance theory. The manufacturer usually gives up control at the time that a product is sold.

Traditionally, the core purpose underlying public nuisance was to assure that public authorities had a legal remedy—injunctive relief—available to terminate conduct of a defendant that was violating a public right and injuring the public safety, health, or welfare.[54] The recovery of damages by a private party sustaining a special injury resulting from defendant's conduct was ancillary to the state's remedy to abate the nuisance. As Prosser noted in 1942, except when public nuisances resulted from a violation of statute or from a defendant's negligence or conduct of abnormally dangerous activities, "the question is not really the nature of the defendant's original conduct but whether he shall be permitted to continue it."[55]

If the principal purpose of the traditional law of public nuisance was to afford courts a means of terminating harmful conduct, the defendant was required to be in possession of the instrumentality causing the public nuisance. A court order directed to the defendant to abate the public nuisance cannot be effective unless the defendant is in control of the instrumentality causing the nuisance. In an earlier case, the federal district court in *City of Manchester v. National Gypsum Co.* had, for similar reasons, rejected a public nuisance claim seeking damages from manufacturers of asbestos insulating materials. The court recognized the traditional essence of the public nuisance action as a means for public authorities to terminate harmful conduct: "Liability for damage caused by a nuisance turns on whether the defendants were in control over the instrumentality alleged to constitute the nuisance, either through ownership or otherwise. If the defendants exercised no control over the instrumentality, then a remedy directed against them is of little use."[56]

Some courts, unlike those of Rhode Island and New Jersey, do not accept, as an element of public nuisance liability, that defendants must control the instrumentality causing the harm at the time when the harm occurs.[57] For example, in the Love Canal litigation, the federal District Court for the Western District of New York held Occidental Chemical Corporation liable for the *"creation* of the 'public health nuisance,'" even though it was not in control of the toxic waste site at the time of the litigation.[58] Further, a comment to the *Restatement (Second) of Torts,* arguably applicable to public nuisance as well as private nuisance, provides that a party that creates a nuisance or participates to a substantial extent in an activity creating a nuisance can be held liable,[59] thus contradicting any requirement of control over the instrumentality causing the nuisance.

If control of the instrumentality is not a required element, the remedies for a public nuisance expand significantly. No longer is a court relegated to ordering a defendant to abate the conduct causing a nuisance. Instead, the court can order the defendant to pay it the costs of abatement—which, although measured differently, look suspiciously like damages—and then can have the government abate the nuisance. It is not unusual for local ordinances to provide that a municipality has the authority to abate a nuisance and then to sue the person responsible for the nuisance for the costs. Similarly, of course, this is the structure of CERCLA.[60]

Public Nuisance as a Land-Based Tort

In a second argument pitting their own interests against those of property owners, defendants also argued in their briefs to the Rhode Island Supreme Court that "the context for the application of the law of public nuisance is site-specific."[61] Most often, they asserted, public nuisance involves "a defendant's use of property in a manner that unreasonably invaded a public space."[62] They argued that "the *situs* requirement for public nuisance is not just an historical accident" but instead overlaps with the requirement that defendants must be in control of the instrumentality causing the harm. Defendants put it thus: "If any nuisance needs to be abated, it must be abated in particular buildings, based upon site-specific conditions."[63]

Again, the Rhode Island Supreme Court agreed with the defendants: "A common feature of public nuisance is the occurrence of a dangerous condition at a specific location." The court similarly noted that "public nuisance typically arises on a defendant's land."[64] This suggestion that public nuisance is limited to harms resulting directly from the defendant's use of its own property obviously is inconsistent with any liability resulting from the mere manufacture or distribution of products. At least two other state supreme courts have made similar observations. In *City of Chicago v. Beretta U.S.A. Corp.*, defendant-manufacturers claimed that "in more than 2,500 reported cases in the over-100-year history of public nuisance law in Illinois, a public nuisance has been found to exist only when one of two circumstances was present: either the defendant's conduct in creating the public nuisance involved the defendant's use of land, or the conduct at issue was in violation of a statute or ordinance." The Illinois Supreme Court essentially accepted this conclusion and declined to expand public nuisance liability "to encompass a third circumstance—the effect of lawful conduct that does not involve the use of land."[65] Similarly, in *In re Lead Paint Litigation*, the New Jersey Supreme Court concluded that "public nuisance has historically been tied to conduct on one's own land or property as it affects the rights of the general public."[66] In contrast, a few courts reject any such restriction on public nuisance liability.[67]

Proximate Causation

The "blame game" as to whether pigment manufacturers or property owners should be responsible for abating the conditions causing childhood lead poisoning also arose in the context of determining which parties' conduct was a proximate cause of the conditions resulting in childhood lead poisoning. Defendants argued that the negligent failure of property owners to maintain their properties so as to prevent lead-based paint hazards constituted an intervening, superseding cause, breaking the chain of proximate causation between the defendants' conduct and the resulting harm. Judge Silverstein rejected this argument and found that the jury reasonably could have found that it was foreseeable to pigment manufacturers that property owners would allow the lead-based paint to deteriorate. Therefore, the property owners' conduct did not constitute a superseding cause, and the manufacturers' conduct remained a proximate cause of the public nuisance.[68]

The Rhode Island Supreme Court, however, held that the lead pigment manufacturers' conduct was not a proximate cause of the alleged public nuisance. The concept of proximate causation was described by the court as follows:

> As a practical matter, legal responsibility must be limited to those causes which are so closely connected with the result and of such significance that the law is justified in imposing liability. Some boundary must be set to liability for the consequences of any act, upon the basis of some social idea of justice or policy.[69]

The court indicated that the test of whether the defendants' conduct was a proximate cause of the public nuisance is "whether the injury is of a type that a reasonable person would see as a likely result of his conduct."[70] In doing so, the court conflated what are often regarded as two separate interpretations of proximate causation: the first, limiting liability on the basis of fairness and social policy factors; the second allowing foreseeability to serve as the sole test of proximate causation. It is probable that at least some courts would hold that the actions of slumlords and other property owners in allowing lead-based paint to deteriorate are foreseeable to the pigment producer.[71] In those circumstances, the manufacturer of the pigment would continue to be liable despite the intervening acts of the landlord in allowing the paint to deteriorate. Obviously, the Rhode Island Supreme Court reached the opposite result under the same articulated test.

As a pragmatic matter, the specific facts in *Lead Industries Ass'n* made it comparatively easy for the court to conclude that the lead pigment manufacturers' conduct was not the proximate cause of the harm, childhood lead poisoning. The manufacturers' conduct occurred at least thirty years ago and, in some cases, more than a century ago. The acts of landlords in failing to properly maintain their residential units were more proximate in time, were in violation of state law, and—because the risks posed by small amounts of lead in a child's bloodstream are so much better understood today than at the time when most of the lead pigment was produced—were arguably more egregious in character. These same facts will not necessarily be present in other public nuisance actions against product manufacturers and other businesses. A harm allegedly resulting from product exposure or other business activity sometimes, but not always, requires irresponsible conduct on the part of others that is more proximate in time than the business's alleged contribution to the harm.

Conflict with the Existing Statutory Framework

In the battle to decide whether the responsibility to abate the conditions causing childhood lead poisoning should rest with pigment manufacturers or with property owners, the Rhode Island courts were not writing on a clean slate. In 1991, the Rhode Island General Assembly had enacted the Lead Poisoning Prevention Act with the express purpose of protecting "the public health and public interest by establishing a comprehensive program to reduce exposure to environmental lead and prevent childhood lead poisoning."[72] The act required property owners to prevent and eliminate lead-based paint hazards and focused on those residences where a child already had been poisoned. As noted earlier, the 1991 act had not been successful in preventing childhood lead poisoning. Indeed, the state's prevention efforts had been less successful than those of other states. In June 2002, the Rhode Island General Assembly overwhelmingly approved legislation substantially strengthening obligations of rental property owners to avoid lead-based paint hazards and, for the first time, imposing an inspection requirement, at least for some housing units, prior to the time when a child was poisoned.[73] The state also adopted extensive regulations to implement the act.[74]

Throughout the decade spanning the Rhode Island litigation, the incidence of childhood lead poisoning in the state declined substantially, with the help of stronger legislation and enforcement actions aimed at the responsibility of landlords to eliminate lead-based paint hazards. Before the filing of Rhode Island's lawsuit against the pigment manufacturers in 1999, more than 11 percent of all Rhode Island children had newly detected elevated blood lead levels (EBLs), but less than two percent had newly detected EBLs in 2005, a 47 percent reduction from the immediately preceding year of 2004.[75]

The recent decision of the Rhode Island Supreme Court and the similar decision from the New Jersey Supreme Court both rest significantly on the courts' understanding that forcing lead pigment manufacturers to abate the presence of lead in residences throughout their respective states would conflict with legislatively enacted statutory schemes for reducing the incidence of childhood lead poisoning. In the Rhode Island opinion, the court stressed that the legislature had enacted a program for preventing childhood lead poisoning, which the court found inconsistent with holding the manufacturers of lead pigment liable. Noting that "the General Assembly has recognized that *landlords* . . . are responsible for maintaining their premises and ensuring that the premises are lead-safe," the court argued that the legislative enactments "did not include an authorization of an action for public nuisance against the manufacturers of lead pigments."[76] The court also made it clear that its decision did "not leave Rhode Islanders without a remedy,"[77] because the existing statutory scheme allowed tenants to seek injunctive relief against landlords who had created lead-based paint hazards by allowing their properties to deteriorate.

If anything, the existence of competing state legislative enactments were even more integral to the New Jersey Supreme Court's dismissal of public nuisance claims against the lead pigment and paint manufacturers. The court stated,

> In examining the Lead Paint Act and its relationship to public nuisance generally, we find its focus on premises owners as the relevant actors to be instructive. The significance is that the presence of lead paint in buildings is only a hazard if it is deteriorating, flaking, or otherwise disturbed. . . . Viewed in this light, we must conclude that the Legislature, consistent with traditional public nuisance concepts, recognized that the appropriate target of the abatement and enforcement scheme must be the premises owner whose conduct has, effectively, created the nuisance.[78]

Later, the New Jersey Supreme Court concluded that an independent claim for public nuisance against a product manufacturer would be inconsistent with a separate and distinct statute, the state's comprehensive and inclusive Product Liability Act.[79] In chapter 9, I address a related, but distinct, issue: do the actions of the attorney general, in a *parens patriae* case where he seeks to address a public health problem in a manner inconsistent with a preexisting legislative scheme, raise constitutional separation of powers concerns?

SOLVING COMPLEX INFRASTRUCTURE PROBLEMS WITHOUT ALL PARTIES BEFORE THE COURT

The blame game between pigment manufacturers and property owners manifested itself in yet another way at the trial court, when the defendant-manufacturers tried to join Rhode Island's state public housing authority and more than three hundred thousand unnamed private property owners as third-party defendants in the litigation.[80] The manufacturers alleged that the property owners, instead of the manufacturers, should be forced to abate the public nuisance and that if the manufacturers were to be held liable, they should be reimbursed, in whole or in part, by the property owners. The third-party complaint alleged that the property owners "failed to maintain their properties, disregarded their duties under federal or state law to keep any lead product on their properties in a safe condition, have ignored other duties as property owners and managers to protect children residing in their properties from lead hazards, or have failed to comply promptly with notices of health violations at their properties."[81] The state accused the manufacturers of wanting to try the case on a "property-by-property" basis, apparently for two reasons. First, this approach would enable the defendants to argue that the state could not establish cause in fact because it could not prove which defendant(s) made the pigment contained in a particular house. Second, focusing on each property individually would facilitate the defendants' efforts to prove that it was the owner's poor property maintenance that had substantially contributed to the conditions that caused childhood lead poisoning.[82] During the next two years, the parties battled over discovery of facts related to specific properties[83] and over whether property owners should be notified of the lawsuit, since their properties arguably could be declared to be part of a public nuisance.[84] In the meantime, in February 2002, Judge Silverstein set aside the defendants' motion and announced that the trial would proceed in phases, with any proceedings against third-party property owners delayed until after a determination of the manufacturers' liability for abatement of the public nuisance. Eventually, Judge Silverstein prohibited the defendants from introducing property-specific evidence at trial.[85]

On one hand, the addition of three hundred thousand third-party defendants to an already complex legal action clearly was impossible. On the other hand, this ruling left unresolved the legitimate question of whether property owners, at least those who failed to maintain their properties and who thus contributed to childhood lead poisoning, should bear some or all of the legal responsibility for the risks posed by lead-based paint hazards.

Only by defining the public nuisance as the presence of lead contained in paint and not as the risks that contributed to childhood lead poisoning did the trial court avoid (or at least defer) this extremely difficult issue. Further, if the courtroom was to be the forum in which accountability for causing childhood lead poisoning was to be resolved, it was unsound, as a matter of both law and public policy, to exclude consideration of the malfeasance and nonfeasance of property owners. Their failure to prevent and to remedy lead-based paint hazards, often (as in Rhode Island since 2002) violates state or municipal law, contributes to causing the vast bulk of such cases, and continues to occur now (not just fifty or one hundred years ago). Following the trial, in which the defendant-manufacturers presented no evidence, the jury foreman, Gerald Lenau told the local newspaper, "Some of us thought a big part of the problem was poor maintenance."[86] Another juror, Kevin Destefanis, said, "Most of us wanted to blame the landlords."[87] Yet the state, in its briefs to the Rhode Island Supreme Court following the judge's entry of judgment on the jury's verdict, claimed that the jury "found that homeowners in Rhode Island were not 'villains' as the Defendants claim."[88] Instead, it argued that the "extreme level of vigilance" by property owners was "another harm that the public ought not have [had] to bear."[89]

THE NATURE OF DEFENDANT'S CONDUCT REQUIRED FOR PUBLIC NUISANCE LIABILITY

Finally, the manufacturers argued throughout the case that any finding of liability on their part required the trial court to find that the public nuisance was intentional and unreasonable, was negligently maintained, or resulted from an abnormally dangerous activity, thereby creating strict liability for any harms caused. In most other jurisdictions, at least until the recent cycle of litigation seeking to hold product manufacturers liable for the costs of abating public health problems, it has been generally understood, as provided in the *Restatement (Second) of Torts*, that the term *public nuisance* describes not a type of tortious conduct leading to a defendant's liability but, instead, a type of harm resulting from a defendant's conduct and for which he may be held liable only if his conduct is otherwise tortious.[90]

As early as the thirteenth century, Bracton concluded that for there to be a public nuisance, a defendant's conduct must be both "injurious and wrongful."[91] Prosser, reviewing the existing case law in 1966, concluded that public nuisance "is always a crime."[92] A comment to the *Restatement (Second) of Torts* is less restrictive but still sets important limitations on liability.

The defendant is held liable for a public nuisance if his interference with the public right was intentional or was unintentional and otherwise action-able under the principles controlling liability for negligent or reckless con-duct or for abnormally dangerous activities. . . . If the interference with the public right is intentional, it must also be unreasonable.[93]

Interesting issues remain, however, as to the meaning of the words *intentional* and *unreasonable* as they are used in one of the alternative require-ments contained in the restatement comment.

In the context of public nuisances allegedly caused by products, the word *intentional* means more than that the manufacturer or distributor in-tends to distribute products. According to section 825 of the *Restatement (Second) of Torts*, "an interference with the public right, is intentional if the actor . . . knows that it is resulting or is substantially certain to result from his conduct."[94] It is possible, in some circumstances, that the requirement of intentional or negligent conduct arguably may be met in the distribution of products. In *City of Cincinnati v. Beretta U.S.A. Corp.*,[95] for example, the court found that the allegations of the city's complaint that defendants "intentionally and recklessly market, distribute and sell handguns that de-fendants know, or reasonably should know, will be obtained by persons with criminal purposes" was sufficient to withstand a motion to dismiss.[96]

Not surprisingly, the Rhode Island Supreme Court, while adopting the "unreasonable interference" provision of section 821B of the *Restatement (Second) of Torts*,[97] did not explicitly adopt comment e's alternative re-quirement of intentional conduct. Other jurisdictions often require the *Restatement*'s formulation of conduct that is intentional and unreasonable, otherwise tortious, or in violation of a specific statute, but the Rhode Island Supreme Court had previously stated that "plaintiffs may recover in nui-sance despite the otherwise nontortious nature of the conduct which cre-ates the injury."[98] In *Lead Industries Ass'n*, the court did attempt to define an "unreasonable interference," even though its discussion of this issue was *dicta* and did not play a direct role in the court's holding. It acknowledged that "what is reasonable . . . is not determined by a simple formula."[99] But the court then proceeded to provide useful guidance on an issue that often has bedeviled courts and commentators.[100]

Whether an interference with a public right is unreasonable will depend upon the activity in question and the magnitude of the interference it cre-ates. Activities carried out in violation of state laws or local ordinances gen-erally have been considered unreasonable if they interfere with a public right. Activities that do not violate the law but that nonetheless create a substantial and continuing interference with a public right also generally have been considered unreasonable.[101]

THE RESPECTIVE BOUNDARIES OF PUBLIC NUISANCE AND PRODUCT LIABILITY

As described previously, the fundamental disagreement between the state and the defendants in the case was whether it was legitimate to characterize the harm complained of as an "enormous public health problem"[102] and therefore to view it "collectively,"[103] rather than as an amalgamation of claims by individual property owners and lead-poisoned children. The manufacturing community argued that this was an attempt to "displace more traditional causes of action (i.e., products liability law)."[104] The state responded that just because manufacturers might be sued under product liability law does not make them immune under other legal claims, such as public nuisance.[105]

The trial court accepted the state's theory that the action against the pigment manufacturers was "*not* a products liability case."[106] According to Judge Silverstein, this obviated the need for the state to satisfy the traditional requirement that in cases against product manufacturers, a particular victim's harm must be found to have been caused by a product manufactured by a specific defendant. Instead, all the state needed to prove was that any particular defendant "caused or substantially contributed"[107] to the collective harm. In doing so, the court found a way to bypass the traditional individualized causation requirement that always had prevented victims of childhood lead poisoning or property owners from recovering against the manufacturers of either lead pigment or lead-based paint. Obviously, the Rhode Island Supreme Court reached the contrary conclusion, similar to the one reached a year earlier by the New Jersey Supreme Court,[108] that public nuisance was not a cause of action available for the sale of an unsafe product.[109]

With rare exceptions, other courts also have rejected the idea that public nuisance law is broad enough in scope to encompass claims based on defendants' manufacture and distribution of lawful products. Instead, states and municipalities are left to pursue more traditional product liability claims in subrogation actions alleging product-caused public health problems. Proving that a product has caused harm to the public health but has not violated a "public right" (as courts have understood that term in the common law of public nuisance) is not enough. Accordingly, the government's subrogation claims against manufacturers of fungible products, such as lead pigment, that cause latent diseases fail without a tort allowing recovery for a collective harm, because the state cannot prove a direct causal link between a particular citizen and a specific manufacturer. In tobacco cases, because the state, as subrogee, "steps into the shoes" of its citizen suffering from cancer or other tobacco-related disease, its recovery is barred by the victim's subjective knowledge of the risks of smoking or by "common knowledge" of such risks among members of the public.

As the Court of Appeals for the Third Circuit explicitly recognized when it rejected a county government's public nuisance claims in an action against gun manufacturers, governments' attempts to use public nuisance law when suing in a *parens patriae* capacity are patently intended to circumvent "the boundary between the well-developed body of product liability law and public nuisance law."[110] The Court of Appeals for the Eighth Circuit commented that to allow recovery for public nuisance "regardless of the defendant's degree of culpability or of the availability of other traditional tort law theories of recovery" would allow nuisance to become "a monster that would devour in one gulp the entire law of tort."[111]

The Rhode Island decision rejecting the application of the public nuisance claim against product manufacturers inflicts a serious, perhaps fatal wound to the idea of using *parens patriae* litigation to solve product-caused public health problems. Public nuisance had represented the most promising of the collective liability theories that would have enabled state and municipal governments to recover against manufacturers despite the obstacles posed to individual victims suffering from product-caused public health problems. The plaintiffs' bar specializing in such litigation probably will try to resurrect such claims as negligent marketing and distribution, which arguably displayed some hints of legal viability in the past. Similarly, the decision of the Wisconsin Supreme Court extending so-called risk contribution theory (a variant of market share liability, discussed in chapter 3) to lead pigment manufacturers probably will inspire courts or legislatures in a few other jurisdictions to adopt a similar approach. Like public nuisance law, risk contribution or market share theories enable the state government to proceed against the manufacturers of fungible products that cause public health problems, even when the government cannot establish a link between a particular victim and a specific manufacturer. When all is said and done, however, the decision of the Rhode Island Supreme Court, even if it is not the final skirmish of the decade-long wave of *parens patriae* litigation against manufacturers, almost certainly is the decisive battle. The lesson, a stark one for many public health advocates, is that if there are to be government solutions to product-caused public health problems, such as childhood lead poisoning, they must originate in the legislature, not the courts.

EVALUATING THE RHODE ISLAND DECISION

The question remains, was the Rhode Island Supreme Court correct in rejecting the state's attempt to hold product manufacturers liable on a public nuisance theory? Any notion that the use of *parens patriae* public nuisance claims against product manufacturers can be confined to a few product-caused public health problems seems unrealistic. In a mass production economy, most product-caused harms are repeated ones. A few products cause harm because of manufacturing defects, that is, defects that make them less safe than the product's intended design.[112] Most often, though, it is an inherent characteristic of a product that causes harm. If a manufacturer makes a product that contributes to harm experienced by a number of consumers, it is easy to characterize the repetitive harms as a public health problem and accordingly as a public nuisance, even if the requirements for liability under products liability law are not met. It is not just tobacco-related illnesses, childhood lead poisoning, or other widespread diseases that arguably fall within the category of public health problems and hence public nuisances. The marketing and distribution practices of handgun manufacturers have been held to be a public nuisance. As seen in chapter 4, automobile safety experts conceptualize the design of automobiles that roll over too easily or slip independently from one gear to the next as a public health problem. The scope of public nuisance law, as conceived by some judges and attorneys general, could encompass virtually any harm resulting from inherent characteristics of any mass-produced product.

Is a Fundamental Change in the Law of Public Nuisance Warranted?

The successful use of public nuisance claims against product manufacturers would have required a significant change in public nuisance law. But this does not necessarily mean that it would have been an unprincipled change in the common law. Occasionally, a series of opinions does represent a major tectonic shift in tort law, such as the opinions pioneering the adoption of strict products liability in the early 1960s or the opinions that abolished contributory negligence as a total bar to recovery a decade later.

Reduced to its essentials, the principle that state attorneys general and members of the mass plaintiffs' bar have been urging courts to accept is that if a manufacturer contributes to harm to many individuals in society, it should pay. This sounds like a very common-sense rule from a lay perspective, albeit one very much at odds with the existing common law. The issue then becomes whether any such change in the law of public nuisance, which would represent one of the most significant breaks with past precedents, is warranted. Should the law of torts allow public nuisance claims by the state against product manufacturers in order to address product-caused public health problems and other repetitive product harms?

As addressed in chapter 3, instrumental approaches to tort law and the formalist version of corrective justice theory offer two contrasting views of the principles that animate tort law. On the one hand, Guido Calabresi's progressive version of law and economics subscribes to the goals of loss distribution and loss minimization. On the other hand, corrective justice perspectives focus on the rights of victims to be compensated for harm they suffer as a result of the wrongdoing of tortfeasors.

PARENS PATRIAE PUBLIC NUISANCE ACTIONS AND INSTRUMENTAL PERSPECTIVES

Loss Minimization

Under instrumental conceptions of tort law, a primary focus is to identify tort rules that will best minimize the costs of harms. The question then becomes whether principles of loss minimization suggest that the tort of public nuisance should be expanded to allow state government to recover from manufacturers the costs resulting from product-caused diseases. Presumably, the prospect of being forced to compensate state governments for the costs resulting from product-caused diseases is likely to encourage manufacturers to be more careful when they distribute products, to test products more completely, and to more adequately warn consumers. Yet two significant sets of factors hint that in terms of loss minimization, the impact of liability rules on manufacturers whose products result in diseases decades later may not be as great as expected.

These same factors often are present in more traditional liability claims against the manufacturers of products produced many years ago, including those filed by individual victims of product-caused harms. My purpose here is not to suggest that these claims, widely accepted as legitimate, should be barred. I tackle only the very narrow issue of whether policy considerations operating in the specific context of *parens patriae* litigation against the manufacturers of products that contribute to public health problems justify profound changes in the traditional understanding of the public nuisance claim. If a common-law system based on precedents means anything, it at least means that the burden of persuasion is greater for those who challenge fundamental principles of the law of public nuisance than it is for those who seek to preserve well-accepted principles governing the law of strict products liability and negligence.

The Time Lag between Product Distribution and the Imposition of Liability. Mass products torts in which the victim manifests a latent disease many years after exposure to the toxic product or even shortly after exposure to the product but many decades after the manufacture and distribution of the product, such as in the case of childhood lead poisoning, pose particular challenges to the idea that tort liability can deter harmful behavior and lead to loss minimization. Obviously, with such a product as paint containing lead pigment, which has not been marketed for residential purposes since 1977, a liability judgment today has no effect on whether that particular product will be distributed in the future. The question remains, however, as to whether liability judgments against lead pigment manufacturers might deter the defendant-manufacturers and other manufacturers from manufacturing and distributing other harmful products in the future.

It is at best questionable whether current adjudications of liability against manufacturers of products that were manufactured and distributed two, three, or even twelve decades ago significantly alter the conduct of today's manufacturers. Corporate managers deciding whether to market a product, how much to invest in product safety testing, and what safety enhancements should be included in the product likely do not give much weight to the prospect of a liability judgment many decades into the future. W. L. F. Felstiner and Peter Siegelman have reached the same general conclusion about the impact of the passage of time on the ability of any accident compensation system to minimize losses: "latent injuries introduce empirical complications that overwhelm the assumptions on which the deterrent effect of tort compensation found in neoclassical economic theory is based."[113]

Those who study corporate behavior conclude that the prospect of liability more than a few years in the future—indeed, the prospect of anything more than a few years in the future—has little impact on corporate decision making. Steven S. Cherensky observes,

> The investment time horizon of utility-maximizing senior corporate managers tends to be short. Senior corporate managers tend to have relatively short tenures in their positions, as such managers often accede to their positions late in their careers and keep these positions for only a few years. More junior managers often move from position to position within the

firms (or worse, from firm to firm). Thus, few managers who invest in long-term research and development projects will be around to reap any benefits from such investments.[114]

The professional interests of corporate managers are probably even more focused on the short term when long-term exposure is only a possibility—as is always the case with a potential legal liability—and not a certainty.[115] Clayton P. Gillete and James E. Krier suggest that for the corporate manager, "certain and demonstrable gains are likely to enhance her standing in the firm more than would the tenuous avoidance of losses."[116]

Further, the tort of public nuisance poses unique challenges to the notion that legal liability for the conduct of manufacturers many decades ago affects the conduct of the defendant-manufacturers and other manufacturers operating today. Many courts have held that provided that a harmful condition is a "continuing" one, as would be the case if the presence of lead pigment on the walls of residential units was found to constitute a public nuisance, the statute of limitations for a public nuisance claim never runs, and the claim is never time-barred.[117] As such, legitimizing the use of public nuisance claims against product manufacturers likely extends the period of time between the manufacture and distribution of a product and any liability judgment resulting from a latent disease or other harm caused by the product.

Despite the arguably rational and profit-maximizing impact of discounting liability judgments a generation or more in the future, it is to be hoped that the corporate decision maker is motivated by other factors, such as a desire to maintain the long-term reputation of his employer or even the altruistic impulse to avoid knowing that one's actions or inactions may result in other human beings suffering from cancer or other latent diseases. Yet principles of profit maximization arguably produce a corporate culture that sometimes renders corporate managers blind to any consequences materializing in the distant future, including legal liabilities resulting from latent diseases. In short, even if the rational, wealth-maximizing decision of a corporate entity would be to eliminate or reduce currently profitable activities that may result in latent disease known to occur only decades after exposure, it is unlikely that corporate managers will make that decision, because of the disparity between their own career interests and the long-term financial interests of their employers.

It is improbable that the imposition of hundreds of millions of dollars in liability under a public nuisance theory, even thirty years down the road, would have no impact on corporations as they evaluate the possible toxic effects of such products, whether such products should be distributed, and how they should be distributed. The present analysis suggests, however, that legislatively or administratively imposed regulation should play a larger role in regulating products that cause latent diseases than in other areas of liability. Regulation of the processes of product safety testing, unlike liability for the results of inadequate testing or research, occurs before or simultaneously with the distribution of the product, not decades later. Thus, the delay between the manufacturer's conduct and the manifestation of the latent disease, facilitated by the failure of public nuisance to time-bar product claims, does not preclude loss minimization that could be achieved through the regulatory processes of legislatures and administrative agencies.

Who Is in the Best Position to Prevent Public Health Problems? If the goal of expanding the boundaries of public nuisance liability is to minimize the incidence of product-caused public health problems, courts also should ask which party is in the best position to prevent such problems. To use Calabresi's terminology, who is the "cheapest cost avoider"?[118] Liberal judges often reflexively conclude, "Manufacturers are usually the 'cheapest cost-avoiders.'"[119]

Many product-caused harms involve causal contributions from multiple actors. Arguably, the careless driver is more morally culpable for the injuries sustained by a passenger in an automobile accident than is the automobile manufacturer who could have designed the car in a way to minimize the injuries to the occupant. Yet courts quite legitimately find that the manufacturer is in the best position to reduce the incidence of enhanced injuries over a range of cases. In other words, the manufacturer is the cheapest cost avoider. In the instance of widespread public health problems caused by product exposure, at least two factors suggest the problematic nature of holding manufacturers liable on this same rationale.

First, a closer examination of the Rhode Island litigation suggests that, at least in some instances, imposing liability on the product manufacturer may "let off the hook" a party in a better position to minimize losses. For example, property owners, not the former manufacturers of lead-based paint, are at least now the cheapest cost avoiders for purposes of minimizing the losses resulting from childhood lead poisoning. As analyzed in chapter 1, most childhood lead poisoning is concentrated in poorly maintained housing. Paint containing lead has been banned for interior use since 1978. Public health officials now more fully appreciate the harm caused by even a small amount of lead in a child's bloodstream. Despite publicity regarding these effects and despite stronger federal, state, and local regulation, landlords continue to allow lead-based paint hazards to proliferate. Yet the net effect of a dramatically expanded definition of public nuisance that would enable a government plaintiff to reimburse property owners for the costs of lead remediation at the expense of paint manufacturers is to reward precisely the parties whose conduct is most responsible for causing childhood lead poisoning.

It is tragically ironic that it is federal, state, and local governments, which have failed so abysmally to fulfill their own responsibilities to prevent the public health problems caused by guns and lead pigment, who now are claiming that product manufacturers were in the best positions to prevent these public health problems and should be held liable. Consider, for example, the damages resulting from handgun violence. According to David Kairys, a key figure in at least two major legal actions brought by municipalities against gun manufacturers, "many state legislatures . . . have taken extraordinary actions to accommodate the industry and the gun lobby."[120] Further, the government obviously possesses the unique capability to deter the criminal's conduct through harsh criminal sanctions, a power not shared by the manufacturer of the handgun. While it is true that both manufacturers and local governments share responsibility for failing to "limit, or require . . . distributors and dealers to limit, the number, purpose or frequency of handgun purchases" and for failing to "monitor or supervise . . . distributors or dealers for practices or policies that facilitate access to handguns for criminal purposes,"[121] it is questionable as to whether the manufacturer is in a better position than the plaintiff-government to reduce the harms flowing from gun violence.

Similarly, state and municipal housing authorities failed to enact or enforce local housing codes that would have prevented the vast bulk of cases of childhood lead poisoning that result from poorly maintained older housing. For decades, federal and state governments required paint that contained highly concentrated lead for public housing projects. There is no evidence that what manufacturers knew of harm resulting from lead exposure was not available to others, including state and municipal governments.

The second problem with using the analysis of cheapest cost avoider in *parens patriae* products litigation is the overwhelming complexity of making such a determination in the case of many public health problems resulting from product exposure. As previously noted, the Rhode Island trial court understandably refused to allow defendants to discover the identities of, much less implead as third-party defendants, the property owners whose failures to maintain their properties had resulted in lead-based paint hazards that dramatically increased the risk of childhood lead poisoning. A common-law court is ill equipped to untangle the causal contributions to widespread public health problems. It is probably for these reasons that courts in Connecticut and California recently have declared the nonjusticiability of complaints alleging that defendant-operators of power plants and automobile manufacturers contributed to the public nuisance of global warming.[122]

These examples of product-caused public health problems—childhood lead poisoning and global warming—do not validate either the conclusion that product manufacturers are necessarily the parties best able to prevent such health problems or the expectation that common-law courts are capable of making these determinations. In short, neither scrutiny of the effects of time delay on loss minimization in cases of latent disease nor the analysis of cheapest cost avoider supports expanding the boundaries of public nuisance liability in *parens patriae* litigation alleging public health problems. Loss minimization does not appear to be a strong argument for overturning preexisting principles of the law of public nuisance, principles that would deny recovery to states in *parens patriae* actions arising from the manufacture and distribution of products. Again, more direct regulation, not litigation, is the answer.

Loss Distribution

The second important instrumental goal for the tort system usually is referred to as "loss distribution." If the costs of accidental harms are distributed widely across many people, they are likely to inflict "less pain" than if they are borne by the original victims.[123] To the individual victim of a product-caused health illness, the disease is likely to be an unexpected, unmitigated disaster. If, however, her damages are compensated by insurance or by the manufacturer—which passes along the costs of the accidents to other consumers—the impact on the victim is reduced.

The factor of loss distribution does not necessarily suggest, however, that the boundaries of public nuisance liability should be expanded to allow actions brought by states and municipalities against businesses for the manufacture and distribution of products. The government is able to spread losses through taxation even more broadly and with fewer transaction costs than manufacturers are able to do through insurance and increased product costs. In other words, in an action brought by a state government against a product manufacturer, the plaintiff, not the defendant-manufacturers, is the superior spreader of losses. Thus, loss distribution, standing in isolation, does not suggest that the boundaries of public nuisance should be expanded in order to accommodate *parens patriae* public nuisance actions seeking compensation for the costs of preventing or treating public health crises.

Administrative Cost Avoidance

Parens patriae actions against product manufacturers, coupled with either the distribution of Medicaid benefits to victims of tobacco-related disease or childhood lead poisoning, or payments by a court-appointed special master to property owners to remediate lead paint hazards, are a highly inefficient means of redressing or preventing the consequences of product-caused diseases. The transaction costs inherent in mass products tort litigation can be huge. A study by the RAND Corporation Institute for Civil Justice shows that victims of asbestos-related disease received 42 percent of the amount spent for litigation resulting from asbestos-related illnesses, with the remaining 58 percent going to attorneys, insurers' expenses and profits, the court system, expert witnesses, and the like.[124] Specific figures of this sort are not available for *parens patriae* product litigation, but lawyers' fees in the state tobacco litigation may have exceeded fifteen billion dollars,[125] and payments of these attorneys' fees have resulted, after the fact, in considerable litigation between the states and their attorneys.[126] In comparison, workers' compensation systems and other compensation systems that do not require an individual claimant to prove the injuring party's tortious conduct have much lower administrative costs than those of the tort system. Admittedly, little evidence is available to quantify the transaction costs involved in seeking and obtaining legislative approval of an administrative, no-fault compensation system and stronger *ex ante* regulation through the political branches.

Nevertheless, there is something disconcerting about the vast discrepancies among transaction costs in these various injury compensation systems and the fact that more than a few plaintiffs' attorneys have walked away from *parens patriae* litigation with hundreds of millions of dollars in fees, having valued their work at more than ten thousand dollars an hour. The tobacco litigation, as shown, resulted in cigarette price increases to raise the funds manufacturers were required to pay under the Master Settlement Agreement. The state governments could have obtained the same funds, far more efficiently, by taxing cigarettes. In short, the inefficiency of addressing product-caused public health crises through the common-law tort system is a strong factor suggesting that the preexisting law of public nuisance should not be overturned to allow its use to address product-caused public health crises.

Parens Patriae *Public Nuisance Actions and Corrective Justice*

Can corrective justice principles justify the dramatic expansion of public nuisance law to enable states to sue product manufacturers for the costs resulting from public health crises caused in part by product exposure? Remember that public nuisance law, as interpreted by Rhode Island and in some of the other states that have enabled such claims to be used in *parens patriae* litigation against product manufacturers, does not require that the state prove that the product manufacturer acted negligently or otherwise tortiously. Further, state and federal governments share blame with manufacturers for causing both tobacco-related illnesses and childhood lead poisoning. For decades, not only did the federal and state governments fail to enact legislation or adopt programs designed to minimize cigarette smoking, but the federal government also aggressively subsidized the tobacco industry.[127] These contributions of federal and state governments to the prevalence of tobacco-related illnesses arguably suggest that corrective justice principles do not support the conclusion that governments should be able to recover from tobacco manufacturers. In view of the records of states and municipalities in persistently failing to enact and enforce effective legislation controlling the distribution of handguns or requiring residential housing to be free of lead-based paint hazards, these governments do not stand on the moral high ground necessary to justify liability on corrective justice grounds when they come to courts as plaintiffs.

CONCLUSION

In 1921, Benjamin Cardozo, later a justice of the U.S. Supreme Court, declared that the judge is "not wholly free . . . to innovate at pleasure" but "is to draw his inspiration from consecrated principles."[128] More recently, Melvin Eisenberg echoed this fundamental principle when he wrote that the legitimacy of the judicial function within a common-law system "depends in large part on the employment of a process of reasoning that begins with existing legal and social standards rather than with those standards the court thinks best."[129] Yet the use of public nuisance law to remedy product-caused public health crises would turn previously accepted fundamental principles of both products liability law and public nuisance law on their heads without a reasoned justification. The common law of torts, of course, has experienced significant change before, though perhaps none that would be so lacking in historical or principled policy underpinnings.

It is easy to understand the plight of the trial court judge facing an underfunded state or municipality and wanting to help her fellow public officials solve a major health problem. Yet even if public policy provides compelling justifications for addressing tobacco-related illnesses, the epidemic of handgun violence, or childhood lead poisoning, the judge, after all, is a judge, not the state's secretary of public health. The court's legitimacy and the authority of its rulings stem from the public's acceptance of a lawmaking process premised on a presumption that precedent will be followed and that departures from precedents will be justified by a professional process of reasoned elaboration. The recent opinions of courts, such as the supreme courts of Rhode Island and New Jersey, refocus our attention on these fundamental principles of the common law. In doing so, they make it very unlikely that the judicial process is the vehicle that will provide solutions for product-caused public health problems.

PART III

A CRITIQUE OF PUBLIC
PRODUCTS LITIGATION

CHAPTER 8

Do Litigation Remedies Cure Product-Caused Public Health Problems?

Viewed superficially, the goal of the *parens patriae* tobacco litigation was to compensate the states for their expenditures resulting from tobacco-related diseases. Retribution also played a role: Ron Motley, one of the lead attorneys in the tobacco litigation, argued, "These gangsters have gotten a free ride for forty years."[1] The most important goal, however, was to change the conduct of the tobacco companies, by imposing an alternative regulatory system through judicial action, bankrupting the companies, or imposing sufficiently severe penalties for certain practices of the companies, particularly the practices of advertising to young people and artificially manipulating the nicotine content of cigarettes. As Graham E. Kelder Jr. and Richard A. Daynard asserted, "the failure of conventional forms of legislative and administrative regulation of tobacco products and the recent shift in the landscape of tobacco litigation indicate that tobacco product liability litigation provides one of the most promising means of controlling the sale and use of tobacco."[2] Daynard, a professor of law who was an early proponent of state litigation against tobacco manufacturers, was an important influence on Mississippi attorney general Michael Moore and others.[3]

The Rhode Island litigation against lead pigment manufacturers was an obvious attempt—even more obvious than the tobacco litigation had been—to use the judicial process to compel a solution to product-caused public health problems. The remedy originally sought by the states in the tobacco litigation was damages to reimburse the states for their past Medicaid expenditures for patients suffering from tobacco-related diseases. In *State v. Lead Industries Ass'n*,[4] however, Rhode Island attorney general (now U.S. senator) Sheldon Whitehouse initially asked the court in the state's complaint for a "judgment ordering the Defendants to detect and abate Lead [*sic*] in all residences, schools, hospitals, and public and private buildings within the State accessible to children." He also asked for "such other extraordinary, declaratory and/or injunctive relief . . . to assure that the State has an effective remedy."[5]

As previously described, the Rhode Island Supreme Court reversed the trial court judgment ordering the defendant-manufacturers to abate the nuisance consisting of lead in premises throughout the state. Prior to the state supreme court's decision, however, Judge Silverstein already had begun to consider the abatement process and had offered a road map for how it would be conducted. Because the Rhode Island lead pigment litigation was the first action in which a jury ordered defendants in effect to solve a massive public health problem, this unique experience has much to teach us about the remedial phases of litigation against manufacturers found to have contributed to complex, multifaceted public health problems.

Following the jury verdict, Judge Silverstein turned to the state to propose a plan for the abatement of the lead nuisance. On September 14, 2007, the Rhode Island attorney general, then Patrick C. Lynch, submitted the state's self-described "ambitious" Lead Nuisance Abatement Plan, detailed in a 126-page proposal that included twenty-four recommendations.[6] Judge Silverstein never issued a remedial decree, because of the reversal of the judgment entered on the jury's verdict finding a public nuisance, but it is probable that the court's remedial order would have substantially resembled the attorney general's proposal. If it had, the trial court's remedial decree would have been the first time that a court had issued such a broad-ranging equitable decree directing manufacturers to solve a product-caused public health problem. The court appeared to tip its hand that it intended to create an autonomous bureaucracy to administer the largest public works program in Rhode Island history and one of the largest in U.S. history. When the attorney general submitted his proposal to the court, he noted the similarities between childhood lead poisoning and public health problems that had been prevalent in an earlier era: "The Rhode Island plan has been crafted to enable . . . lead poisoning from exposure to lead paint . . . to join typhoid, cholera and tuberculosis as housing-associated diseases of the past."[7]

In this chapter, I assess the efficacy of the end products of *parens patriae* litigation in solving product-caused health problems. I first evaluate the negotiated outcome of the tobacco litigation against the goals originally set by those who promoted or brought the litigation. How well has the Master Settlement Agreement achieved Kelder and Daynard's regulatory goals, as well as the retributive goals of Motley? I also evaluate whether the settlement can be justified as a means of compensating the state for either past tobacco-related expenditures or current and future efforts to prevent smoking. I then describe the Rhode Island attorney general's proposed abatement plan in the lead pigment litigation and speculate both as to whether it would have been successful in eliminating childhood lead poisoning and as to how it would have altered the traditional understanding of what it is that courts do. Finally, I consider the argument, made by several notable scholars, that regulatory public health litigation can be justified as a means of influencing the regulatory policies of legislatures and administrative agencies.

Two factors distinguish the tobacco litigation from the lead pigment litigation. First, the tobacco litigation was resolved through settlement, while it appears that the Rhode Island litigation would have resulted in a court-imposed judicial decree. Second, because cigarettes and other tobacco products continue to be manufactured and distributed, reducing the use of such products is the primary means of preventing tobacco-related diseases, but because lead-based paint has not been sold for decades yet remains in place in residences and schools, where it continues to pose a risk, the prevention of childhood lead poisoning requires more comprehensive and challenging measures than merely reducing or prohibiting the use of a product.

THE TOBACCO SETTLEMENT

The Global Settlement Agreement

As described in chapter 6, the agreement originally reached between the tobacco companies and the states, the Global Settlement Agreement, fell apart because of the unwillingness of Congress to enact implementing legislation. If the agreement had become effective, it would have transformed the way the tobacco industry operated in the United States.[8] The Global Settlement Agreement substantially limited the nature of tobacco advertising and promotion, banning both such advertising icons as Joe Camel and the Marlboro Man and outdoor advertising of tobacco products. The companies agreed to the provisions of the proposed FDA rules limiting the sale of tobacco products to youth. Cigarette packages would have contained new, stronger rotating warnings about the health consequences of smoking. The agreement restricted smoking in the workplace and in indoor public facilities. It also imposed substantial restrictions on tobacco lobbyists.

Perhaps most important, the agreement acknowledged the right of the FDA to require "the gradual reduction, but not the elimination of nicotine yields, and the possible elimination of . . . other harmful components of the tobacco product."[9] Even under this original settlement, however, the FDA could act in this manner only if it could show that the lowering of nicotine levels or the banning of specified ingredients would substantially reduce health risks, be technologically feasible, and not result in a contraband market. Some representatives of the tobacco industry believed that it would be difficult for the FDA to meet these conditions, and public health critics were very critical of these limitations on the FDA's regulatory authority.

As a part of the originally proposed agreement, the tobacco companies agreed to pay $368.5 billion over twenty-five years. The terms specifically provided that these payments were to be reflected in the cost of cigarettes, which the parties anticipated would discourage smoking by youth. Of course, revenues from increased prices for cigarettes also meant that the companies would not really be paying the damages. Under the so-called look-back provisions of the agreement, the companies also would make substantial additional payments if periodic targets for reducing youth smoking were not achieved.

In terms of liability, the Global Settlement Agreement would have settled all state actions, class actions, other "aggregate actions" (e.g., those involving consolidation of claims), and individual actions alleging nicotine dependence or addiction. It preserved other actions brought by individual victims of tobacco-related diseases, but it left intact all the traditional obstacles to proving liability in such actions. Significantly, it also prohibited punitive damages. Aggregate compensatory damages paid to individuals were capped on an annual basis, and any amounts paid toward those judgments reduced the tobacco companies' required annual payments to the states. What the tobacco companies gained through the settlement were predictable caps on liability and the ability to move forward to achieve profitability in arenas other than the domestic cigarette market—namely, cigarette sales in foreign markets and other lines of business in the United States.

The Settlement Falls Apart

Even though the Global Settlement Agreement was negotiated in private by representatives of the tobacco companies, on one side, and those speaking for the state attorneys general and mass plaintiffs' attorneys, on the other side, it obviously represented a political resolution of a major public health problem. The agreement, however, required congressional action because it altered the legal rights and responsibilities of several groups that were not parties to the agreement, including individual victims of tobacco-related illnesses, individuals who might have been represented in future class actions, and tobacco companies that were not parties to the agreement. That enactment of congressional legislation never happened, and the Global Settlement Agreement was never implemented.

Following the announcement of the settlement agreement, many public health advocates blasted it as a sweetheart deal for the tobacco companies. Both David Kessler, who had recently resigned as FDA commissioner, and former surgeon general C. Everett Koop argued that the settlement would weaken the agency's control over tobacco products, at a time when the courts had not yet rejected the FDA's attempt to establish regulatory authority over cigarettes.[10] When Attorney General Moore met with representatives of the American Cancer Society and antitobacco advocates, they astutely asked him, "Who gave you the right to make health policy for the country?"[11] Tobacco farmers opposed the settlement because it contained nothing to protect their interests.[12] Trial lawyers—other than those few actively involved in the state litigation and *Castano* class actions, who stood to profit handsomely from the settlement—opposed it because it restricted the liability of the tobacco companies.

Led by Senator John McCain, a bipartisan group of senators introduced legislation that, had it been enacted, would have been much tougher on tobacco companies than the settlement agreement itself. The legislation would have required payments by tobacco companies of $516 billion, imposed more restrictions on advertising and promotion, and granted the FDA greater regulatory authority.[13] The tobacco industry believed that the McCain bill pushed it too far, and in April 1998, it indicated that it would fight the bill and withdraw from the settlement.[14] After an extensive lobbying and advertising campaign by the industry, the bill died on the Senate floor on June 17, 1998. Christine Gregoire, then Washington State attorney general and one of the lead negotiators for the state attorneys general, later reflected that the Global Settlement Agreement had been "a golden opportunity for major reform."[15]

The Master Settlement Agreement

As Congress deliberated the legislation called for by the global settlement, the four states with approaching trial dates in their *parens patriae* actions—Mississippi, Florida, Texas, and Minnesota—each settled for a sizable amount measured in billions of dollars; in some cases these settlements also included weaker versions of the Global Settlement Agreement's restrictions on tobacco companies' marketing to youth.[16] Meanwhile, Attorney General Gregoire and one of the lead plaintiffs' attorneys, Joe Rice, pushed for an alternative settlement of the *parens patriae* actions that would not require the approval of Congress.[17]

The respective bargaining powers of the states and the tobacco companies in the new round of negotiations differed greatly from those leading to the Global Settlement Agreement. Both the failure of Congress to act and the then-recent decision of the Court of Appeals for the Fourth Circuit in *Brown & Williamson Tobacco Corp. v. FDA*,[18] which held that the FDA lacked regulatory jurisdiction over tobacco under existing law, denied negotiators the ability to reach an agreement conferring the FDA such jurisdiction. In the face of this reality, coupled with the inability to confer liability protection on a nationwide basis without congressional approval, the negotiations increasingly focused on money. Certainly, money was an important factor to both fiscally strapped state governments and their lawyers, who were to be paid on a contingent fee basis. On November 16, 1998, the parties reached agreement on the so-called Master Settlement Agreement.[19]

The Master Settlement Agreement (MSA) required the tobacco companies to pay $206 billion, most of it in annual installments over a twenty-five-year period. It restricted the companies' advertising and promotion activities, but not as vigorously as the Global Settlement Agreement would have. Cartoon characters, such as Joe Camel, were forced into retirement as advertising spokespersons for tobacco companies. The signatory companies agreed to forgo brand-name advertising of events targeted to youth, including most concerts and specified athletic events. Outdoor advertising and transit advertising were prohibited. Wearing apparel advertising tobacco products and the distribution of free samples in locations accessible to youth became a thing of the past. Although many of these restrictions were designed to limit advertising to youth, the MSA, unlike its predecessor, did not directly limit sales to youth, nor did it include look-back provisions that would have imposed penalties on manufacturers if youth smoking did not decrease. Most important, the MSA did not grant the FDA regulatory authority over tobacco products. It did contain restrictions on lobbying on behalf of tobacco companies.

THE SETTLEMENT OF *PARENS PATRIAE* LITIGATION AS LAWMAKING

The history of the negotiation of the original Global Settlement Agreement, its rejection by Congress, and the subsequent agreement on the substantially weakened Master Settlement Agreement carries important lessons for those who see *parens patriae* litigation as a superior means of lawmaking compared to the ordinary legislative process, regardless of how corrupted the legislative process may be by political contributions and the lobbying of manufacturers. The negotiations leading to the Global Settlement Agreement and the Master Settlement Agreement can hardly be described as exemplars of democratic process. The bargaining began in earnest when Senate majority leader Trent Lott, concerned that the appearance of Republicans' strong ties to the tobacco industry hurt them politically, suggested to his brother-in-law Dickie Scruggs, a private attorney working with Mississippi attorney general Michael Moore, that he should settle the case. Lott offered to put him in touch with two Republican operatives with close ties to the tobacco industry.[20] Over the next two-and-one-half years, negotiators for both sides met many times. A small number of representatives from the nation's state attorneys general and several lawyers from mass plaintiffs' firms handled the real bargaining. On occasion, the states were represented by a single private plaintiffs' attorney with no state attorneys general present. According to Mollenkamp, there was tension between the attorneys general, who were pushing for reimbursement of state expenditures resulting from tobacco-related illnesses and programs to reduce smoking among youth, and the private attorneys, who were more interested in smoking cessation programs and large fees.[21]

The need for congressional approval of legislation implementing the Global Settlement Agreement, the original agreement, provided the first real democratic access to the lawmaking process during the tobacco negotiations. As previously noted, the first to attack the Global Settlement Agreement were public health advocates (e.g., former FDA commissioner David Kessler), not the tobacco companies. Antitobacco activists complained about being excluded from the negotiating sessions where the deals were cut. This public participation caused Congress to strengthen the requirements imposed on the tobacco manufacturers to the point where the deal became unacceptable to them. Ironically, the outcome was the Master Settlement Agreement, a substantially weakened deal that did not require congressional approval. In other words, the MSA, which became the nation's blueprint for reducing the incidence of the nation's most pervasive and important public health crisis, was adopted without the opportunity for either public comment or the accountability that, at least in theory, characterizes the legislative process. The public health advocates and public interest lawyers who saw *parens patriae* litigation as an alternative to a legislative process characterized by excessive influence from tobacco companies had created a closed-door process where there were virtually no checks on the special interests divvying up the pie.

TOBACCO-RELATED DISEASE PREVENTION AND OTHER GOALS OF THE MASTER SETTLEMENT AGREEMENT

The Master Settlement Agreement is by far the most important negotiated resolution resulting from *parens patriae* litigation to date. A logical place to begin evaluating product regulation through *parens patriae* litigation is to assess whether the MSA in fact was effective in achieving its proponents' objectives and other traditional goals of tort litigation. First and most important, how effective is the MSA likely to be in reducing smoking and tobacco-related diseases?[22] The second set of criteria for evaluating the MSA focus on compensation and loss distribution. Have the funds received by the states adequately compensated the states for their expenditures necessitated by tobacco-related illnesses? How effective has the combination of *parens patriae* litigation and state Medicaid programs been in compensating victims of tobacco-related diseases for the harms they experienced directly? The final issue concerning compensation and loss distribution is whether the MSA distributed the losses resulting from tobacco-related diseases in a desirable manner. The third set of criteria asks whether the MSA achieved Ron Motley's retributive goal; that is, did "the gangsters" pay? Fourth and finally, in an era in which government is so often criticized, justifiably or not, for being inefficient, does the MSA attempt to accomplish these other objectives in an efficient and cost-effective manner?

An obvious beginning point in assessing whether the MSA provided an effective policy for reducing the incidence of smoking is to look at trends in smoking rates since its adoption, though such analysis turns out not to be terribly helpful. Cigarette smoking among American adults declined steadily from 1997 through 2004, after having remained essentially unchanged from 1990 to 1997.[23] Of course, it is impossible to attribute this decline to any one factor, such as the provisions regulating cigarette advertising and promotion contained in the MSA. Further, the decline in smoking rates appears to have stalled between 2004 and 2007. Perhaps most discouragingly, 23.9 percent of young adults between the ages of 18 and 24 smoked in 2007, a rate higher than for any other age-group. A 2008 survey showed that smoking among teenagers declined in 2008, but smoking rates for this group were still higher than in the early 1990s.[24]

A recent publication of the Centers for Disease Control and Prevention suggests that the lack of decrease in cigarette smoking during the past several years may be attributable in part to an almost doubling of the tobacco industry's promotional expenditures, representing a shift to promotions still allowed under the MSA, including coupons, "buy one, get one free" offers, and incentives for retailers to place their products in conspicuous locations. In addition, the funding of state and local efforts at tobacco control and prevention between 2004 and 2007 has decreased more than 20 percent.[25] Given the focus on restricting specific sorts of advertising and marketing contained in the MSA, the increase in promotional expenditures is particularly troubling.

Perhaps the most important consequence of the MSA was that the states and the tobacco companies became financial partners, arguably inhibiting the cash-strapped states from adopting and implementing effective antismoking policies. Christine Gregoire, a lead negotiator of the MSA, acknowledged, "The money in the tobacco settlement is as addictive to states as the nicotine in cigarettes is to smokers."[26] Daynard is even harsher in his assessment: "Under the agreement, there's a multi-state permanent lobby for business-as-usual."[27]

Many states "securitized" the expected stream of revenues from the tobacco companies' annual payments required by the MSA.[28] To get larger amounts of cash for immediate needs, about half of the states sold bonds to investors that were secured by the expected revenue stream to be generated by the payments from tobacco companies under the MSA. Leading antitobacco activist Stanley Glantz observed, "Probably the tobacco industry will win in the long run, largely because of the securitization of the money putting pressure on the states to keep consumption up to get their bonds paid off."[29]

THE CARTELIZATION OF THE TOBACCO INDUSTRY

In view of the comments of Ron Motley and others suggesting retributive objectives for the tobacco litigation, it is fair to ask whether the Master Settlement Agreement punished the manufacturers for their past conduct. I consider the issue of retribution before I take up issues related to compensation and loss distribution, because to the extent that many proponents of the tobacco litigation believed that the well-being of tobacco manufacturers was inversely correlated with the reduction of tobacco-related diseases, the issues of disease prevention and retribution overlapped, and the line between them became blurred.

Mark Curriden, who covered the tobacco litigation for a decade as a journalist, recently concluded that a decade later, "it's business as usual."[30]

In an analysis published in 2004, F. A. Sloan, C. A. Mathews, and J. G. Trogdon reached these perhaps surprising conclusions:

1. The return on investment for the tobacco industry, an important measure of industry profitability, exceeded returns on investment in other industries during the period 1999 through 2002.
2. During the same period of time, domestic tobacco revenues were greater than those before the Master Settlement Agreement.
3. Total advertising expenditures for the tobacco industry increased at a higher rate during 1999–2002 than they had during 1990–1998, the period preceding the MSA.[31]

In addition, tobacco companies dramatically increased sales of cigarettes in the international market, particularly in developing countries.[32] Finally, the leading tobacco companies of the pre-MSA era often purchased or combined with nontobacco businesses, such as food or brewing companies, to reduce the impact of any downturn in the tobacco industry, whether it resulted from reduced consumer demand or additional litigation.

The MSA enabled the largest tobacco companies to raise the price of a package of cigarettes far more than was necessary to meet their financial commitments under the agreement, by protecting them from price competition from nonsignatory tobacco companies. The price of a pack of cigarettes increased by forty-five cents following the approval of the MSA and by a total of seventy-six cents within a few years, far more than the approximately nineteen cents per pack estimated to be necessary to make the tobacco companies' payments under the agreement.[33] The MSA essentially intended to create a state-legitimized cartel that protected the four major tobacco manufacturers that signed the agreement from price competition from new entrants into the marketplace for cigarettes.[34] Other than the four major companies that had negotiated the agreement, other manufacturers were invited to become signatories to the agreement. If they signed, they were not liable for payments under the MSA unless their future revenues exceeded 125 percent of what they had been before the enactment of the MSA. In other words, those other companies that agreed to comply with the MSA were allowed a modest increase in revenues before becoming liable for payments to the states under the agreement, but only if they capped their growth at a limit that did not threaten the four industry behemoths.

The remaining risk to the initial four signatories was that new competitors, not obligated under the MSA to make annual payments, would enter the market and attempt to sell cheaper or "discount" cigarettes that did not reflect the cost of the payments to the states. Here, the MSA protected the major companies by recommending that each state pass model legislation that would require future entrants into the cigarette market to make payments to the state in order to sell cigarettes. These payments, however, were categorized as "escrow payments." On one hand, these payments were to be returned to the nonparticipating manufacturer after twenty-five years if they were not needed to satisfy any judgment obtained by the state in a claim similar to those settled by the MSA. On the other hand, these payments, unlike the required annual payments of the major manufacturers, were not tax deductible, putting the discount or nonparticipating manufacturers at a financial disadvantage. The major tobacco companies that signed the MSA had protected themselves from price competition from new cigarette manufacturers that were not obligated to make payments under the agreement. Further, because there were only four companies signing the agreement, it became easier for these companies to collude in raising cigarette prices more than was necessary to finance their obligations under the agreement. Accordingly, the MSA appeared to facilitate higher profit margins for these four tobacco manufacturers in the domestic cigarette market.

The major companies' efforts to reduce competition were not entirely effective. The market share of nonsignatory companies increased from 0.37 percent before the MSA to more than 8 percent by 2003.[35] Even in this situation, however, the MSA protected the participating tobacco manufacturers. It provided that if the settling manufacturers lost market share to nonparticipating manufacturers due to compliance with the MSA, their annual payments under the agreement would be reduced and in some cases eliminated.[36] In fact, as the sale of discount cigarettes by nonparticipating manufacturers rose, Philip Morris and other manufacturers threatened to withhold a significant portion of the baseline payments to the states.[37]

States sometimes have found it difficult to enforce the provisions of the MSA governing nonparticipating manufacturers.[38] Simultaneously, the legitimacy of the state-legitimized cartel created by the MSA has faced challenges from nonparticipating manufacturers, on antitrust and constitutional grounds under federal law and on antitrust and unfair competition grounds under state law. None of these legal challenges has yet reached the U.S. Supreme Court, but a majority of the federal appellate courts hearing such challenges have upheld the anticompetitive provisions of the MSA.[39] In *Freedom Holdings, Inc. v. Spitzer*,[40] however, the Court of Appeals for the Second Circuit reversed the trial court's dismissal of an antitrust challenge to the MSA. In its opinion, the court of appeals quoted from the testimony of Robert Pitofsky, chair of the Federal Trade Commission at the time of the original Global Settlement Agreement and Senator McCain's resulting legislation.

> This provision, as it reads here, says that the tobacco executives can get together in a room and agree on what the price of a pack of cigarettes is. It could be a price much higher than the cost of the annual payments. That seems to me not sensible.[41]

The court also expressed a strong preference for direct regulatory legislation—rather than relying on tobacco companies, protected from antitrust actions—to increase the price of cigarettes and thereby discourage smoking, particularly among youth: "Given the broad police powers of the state to regulate the marketing of dangerous commodities, effective public health measures other than affording tobacco manufacturers a cartel are ubiquitous and far more obvious than the complex market-share arrangements of the MSA."[42]

So far, the U.S. Supreme Court has declined the opportunity to resolve the dispute among the circuits on the issue of whether or not the MSA and the implementation of its provisions violate federal antitrust law. If the provisions of the MSA are upheld when challenged on antitrust grounds (a seemingly probable result), it will become clear that the results of the negotiations concluding the *parens patriae* tobacco litigation did not punish the manufacturers that many public health advocates refer to as "merchants of death." Instead, it is "business as usual"—or, perhaps more accurately, "business better than usual"—for an industry in which the four largest players now operate with advantages that otherwise would be illegal under federal law.

THE MSA AND COMPENSATORY OBJECTIVES

The State of Mississippi's complaint against the tobacco manufacturers alleged that it had paid significant health care expenses for thousands of its citizens who suffered from tobacco-related illnesses.[43] Yet W. Kip Viscusi's convincing financial analysis proves otherwise.[44] Viscusi argues, for example, that one of the consequences of the shortened life expectancies of smokers is that when their citizens smoke, state and federal governments save both Social Security and Medicare costs that more than offset any increased medical expenses they pay. In short, says Viscusi, "the striking economic result is that cigarettes are self-financing for every state and for the federal government when viewed from a variety of insurance cost perspectives."[45] Obviously, Viscusi is not arguing that the patients themselves and their families are better off; he is arguing that state and federal governments are, though only in financial terms. Nevertheless, his analysis is remarkable. When all state expenditures and state receipts are included in the analysis, tobacco-related illnesses and deaths actually save the states money.

Because the parties to the Master Settlement Agreement thought they were reimbursing the states for past expenditures for tobacco-related diseases, there was most often little concern about whether settlement proceeds were spent for programs of smoking control or for unrelated expenditures. Yet, as Viscusi demonstrates, the states' receipts from payments under the MSA more than compensated them for past expenditures necessitated by tobacco-related diseases. Under these circumstances, it might be expected that states would spend a sizable portion of the proceeds from the MSA for programs in tobacco prevention. The Centers for Disease Control and Prevention, for example, recommended that a state spend an amount equal to at least 20 percent of its proceeds from tobacco settlement on programs for smoking prevention,[46] but states reportedly spent less than one-quarter of the recommended amount.[47] In some cases, the funds were spent on such items as a NASCAR speedway, golf courses, and, perhaps most ironically, a tobacco warehouse.[48]

Higher cigarette prices did serve another function: discouraging smoking, particularly among the young. At the same time, it is disconcerting to realize that both cigarette excise taxes and the higher cigarette prices resulting from the MSA not only discourage smoking but also are highly regressive forms of taxation to solve state budget crises and finance golf courses. People from lower-income groups, in the aggregate, tend to smoke far more than those from higher-income groups.[49] As a result, they not only pay a higher proportion of their incomes as cigarette taxes and increased cigarette prices; they also pay more in absolute dollars.

Even more fundamentally, if *parens patriae* litigation against the manufacturers of cigarettes is viewed as a means of circumventing the inability of the individual lung cancer victim to recover from the tobacco manufacturer because of defenses resulting from her own conduct in smoking, *parens patriae* litigation is, at best, an imperfect and incomplete alternative. First, for the majority of its smoking citizens, whose tobacco-related medical expenses are paid by private insurers or other sources and not by the state, the state recovers nothing. Second, any compensation received by the state resulting from a victim's tobacco-related disease includes only compensation for medical expenses and nothing for lost wages, noneconomic damages, or other items of damages. Even in the case of the Medicaid recipient, the cancer victim himself receives no compensation other than the state's payment of medical bills, which the Medicaid program would have paid even in the absence of the *parens patriae* litigation. Following the MSA, numerous Medicaid recipients sued to receive a portion of the benefits, and these actions were unsuccessful.[50]

THE TRANSACTION COSTS AND EFFICIENCY OF
PARENS PATRIAE LITIGATION

Parens patriae litigation against product manufacturers seeks to use the judicial system to solve public health problems. It is a remarkably inefficient process for doing so. When all was said and done, the result of the MSA was that participating manufacturers raised the price of cigarettes in order to pay the states. In other words, the result of the MSA was to impose the equivalent of an excise tax on cigarettes. Yet, in addition to the delay of more than four years to resolve the litigation, the privately retained plaintiffs' firms ended up receiving more than $15.4 billion dollars in compensation, paid by the tobacco companies as agreed to in the MSA.[51] Private counsel representing the state of Florida, for example, received $3.43 billion. According to Viscusi, the fee received by private counsel representing the state of Texas translated to $150,000 per hour.[52] Frequently, there had been long-standing political and even personal relationships between the state attorneys general and the private lawyers they employed. When a few states balked at payment of the previously agreed-on fees, these disputes were submitted to arbitration pursuant to the terms of the MSA and sometimes litigated.[53] The increased price of cigarettes obviously also included a surcharge to pay for these fees. An excise tax on cigarettes would have avoided such substantial transaction costs. Taxation would have been far more cost-effective and efficient than *parens patriae* litigation.

THE PURPORTED SUPERIORITY OF *PARENS PATRIAE* LITIGATION
AS A MEANS OF REGULATING TOBACCO

To its proponents, state tobacco litigation promised superior outcomes contrasted with the legislative process corrupted by tobacco company lobbyists and campaign contributions. But by the time the MSA was negotiated, they complained about the backroom deals between the manufacturers and the attorneys general that excluded them in a way that the legislative and administrative processes never could have.

The ends did not justify the means. The Master Settlement Agreement, a de facto form of national legislation, did not offer an effective tobacco control strategy. It is not surprising, given those who negotiated it, that what it primarily did offer was dollars. The plaintiffs' attorneys who represented the states and were among those who negotiated on their behalf ended up obscenely wealthy. States, many of them short of cash as a result of strong political opposition to taxes, solved their budgetary problems, at least temporarily, at the expense of addicted smokers. A settlement couched in terms of the states receiving billions of dollars and restrictions on the ability of tobacco manufacturers to promote and advertise their products to youth, coupled with claims that the evil tobacco manufacturers had been forced to pay for their wrongdoing, had popular political appeal. It is no accident that the abbreviation *AG* used for "attorney general" is sometimes said to stand for "aspiring governor."

The tobacco companies certainly benefited from the MSA and thrived in its aftermath. Manufacturers and their counsel will almost always know more about their industry and how to devise negotiated agreements that appear to regulate but in fact appease. In short, all those present at the bargaining table did well. The public fared less well. None of these influences is unique to the tobacco litigation. Instead, each of them will likely recur in the settlement of most, if not all, *parens patriae* actions seeking resolution of product-caused public health problems.

Some will claim that the fault in the tobacco litigation remains with a Congress that is too beholden to manufacturers' lobbying and campaign contributions. That explanation suffices for the failure of the original Global Settlement Agreement to be legislatively implemented. It does not, however, explain the shortcomings of the Master Settlement Agreement as a national blueprint for tobacco control and prevention. That debacle, instead, lies at the hands of a closed negotiating process that predictably served the interests of those directly involved, while it sold short the public interest. Yet plaintiffs' lawyers and state attorneys general continued to promote *parens patriae* litigation as a means to force product manufacturers to the bargaining table in order to impose alternative regulatory frameworks.

THE JUDICIAL DECREE AS A SOLUTION FOR PUBLIC HEALTH PROBLEMS: RHODE ISLAND'S PROPOSED LEAD NUISANCE ABATEMENT PLAN

I now turn to a description of the State of Rhode Island's proposed remedial plan in the lead pigment litigation. As noted previously, Judge Silverstein had made it clear that his ultimate remedial decree likely would closely track the state's proposal. Although it obviously is impossible to report the success or failure of any judicial plan that might have been adopted if the Rhode Island Supreme Court had not reversed the trial court judgment in favor of the state, a mere summary of the attorney general's proposal strongly suggests that a judicial decree resembling that proposal would have exceeded the institutional capacity of the court, as well as the appropriate functions of a court as they traditionally have been understood.

The attorney general's proposed abatement plan acknowledged that it was "impractical" to remove all lead paint from Rhode Island residences and schools,[54] but it nevertheless called for an effort that would have required lead remediation in approximately 240,000 Rhode Island residences and 758 schools and child care centers,[55] at an estimated total cost to the defendants—and presumably their insurers—of more than $2.4 billion.[56] The plan called for more than ten thousand trained and skilled workers to spend approximately 8,013,000 "labor days" over a period of several years to abate the nuisance.[57] In all except a few instances, families who lived in the units where control of lead hazards was ongoing would have been relocated at the defendants' expense.[58] The plan acknowledged that abatement usually takes two to four weeks in a single-family unit and four to eight weeks in a building with three or more housing units, yet it boldly and optimistically (but probably unrealistically) predicted that those times could have been reduced to three and a half to five days and seven to ten days, respectively. Of course, if these projections were not met, the cost of abatement would have risen considerably. The plan anticipated that occupants would have been compensated for their meals and transportation during their displacements.

Despite the fact that the jury found the "cumulative presence" of lead throughout the state to be a public nuisance, the proposed abatement plan did not call for the removal or abatement of all lead paint.[59] The attorney general admitted, "So-called 'lead-free' housing is not practical, measurable, or achievable and no existing program has ever incorporated such a standard as a general measure."[60] Instead, the objective was to make "the bulk of Rhode Island's housing units and other buildings, which include schools, day care centers and other facilities children may be in or visit *safe* for children for the expected life of the building."[61] This limited remedial objective, presumably deemed sufficient by the attorney general to abate the public nuisance, appears to be at odds with the state's core argument throughout the litigation that it was the presence of the lead pigment, in and of itself, that comprised the public nuisance. The state earlier had re-

jected the defendants' argument that it was the failure of the property owners to maintain paint containing lead in a safe manner that accounted for
the public health threat and caused any alleged public nuisance. The abatement plan then described at great lengths the steps that could be taken to
accomplish the attorney general's articulated goals, as well as the costs and
benefits of each set of measures.

The plan also allowed the possibility that lead-based paint hazards in
some units would not be eliminated. Consider what the plan anticipated
would happen if either the owner or the occupants of a property refused to
participate when the monitoring authority offered abatement, most likely
because they did not want to relocate. The attorney general's plan offered,
as one approach, that the owner, the occupants, or both might be ordered
to participate, in much the same way that a court might order the parents
of a child to allow a vaccination of a child.[62] The scope of thousands or tens
of thousands of evictions was yet one more aspect that suggested the plan's
impracticality. The plan also suggested that "the property could be bypassed completely for abatement."[63] The attorney general's plan would thus
have failed to entirely eliminate lead-based paint or even lead-based paint
hazards in houses and other buildings occupied by children. Even if the
abatement effort achieved all of its massive and seemingly overly ambitious
goals, the remaining cumulative presence of lead in buildings throughout
Rhode Island might, when viewed from the perspectives of the state and
the trial court, still have posed a risk of childhood lead poisoning. According to the trial court's own standards, the public nuisance would not have
been eliminated.

At the same time, the attorney general's plan exposed the reality that
lead-based paint hazards simply constitute a subset of a wider array of
problems in older housing stock that property owners have allowed to deteriorate. For example, the plan envisioned that in many instances, the
oversight authorities would have been required to repair structural
deficiencies known to facilitate the development of lead-based paint hazards.[64] Leaky roofs and leaky faucets, missing downspouts and gutters,
holes in walls, broken windows, and improper grading of the lawn all cause
paint to deteriorate and hence result in lead-based paint hazards. These
problems, claimed the attorney general's report, would have to be cor-
rected—in the case of a leaking roof, for instance, by replacing a portion of
the roof. At the same time as windows were to be replaced to avoid the lead
dust created when old windows are raised and lowered, it only made sense,
according to the proposal, to install "Energy Star label" windows.[65] Therefore, the attorney general's abatement plan anticipated that the manufac-

turers of lead pigment would have paid for a wide variety of structural improvements that, while beneficial, would have solved housing problems that had nothing to do with the presence of lead pigment.[66] Of course, we do not know whether the trial court, in its supervisory role over the abatement process, would have allowed such expenditures. Unlike the tobacco settlement, inherent in the remedial phases of the Rhode Island lead pigment litigation, as envisioned by both the attorney general and the trial court, was judicial supervision over the use of the funds received from the defendants.

In any event, the abatement plan anticipated that a Rhode Island trial court judge was to supervise a public works project funded by the lead pigment manufacturers that would inspect the state's approximately 240,000 older residences and, unit by unit, decide on abatement plans that might require structural renovations, coordinate with owner preferences, relocate residents during the meantime, order recalcitrant occupants out of their homes, inspect completed abatements, and pay appropriate inspectors and contractors. All of this was to have been accomplished, for a quarter million homes, within four years and within a $2.4 billion budget. It is likely, of course, that the court would have needed to address issues raised by homeowners and tenants unhappy about being forced from their homes. Presumably, the defendant-manufacturers would have returned to court frequently to object to plans for which they apparently would have been ordered to write a blank check. No court, even in cases of constitutionally mandated school desegregation or prison reform, has tackled a project of this scope and complexity. For that matter, no federal or state housing authority has engaged in a comparable enterprise, regardless of what they should have done to eliminate childhood lead poisoning. The proposal was a daunting one and was almost certainly unrealistic.

THE RELATIONSHIP BETWEEN PUBLIC PRODUCTS LITIGATION AND POLITICAL REFORM

Proponents of *parens patriae* and other public health litigation against product manufacturers often argue, as scholar Lynn Mather has, that the most important role played by such litigation is as "the centerpiece of an overall political strategy."[67] Similarly, William Haltom and Michael McCann remind us that "the penchant for rights-based litigation must be understood in relationship to the larger institutional and cultural features of U.S. politics."[68] Timothy D. Lytton recently identified several ways in which public health litigation may influence regulatory policy, including "framing issues in terms of institutional failure and the need for institutional reform; . . . generating policy-relevant information; . . . [and] placing issues on the agendas of policy-making institutions."[69]

Haltom and McCann's investigation of media coverage of the states' *parens patriae* tobacco litigation leads them to conclude that the litigation enabled proponents of tobacco control to change the policy debate from one framed by a public perception of individual smokers' responsibilities and greedy private trial lawyers to one where "noble public prosecutors who are 'above politics'" placed the causal and moral responsibility on the tobacco companies.[70] Because of the state attorney general's critical role in the litigation and because of the well-documented evidence of corporate duplicity, the *parens patriae* litigation came to be framed as a "quasi-criminal" proceeding.[71] According to Haltom and McCann's research, after 1998, press coverage of tobacco litigation reflected a less unfavorable view of plaintiffs' attorneys and increasingly focused on corporate irresponsibility. They conclude that the states' tobacco litigation helped to pave the way for public acceptance of both the restrictions contained in the MSA and stronger local antismoking ordinances enacted after 1998.[72]

Other scholars contend that public interest litigation impedes legislative reform instead of facilitating it. In his classic study *The Hollow Hope: Can Courts Bring About Social Change?* Gerald N. Rosenberg concludes "that courts act as 'fly-paper' for social reformers who succumb to the 'lure of litigation.'"[73] He explains, "Courts also limit change by deflecting claims from substantive political battles, where success is possible, to harmless legal ones where it is not. Even when major cases are won, the achievement is often more symbolic than real."

On a related issue, Wendy Wagner writes that public interest litigation of any type, including *parens patriae* actions, serves an important additional role because "the tort system can be more effective than the regulatory system in accessing the various types of information needed to inform regulatory decisions."[74] Wagner notes that much of the information regarding product risks is uniquely in the hands of the regulated manufacturers. Plaintiffs' attorneys, working on a contingent fee basis, are far more motivated to discover the manufacturer's knowledge of risks than are regulators, who may be deterred by political considerations. Further, because a trial unfolds as the telling of a narrative to a jury, information about product risks is more accessible to the public through the judicial process than it is within the complex, highly specialized regulatory maze.

For example, until information came to light, at least indirectly, through the litigation process, Congress and the FDA were not fully aware that tobacco manufacturers had manipulated the nicotine content of cigarettes for decades before the 1990s. Recall that Merrill Williams had worked with a Louisville law firm representing defendants in tobacco litigation and had compiled more than four thousand internal documents that were highly incriminating to the companies. After he was laid off from his job, he turned over these critical stolen documents to attorneys who had sued tobacco companies in the past. Eventually, these files worked their way into the hands of the privately retained attorneys representing the State of Mississippi. These documents might have eventually become public without the tobacco litigation, perhaps when antitobacco activists later placed them on a website or shared them with the media, but maybe not. In theory, of course, legislatures and administrative agencies, sometimes assisted by the press, should be able to obtain information about product risks. In reality, as Wagner points out, this often does not occur.[75]

The scholarship of Haltom and McCann and of Wagner seems persuasive. It is true that the states' lawsuits occurred only after FDA commissioner David Kessler's decision to regulate tobacco as a drug, television reports that tobacco companies had "spiked" the level of nicotine in drugs, the release of the incriminating documents stolen from Brown & Williamson, and the congressional hearings in which executives of tobacco companies embarrassed themselves by testifying that nicotine was not addictive. By 1994, tobacco control clearly was on the public agenda, and the public already was beginning to frame the tobacco epidemic as the result of the tobacco industry's perfidy. Nevertheless, Haltom and McCann's research suggests that although it is difficult to untangle the relationships among litigation, regulatory and legislative proceedings, and media publicity during the mid-1990s, the *parens patriae* litigation had a distinctive effect in shaping public opinion and the political debate. Of course, since the MSA, no stronger federal legislation for tobacco control or regulation were enacted for eleven years, but this failure can be laid at the feet of a Republican administration and a Republican Congress that were pervasively antiregulatory in economic matters.

Rhode Island's pigment litigation may have had a stronger effect, albeit perhaps an ironic impact from the perspective of property owners, in leading to legislative reform in that state. In 2002, midway through the litigation, the Rhode Island General Assembly passed legislation that, for the first time, imposed strong requirements on landlords to implement interim controls.[76] With state litigation pending against the pigment manufacturers, it became impossible for landlords to argue that they should not be the only party bearing the costs of ending childhood lead poisoning. In April 2007, however, before the Rhode Island Supreme Court's reversal of the trial court's judgment, but also prior to the time when Rhode Island might have expected to receive any funds as a result of the litigation, Governor Donald L. Carcieri proposed cutting the state's budget for prevention of lead poisoning, justifying the reductions on the basis that he expected the money to be replaced by funds from the lawsuit.[77]

CONCLUSION

For public interest advocates and public health officials, the lessons of both the states' tobacco litigation and the Rhode Island pigment litigation are likely to be disappointing for the most part and hard to accept. Even when tobacco companies arguably were pressured by the prospects of huge liabilities as a result of the *parens patraie* litigation filed by most American states, they were able to negotiate a settlement that served their interests well and one that many health experts now acknowledge to have been a mistake. Similarly, the decision of the Rhode Island Supreme Court to reverse the trial court judgment ordering the pigment manufacturers to abate the nuisance causing childhood lead poisoning in Rhode Island was a likely dead end for judicial attempts to end childhood lead poisoning. Even if the trial court's judgment had been affirmed, there was little to suggest that the court's remedial decree would have been either effective or appropriate.

What the states' *parens patriae* litigation did accomplish was to enable antitobacco reformers to reframe both the public's perception of responsibility for tobacco-related diseases and the debate within the political processes. The full effect of this reframing of the public debate may not yet have fully impacted the federal regulatory processes. In contrast, property owners' expectations that lead pigment manufacturers would end up paying the costs of lead abatement may ironically have facilitated the enactment of Rhode Island legislation requiring property owners to implement interim controls and pay for it themselves.

Impersonating the Legislature: State Attorneys General and *Parens Patriae* Products Litigation

Litigation filed against product manufacturers by state attorneys general attempts to change the structure of either product regulation or government programs seeking to prevent product-caused public health problems. For example, tobacco manufacturers today operate under a set of detailed regulations governing many aspects of their operations, including advertising directed toward young people, and these regulations are strikingly similar to proposals previously rejected by Congress.[1] Federal regulators and state legislators, however, did not devise this regulatory regime. The new regulations resulted when state attorneys general and their partners, a handful of plaintiffs' firms specializing in mass products litigation, brought manufacturers to the bargaining table by filing lawsuits asserting novel substantive claims, such as public nuisance.

It is true that the new regulatory schemes were imposed on the manufacturers of cigarettes through settlement, not through judicial action. However, while a product manufacturer might rationally decide to "roll the dice" in the first several rounds of litigation brought by individual claimants[2] and wait for a pattern of liability or no liability to emerge, few manufacturers are willing to risk trial when the plaintiff is a state that may collect billions of dollars as a result of harms allegedly suffered by millions of its residents, much less a consortium of states operating in concert.[3] The initiation of litigation by a state attorney general against the manufacturer of a mass product is, in its own right, a far weightier matter than simply the initiation of a more typical civil lawsuit. In and of itself, it represents the exercise of regulatory power.

Not all actions brought by state governments against product manufacturers settle, but the influence of the state attorney general potentially remains decisive. Consider the recent litigation brought by the Rhode Island attorney general against the manufacturers of lead pigment. Rhode Island's legislatively enacted Lead Poisoning Prevention Act[4] had declared that a property owner's failure to prevent lead-based paint hazards constituted a public nuisance.[5] Yet the trial court's judgment envisioned the adoption of the attorney general's complex regulatory proposal, which would have shifted financial responsibility to eliminate or reduce lead-based paint hazards in hundreds of thousands of Rhode Island residences from property owners to the pigment manufacturers.[6]

Both the Rhode Island Supreme Court and the New Jersey Supreme Court, traditionally one of the most influential and pro-plaintiff courts in the country on issues related to products liability, recently have explicitly noted, in their opinions dismissing government actions against lead pigment or paint manufacturers, the antidemocratic character of executive branch officials and courts seeking to use common-law claims to remedy a public health problem previously addressed in a different and inconsistent manner by the state legislature. In *In re Lead Paint Litigation,*[7] the New Jersey court dismissed multiple actions filed against manufacturers of lead paint and lead pigment by municipalities seeking to hold defendants responsible for financing the costs of efforts at preventing lead poisoning. The court extensively reviewed federal and state legislative efforts to reduce the incidence of childhood lead poisoning and acknowledged that "under the [New Jersey] Lead Paint Act, responsibility for the costs of abatement rests largely on the property owners."[8] The court found that the state statute, which declared lead-based paint contained within residential dwelling units "'to be a public nuisance,'"[9] was inconsistent with any attempt to hold a product manufacturer liable for the costs of abating the same problem.[10] Similarly, the Rhode Island Supreme Court noted that the state legislature's "statutory schemes . . . reflect the General Assembly's chosen means of responding to the state's childhood lead poisoning problem." The court further stated that "the General Assembly has recognized that landlords, who are in control of the lead pigment at the time it becomes hazardous, are responsible for maintaining their premises and ensuring that the premises are lead-safe."[11] In short, each court acknowledged that it would be inappropriate for it to impose a regulatory scheme inconsistent with the one established by its respective state legislature to address the public health problems caused by childhood lead poisoning.

Any tort claim typically—and in my opinion, legitimately—has a regulatory effect. Recent state *parens patriae* actions against product manufacturers, however, are distinguishable from ordinary tort claims. Traditionally, when residents of a state are harmed by the arguably tortious conduct of other private parties, the state serves as the umpire of the litigation, not as the champion of one set of contestants. Here, in contrast, the attorney general sues, at least indirectly, on behalf of victims of tobacco-related diseases or owners of properties that are covered with paint containing lead pigment. More important, such *parens patriae* litigation has as its primary goal the creation of a new regulatory regime governing an industry or establishing comprehensive statewide programs to address a public health problem without going through the usual legislative or regulatory processes.

NEW REGULATORY REGIMES IN THE FACE OF *EX ANTE* LEGISLATIVE STRUCTURES

In assessing the legitimacy of the attorney general's *parens patriae* action against a product manufacturer, the real issue usually is not that the legislature has failed entirely to regulate a product or to address a public health problem. Rather, it is that the legislative response is not deemed optimal by the state attorney general. He and his political allies within the public interest community and the plaintiffs' bar specializing in mass products litigation recognize, as the late Gary Schwartz noted more than a decade ago, that, generally speaking, manufacturers are better able to influence state legislatures through lobbying and campaign fund-raising than are consumers.[12]

Despite its flaws and gaps, the negotiated resolution of the states' tobacco litigation represented a comprehensive, new regulatory scheme governing the advertising and promotion of tobacco products.[13] As described in chapter 5, Congress, state legislatures, and the administrative agencies they created had extensively regulated the promotion and marketing of tobacco products prior to the Master Settlement Agreement (MSA). Yet the MSA, negotiated by the state attorneys general and resulting from litigation initiated by them, led to an entirely new regime that often governed the same conduct. State attorneys general and plaintiffs' attorneys specializing in mass products litigation ended up playing the regulatory roles traditionally handled by Congress, state legislatures, the federal Food and Drug Administration, the Federal Trade Commission, and the Federal Communications Commission.

The attorney general's proposal for the remedial phases of the Rhode Island pigment litigation explained how the lead remediation was to be accomplished, in far greater detail than any set of state statutory or regulatory enactments governing prevention of lead poisoning anywhere in the United States. For example, in its discussion of "building component replacement" (i.e., the removal of doors, windows, and trim), one very lengthy paragraph of the attorney general's proposal instructed, "Using a garden sprayer or atomizer, lightly mist the component to be removed with water."[14] The proposal contained scores of pages of instructions at a similar level of specificity. Unlike judicial decrees on school desegregation or prison reform, these instructions were directed not to public officials but to tens of thousands of private individuals. Some of the directives appeared to be unrelated to the core objective of the public nuisance caused by lead pigment. For example, the attorney general's plan required that windows replaced as a part of the lead remediation "should carry the Energy Star label."[15] Although such a choice might reflect a wise decision by an individual consumer or even an appropriate choice by an administrative agency, it seems out of place as a judicially mandated remedy addressing childhood lead poisoning. Further, the plan anticipated that changes to the detailed abatement procedures for any of the 240,000 residences could be "granted on a case by case basis due to unique or compelling needs, e.g., clapboard siding is preferred by the owner instead of vinyl siding."[16]

The plan envisioned "pilot projects" in selected neighborhoods and cities,[17] studies to determine the extent of lead hazards in other buildings,[18] and administrative flexibility as the effort moved forward.[19] In form at least, the proposal more closely resembled the approach of a legislative or administrative agency or even that of a public interest think tank, rather than one typically proposed by a litigant to a court. Obviously, the attorney general's proposed regulatory regime for eliminating childhood lead poisoning in Rhode Island has been rendered moot by the Rhode Island Supreme Court's reversal of the trial court judgment ordering the defendants to abate the public nuisance. Nevertheless, the plan, considered in the context of the trial court judge's expressed intent to rely heavily on the attorney general's recommendations, offers a revealing picture of what judicial decrees attempting to address public health problems caused by products might look like in similar litigation in the future.

The regulatory scheme that the Rhode Island attorney general sought to implement through his litigation against the manufacturers of lead pigment was in tension with the one previously adopted by the state legislature and described in chapter 5. The attorney general would have placed the financial responsibility for addressing lead-based paint hazards on manufacturers of lead pigment, not on property owners. Further, the litigation initiated by the attorney general had already resulted in the trial court's determination, although later reversed, that the presence of lead itself in residential premises throughout the state constituted a public nuisance and that product manufacturers were responsible for abating this nuisance.[20] This obligation is a substantially more demanding one than the legislature's determination that owners of rental properties should undertake far more limited, so-called interim control measures to eliminate lead-based paint hazards so as to substantially reduce the risk of childhood lead poisoning.

THE DOMINANT ROLE OF THE ATTORNEY GENERAL IN THE NEW REGULATORY REGIMES

If the issue is viewed superficially, the attorney general clearly possesses the power, under either the state constitution or statutes, to initiate claims on behalf of the state's interests and to reach a negotiated resolution of the issues in dispute. The recent filings of claims against product manufacturers, however, are far more than the mere filing of lawsuits. Litigation driven by the attorney general against product manufacturers represents a major shift in regulatory power from the legislative branch and administrative agencies to the attorney general, in a manner that could not have been foreseen even fifteen years ago.

Four factors explain, in large part, why the attorney general is able to dominate the content of the regulatory regimes imposed by *parens patriae* litigation. First, as described in chapter 6, some courts have dramatically expanded the doctrine of *parens patriae* to enable states to recover damages or the costs of abatement for harms suffered in the first instance by individual citizens of the states. Second, intertwined with the first factor is the emergence in at least a few jurisdictions of new, vaguely defined causes of action (notably public nuisance) that enable the attorney general to pursue regulatory litigation even if the individual victims who directly suffered the harms could not themselves successfully sue the manufacturer. These two developments give the state attorney general, the person who decides whether to file claims on behalf of the state, enormous leeway in selecting product manufacturers as litigation targets.

It might be argued that the attorney general, as counsel for the state, has the same right to file any nonfrivolous claim as would the attorney for any private litigant, but this argument ignores the distinctive role of the attorney general in *parens patriae* litigation. The American Bar Association's ethical rules governing the prosecutor recognize that he "has the responsibility of a minister of justice and not simply that of an advocate."[21] Both professional ethics codes and judicial opinions establish that this special obligation applies to his role in civil litigation as well as in criminal proceedings.[22] As the U.S. Supreme Court has noted, the government attorney "is the representative not of an ordinary party to a controversy, but of a sovereignty."[23]

Most of the time, the decision of whether the state, acting as *parens patriae*, will decide to sue or not to sue those firms that produce any particular product lies solely within the discretion of the state attorney general. Speaking about the attorney general's decision on whether to initiate government-sponsored litigation, Eliot Spitzer, formerly the attorney general of New York and later the governor, stated, "The likelihood that I would consult the legislature in this process is rather slight."[24] These discretionary decisions of state attorneys general regarding which manufacturing industries to target obviously represent a critical aspect of product regulation in today's economy and a major shift in the allocation of powers among the coordinate branches of government.

The Attorney General's Superior Bargaining Power in Settlement Negotiations

The third factor that enables the attorney general to dominate *parens patriae* products litigation is that defendant-manufacturers facing potential liability for the amalgamation of tens of thousands or even millions of individual claims if they "roll the dice" will most often probably decide that settlement is a more prudent option. The state's bargaining power is considerably greater, perhaps unconscionably so, as compared with that of any individual litigant in more typical litigation. Judge Richard Posner once described the "intense pressure to settle" that faces an industry in aggregate litigation when a single jury holds "the fate of an industry in the palm of its hand"; he contrasted it with the "decentralized process of multiple trials involving different juries, and different standards of liability, in different jurisdictions."[25] This bargaining leverage is even greater when the product involved has caused widespread health problems affecting millions of people. Further, in the tobacco litigation, more than forty states negotiated as a single block. The unacceptable risks of "betting the industry" on one or two jury verdicts places the defendant-manufacturers' counsel in an untenable negotiating posture. At some point in settlement negotiations, the attorney general, who possesses dominant bargaining power on behalf of the state, more closely resembles a regulator than he does an attorney or negotiator in more typical settlement negotiations.

The Pragmatism of Judicial Deference to the Attorney General's Regulatory Role

The fourth factor that may buttress the attorney general's dominance in public products litigation is the possibility that when the court issues a complicated remedial decree, it is likely to defer to the attorney general's recommendations. In many cases, it would be naive to expect that when such litigation does not settle and instead extends to trial, trial and appellate court judges reliably provide an effective check on the attorney general's attempt to expand the boundaries of common-law torts and to resist the appeal to impose a new regulatory scheme on product manufacturers. Realistically, whether this happens likely depends on whether the trial court and appellate court judges view their respective judicial roles as ones sometimes characterized as "activist" or as ones that are more restrained. Some judges are more attuned than others to using the powers of the court to seek to solve massive social problems. It is important to remember, however, that in a *parens patriae* action against product manufacturers, which no doubt alleges statewide harms, the attorney general often can "forum shop" to identify a trial court judge likely to be sympathetic to his cause. To the extent that a national plaintiffs' firm specializing in mass product torts selects the state in which to file an initial legal action against a particular industry, it obviously will choose one in which the state supreme court is likely to be favorable.

Even ignoring forum shopping by the attorney general and his litigation partners, judges may fail to provide an effective check on the state attorney general. Consider a state supreme court evaluating an appeal from a trial court's judgment ordering out-of-state defendant-manufacturers to pay the court or the plaintiff-state billions of dollars to address a product-caused public health problem. In a somewhat similar situation, at least one state supreme court justice has openly acknowledged that it was impossible for him to ignore the economic plight of the residents of his own state: "Trying unilaterally to make the American tort system more rational . . . will only punish our residents severely."[26] Thus, even if the state's *parens patriae* action against a product manufacturer is fully litigated and appealed, it is possible that courts would be reluctant to check the attorney general's discretionary choices. This judicial acquiescence obviously augments the attorney general's power.

In summary, in state litigation against product manufacturers, the attorney general is far more than an attorney representing a litigant, even one as powerful as the state. Novel interpretations of the scope of the state's *parens patriae* standing in some jurisdictions allow the attorney general to amalgamate the damages resulting from millions of separate harms suffered directly by the residents of his state. If coupled with expansive reinventions of historically limited causes of action, such as public nuisance, *parens patriae* standing enables the attorney general to circumvent the individual victim's inability to prove causation, as well as affirmative defenses that typically prevent the individual victim's recovery. Together, expansive interpretations of standing and public nuisance have the potential to give the attorney general the power and discretion to sue virtually any manufacturer of a mass product that repeatedly has caused harm. At the end of the litigation process, the attorney general possesses overwhelming bargaining power because of the immense scope of damages that otherwise would result from the amalgamation of claims. If the manufacturers refuse to settle, however, pragmatic realities, at least in many jurisdictions, suggest that the courts may not effectively check the attorney general's attempt to regulate the defendant's industry. The attorney general speaks for the sovereign as a powerful product regulator.

APPROPRIATE INSTITUTIONAL BOUNDARIES AND *PARENS PATRIAE* LITIGATION DRIVEN BY THE ATTORNEY GENERAL

Competency

In his insightful essay "The Limits of the Law," Peter H. Schuck describes three separate critiques of the "growing ambitions for law,"[27] two of which are relevant here in assessing the boundaries between, on the one hand, regulatory regimes established by legislatures and the administrative agencies they create to address specific social problems and, on the other, comparable regimes arising from *parens patriae* litigation driven by the attorney general. He distinguishes between "the functionalist critique" and the "illegitimacy critique." At issue, on functional grounds, is whether the attorney general (working within the judicial process) or the state legislature is best equipped to address, in an effective and fair manner, the difficult social problems resulting from the use of such products as tobacco and lead pigment. Which can better create either a new regulatory framework for products or comprehensive programs to address the public health problems such products cause?

Schuck suggests that "judges are institutionally ill equipped to obtain and integrate the information and values needed for sophisticated policy analysis."[28] As chapter 8 illustrated in the context of tobacco and lead pigment litigation, even if it is assumed that judges and juries are capable of conducting the polycentric analysis required by the liability phase of complex products cases, it is unlikely that judges are competent to devise and implement remedies to address complicated public health problems. Of course, for decades, federal courts tackled difficult social problems when they found that school systems, prisons, mental health facilities, and similar state institutions had been operated in an unconstitutional manner. Even in this arena of judicial activism that is arguably justified constitutionally, many commentators suggest that the judicial process cannot effectively institute structural reforms.[29] Yet judicial supervision of repairs and renovations in hundreds of thousands of Rhode Island homes would have been beyond the scope of even the broadest and most complex constitutionally mandated cases of school desegregation or prison reform. The trial court judge in the Rhode Island pigment litigation himself acknowledged both his own lack of public health expertise and the difficulty of using the traditional judicial process to solve polycentric problems.[30]

Legitimacy

I now turn to using Schuck's criterion of "legitimacy" to assess *parens patriae* litigation driven by the attorney general. Legitimacy, according to Schuck, begins, at a minimum, with the constitutional allocation of powers, but it also includes other factors affecting public respect for the law, such as the necessity of "decision procedures" that are "minimally fair, accurate, and participatory." Further, as Schuck acknowledges, the law's legitimacy is "closely linked" with its effectiveness.[31]

The deliberative legislative process, with its assortment of checks and balances (either constitutionally mandated or arising from legislative tradition), helps to legitimate the lawmaking process. James A. Henderson Jr. recently concluded, "It is commonly understood that, in a representative democracy, macro-economic regulation is accomplished most appropriately by elected officials and their lawful delegates."[32] The same could certainly be said about whether the more appropriate government entities to solve widespread and complex public health problems are the state attorneys general or legislatures and the state departments of health and housing that they authorize. The legislative process provides, theoretically and—to a greater or lesser extent—realistically, an opportunity for all parties to be heard and for their experts to testify. In contrast to the attorney general's decision making, this process is a comparatively open one.

In the analysis that follows, I examine how the allocation of powers among the three coordinate branches of government, as reflected in state and federal constitutions, might apply to state public health litigation against product manufacturers. In my opinion, *parens patriae* litigation driven by the attorney general blurs the allocation of powers among the coordinate branches of government as articulated in the constitutions and as traditionally understood. Such litigation violates the spirit of our constitutional frameworks in a manner that ordinary tort actions, even if they also have regulatory effects, do not. Having said that, my guess is that courts will not dismiss *parens patriae* actions on this basis but will allow the legislative and executive branches to sort out their own issues. Nevertheless, precedents concerned with separation of powers provide a powerful measuring rod against which to assess the legitimacy of product litigation initiated by the attorney general.

When the attorney general pursues a detailed regulatory framework to govern an industry, he infringes on the governmental powers allocated to the legislature by the state constitution and to the administrative agencies specifically authorized by the legislature to regulate a particular industry. This is true regardless of whether his proposal results in a negotiated consent decree, as in the tobacco litigation, or in the judicial adoption of the attorney general's recommendations in a court decree. While an administrative agency, traditionally regarded as part of the executive branch, frequently adopts rules similar to legislation and renders adjudicatory rulings with regulatory policy implications that extend far beyond a dispute with any particular regulated party, it does so because the legislature has commissioned the agency to implement its regulatory vision within the parameters the legislature has set. In contrast, the attorney general often acts alone.

Neither the states' tobacco litigation nor Rhode Island's lead pigment litigation can be legitimated as an example of "ordinary" tort action where one party seeks compensation from another whose tortious conduct caused the claimant's harm. The regulatory frameworks established by the MSA in the tobacco litigation and the remedial decree unsuccessfully pursued by the Rhode Island attorney general in the pigment litigation were entirely different creatures from a mere string of judgments holding the defendant liable for its tortious conduct. State attorneys general, powerful officials within the executive branches of their respective states, had consciously intended to govern industries through detailed regulatory frameworks. Compensation for victims of the harms was of secondary consequence.

This realistic appraisal of what happens in the resolution of state *parens patriae* litigation does not fit neatly into existing constitutional doctrinal pigeonholes governing the allocation of powers among the three coordinate branches of government—the executive, legislative, and judicial. To the extent that judicially devised remedial decrees in state-sponsored products litigation resemble or even supplant legislative regulation, manufacturers may argue that the remedial issues presented to the court are nonjusticiable because they represent political questions. My analysis, however, suggests that the governmental entity that de facto creates the regulatory framework is not the court but the attorney general, a member of the executive branch, even when the court ultimately stamps the regulations governing industry with its imprimatur. If the constitutional issues are viewed through the lens of whether the attorney general has assumed powers that belong to the legislature, the appropriate doctrinal category becomes separation of powers, a matter closely intertwined with that of justiciability.[33]

JUSTICIABILITY AND COMPREHENSIVE DECREES OF PRODUCT REGULATION

Not surprisingly, when state attorneys general file *parens patriae* actions seeking to impose stronger product or environmental regulation, business interests assert that such actions are political questions, committed to the political branches of government, and therefore are nonjusticiable. The federal regulatory climate during the recent administration of George W. Bush was more probusiness and antiregulatory than that at any time in recent memory. To many who were committed to addressing public health problems, the judicial branch appeared to be the last hope for what they perceived to be sound environmental and product regulation.

In deciding whether matters are nonjusticiable because the issues the parties seek to litigate pose political questions, courts generally rely heavily on the analytical framework established by the U.S. Supreme Court in 1962 in its seminal case addressing political questions, *Baker v. Carr.*

Prominent on the surface of any case held to involve a political question is found a textually demonstrable constitutional commitment of the issue to a coordinate political department; or a lack of judicially discoverable and manageable standards for resolving it; or the impossibility of deciding without an initial policy determination of a kind clearly for nonjudicial discretion; or the impossibility of a court's undertaking independent resolution without expressing lack of the respect due coordinate branches of government; or an unusual need for unquestioning adherence to a political decision already made; or the potentiality of embarrassment from multifarious pronouncements by various departments on one question.[34]

The *Baker* court stated that dismissal of a case on the basis of a political question may be appropriate even if only one of the factors listed in this framework is "inextricable" from the case.[35] More recently, a plurality of the Supreme Court described each of the six factors as "independent tests."[36]

In 2005, in *Connecticut v. American Electric Power Co.*,[37] a federal district court dismissed, on political question grounds, a suit brought by a number of northeastern states against several major power companies, alleging that their emissions of carbon dioxide into the atmosphere had contributed to the public nuisance of global warming. The court concluded that because of the balancing of policy interests necessary to resolve the case, it faced "the impossibility of deciding without an initial policy determination of a kind clearly for nonjudicial discretion."[38] The court also noted the complexity of this initial policy determination. Similarly, when the attorney general of California initiated regulatory litigation alleging that automobile manufacturers had contributed to the public nuisance of global warming by producing automobiles that emitted carbon dioxide, a federal district court dismissed the complaint because it raised nonjusticiable political questions.[39] The California federal district court, like the Connecticut court, concluded that it would be required to make an initial policy determination, even though the complaint in the California litigation asked for only damages and not equitable relief.[40] Further, the court reasoned that there were no "judicially discoverable or manageable standards by which to properly adjudicate" the plaintiff's claim.[41] The court stated that it was "left without a manageable method of discerning the entities that are creating and contributing to the alleged nuisance," because there were "multiple worldwide sources of atmospheric warming across myriad industries and multiple countries."[42]

State *parens patriae* litigation against product manufacturers appears to raise troubling justiciability issues—at least in the remedial phases of litigation, when the plaintiff state seeks a complex regulatory regime, such as that proposed by the Rhode Island attorney general in the lead pigment litigation. Because the products have left the manufacturers' possession, the nuisance cannot be abated by a simple judicial order to defendants to abate their conduct. Instead, if widespread public health problems are to be remedied, a massive governmental effort is necessary. The question is whether this can be accomplished by a court or whether, instead, such a public health situation requires a response from the legislature and appropriate administrative officers. The myriad policy decisions necessary to remediate lead-based paint hazards throughout the state may be at the core of the issues that the Supreme Court regards as political questions.

Admittedly, many of these same concerns arise in any public interest litigation asserting that governmental entities, such as prisons or school districts, have violated constitutional rights. Courts often have found such disputes to be justiciable. The purview of judicial legitimacy arguably should be drawn more narrowly, however, when the defendants are private parties and not the state and when the plaintiff's action depends not on the Constitution but solely on the common law—and a questionable expansion of common-law principles at that.

SEPARATION OF POWERS AND REGULATORY LITIGATION SPONSORED BY THE ATTORNEY GENERAL

Viewing complex judicial decrees that attempt to implement solutions to public health problems through the lens of justiciability—that is, determining whether the matter is suitable for judicial resolution or should be left to the legislative branch—risks obscuring the dominant role of the state attorney general in *parens patriae* litigation against product manufacturers. In litigation against the manufacturers of cigarettes and lead pigment, as well as other products, state attorneys general have taken on for themselves the power to initiate and pursue regulatory litigation against product manufacturers. In doing so, they have expropriated functions traditionally handled in the constitutional framework by the legislative branch and administrative agencies specifically tasked by the legislature to regulate particular products. The attorney general determines whether a manufacturer's conduct impairs the public interest. He seeks regulation of products that is both extensive and detailed. His actions result in increases in the costs of products, such as cigarettes, which constitute de facto tax increases. These are functions traditionally allocated to the legislative branch. Thus, viewed realistically, the conflict in product-related regulatory litigation initiated by the attorney general is not so much between the legislative and the judicial branches as it is between the legislature and the attorney general, a member of the executive branch.

Separation of powers analysis usually concerns the allocation of powers among the coordinate branches of government within a single sovereign, that is, either the federal government or the state government. Because the regulation of many products, including cigarettes, is primarily a function of the federal government during the post–New Deal era, regulatory litigation initiated by the state attorneys general raises intertwined issues of separation of powers and federalism, sometimes including questions of federal preemption.[43] If anything, the assertion of product-regulatory authority by the state attorney general in an area where the U.S. Congress already has exercised its own authority under the U.S. Constitution's commerce clause is doubly upsetting to our constitutional framework. Further, the attorney-general-driven state litigation often—in a nationwide economy—affects both consumers and producers beyond the state's own borders. In litigation sponsored by the state attorney general against lead pigment manufacturers, the intertwining of separation of powers and federalism issues, although still present, plays a less significant role than it does in the regulation of tobacco products, because most legislation for the prevention of childhood lead poisoning is enacted by the state. Here I analyze the attorney general's regulatory powers only as a matter of the allocation of powers between the attorney general and the legislature—whether federal or state—and its authorized agencies.

State Governments and Allocation of Powers

The state attorney general typically is regarded as a member of the executive branch of the government, even when he does not answer to the governor and is independently elected. Strictly speaking, of course, the state attorney general usually is authorized to file suits on behalf of the state.[44] However, in some instances, the attorney general's use of power, particularly when he acts in concert with his peers to control the remedial outcome of public products litigation, suggests that he may be impeding on powers traditionally regarded as constitutionally delegated to a coordinate branch of government, the legislature.[45]

Separation of powers principles call for the diffusion of power among the legislative, executive, and judicial branches of government. Approximately forty state constitutions explicitly provide that separation of powers principles apply to their state governments.[46] Even in the absence of such explicit provisions, many scholars find separation of powers implicit within the structure of state government.[47] Additionally, Laurence Tribe argues that the language of the U.S. Constitution implies that separation of powers principles pertain to the states.[48] For instance, as Tribe notes, the guarantee clause of the Constitution assumes that states will have distinct legislatures and executives.[49]

Even though the separation of powers principle typically apply to both the federal and state governments, there are significant differences between how state constitutions and the federal constitution allocate powers among the coordinate branches of government. Most important, unlike congressional powers, which are limited to the set of enumerated powers listed in the U.S. Constitution, the powers of the state legislature are plenary (assuming that the legislature is otherwise constitutional) in the absence of constitutional provisions that either limit legislative powers or grant powers to the executive or judicial branch.[50] In other words, all powers not explicitly allocated to the executive or judicial branch by the state constitution are reserved to the legislature.[51]

Applying Federal Separation of Powers Analysis by Analogy

State supreme courts seldom have been asked to address separation of powers issues. When they have, they often have borrowed from the U.S. Supreme Court's analysis of the federal separation of powers. Therefore, in this section, I consider how separation of powers doctrine, as understood by the U.S. Supreme Court, informs separation of powers analysis at the state level.[52]

Courts and commentators often use the opinions of Supreme Court Justices Hugo Black and Robert Jackson in *Youngstown Sheet & Tube Co. v. Sawyer*[53] to structure separation of powers analysis, despite the academic controversy the case has spawned in recent decades.[54] Contemporary scholars characterize Justice Black's opinion of the Court as a "formalist" approach to constitutional interpretation, because it requires that the exercise of power by any coordinate branch of the government be justified by a firm textual basis in the Constitution.[55] In 1952, in *Youngstown*, Justice Black held that President Truman lacked the constitutional power to direct his secretary of commerce to seize the nation's steel mills to avert their shutdown as a result of a labor strike during the Korean War.[56] The Court found the constitutional analysis remarkably simple: neither any statute nor any provision of the Constitution granted the president such power, either expressly or impliedly.[57] Notably, Congress previously had rejected the use of plant seizure to prevent labor stoppages.[58]

Under Justice Black's approach, determining whether the attorney general violates separation of powers when he initiates regulatory litigation against product manufacturers ultimately depends on how such litigation is interpreted. If the issue is viewed superficially, the attorney general almost always possesses the power, under either the state constitution or statutes, to initiate claims on behalf of the state's interests. As previously explained, however, the recent examples of the filings of regulatory litigation are far more than the mere filings of lawsuits. The appropriation of regulatory powers belonging to the legislative branch suggests that the attorney general, as a member of the executive branch, violates separation of powers principles as Justice Black and other formalists conceive them.

The alternative to the formalist approach to separation of powers is the so-called functionalist approach. Functionalism, a less rule-bound, more policy-oriented approach to separation of powers, seeks an appropriate balance of power among the three coordinate branches.[59] The concurring opinion of Justice Jackson in *Youngstown*[60] follows a functionalist approach and has been regarded for decades as a starting point for assessing the legitimacy of the exercise of executive power. According to Justice Jackson, the president's power under the Constitution is greatest when Congress has authorized his actions, either expressly or impliedly. In this situation, his actions are justified by the combined powers constitutionally granted to the executive and legislative branches. Conversely, the president's power is "at its lowest ebb" when he "takes measures incompatible with the expressed or implied will of Congress," because "he can rely only upon his own constitutional powers minus any constitutional powers of Congress over the matter."[61] In a middle category, Jackson's "zone of twilight,"[62] the constitutional analysis becomes most murky. Here, either the president and Congress have concurrent authority, or the allocation of powers is ill defined. According to Justice Jackson, in this sphere, "congressional inertia, indifference or quiescence may sometimes, at least as a practical matter, enable, if not invite, measures on independent presidential responsibility."[63] Nevertheless, in *Youngstown*, Justice Jackson concurred in the Court's judgment striking down the seizure of the steel plants even during a time of war.

Justice Jackson's Analytical Framework and Regulatory Product Claims Initiated by the Attorney General

Regulatory Product Claims with Legislative Approval

Employing Justice Jackson's analytical framework, an attorney general who brings such a claim against a manufacturer with the express or implied approval of the legislature operates at the highest level of authority. He possesses both the powers granted to him in the state constitution and those delegated by statute. For example, during the tobacco litigation, the state legislatures of Florida, Maryland, and Vermont all passed legislation that implicitly authorized their states' *parens patriae* litigation.[64] In each of these states, there is little question that the attorney general possessed the constitutional authority to file litigation against the tobacco manufacturers, even if the attorney general's intent was to impose a new regulatory regime, which, in fact, is what transpired. The attorney general's authority also is strengthened and presumably legitimized when the legislature enacts statutory provisions declaring that the manufacture or distribution of specified unlawful products constitutes a public nuisance.[65]

Regulatory Product Claims in the Face of Legislative Disapproval

Justice Jackson's opinion makes it clear that a state legislature has the constitutional power to prevent the attorney general from filing a *parens patriae* action against either the manufacturer of a specific product or the manufacturers of all products. For example, during the cycle of litigation against firearm manufacturers, a number of state legislatures prohibited the filing of such actions by the state attorney general.[66] Even in the absence of legislation explicitly prohibiting *parens patriae* regulatory claims by the attorney general, the legislative response to such social problems as childhood lead poisoning or tobacco-related illnesses may implicitly signal a rejection of the approach the attorney general seeks to implement in the product regulatory litigation.

Consider, for example, the situation in Rhode Island when the state legislature enacted a comprehensive regulatory scheme designed to prevent childhood lead poisoning by placing the burden of eliminating lead-paint hazards on property owners and mandating that they undertake specified measures to render residential properties "lead-safe." Despite this legislative scheme, the state attorney general filed a *parens patriae* regulatory action that sought to hold manufacturers of lead pigment, not property owners, financially responsible and that imposed more demanding standards for remediation of lead-based paint hazards. In this situation, it appears that the legislature implicitly rejected some of the fundamental goals of the state attorney general. As previously noted, the Rhode Island Supreme Court, as well as the New Jersey Supreme Court, relied heavily on legislative enactments at odds with the state's or municipalities' litigation when they dismissed government actions against pigment and paint manufacturers.

Regulatory Products Claims and Legislative Silence

The issue of whether the attorney general is authorized to act alone to initiate a regulatory civil action in the absence of either legislative approval or disapproval, whether expressed explicitly or impliedly, is more difficult under Justice Jackson's analysis. If the legislature has been truly silent, as contrasted with the situation where the legislature merely has failed to respond with the aggressive stance preferred by public interest advocates, the resolution of this question is a difficult one, which must be informed by one's evaluation of the respective institutional competencies and fairness of the legislature and the office of the attorney general.

Public interest advocates and mass plaintiffs' attorneys argue that the state attorney general is justified in acting on his own to sue tobacco and lead paint manufacturers because the legislature has failed to regulate those industries. They further argue that if the legislature believes the attorney general is impinging on its authority, it can always enact legislation preventing such litigation, either generally or against a specific industry. This latter argument is suggested by the U.S. Supreme Court's reasoning in 1981 in *Dames & Moore v. Regan*,[67] where the Court held that Congress had implicitly approved the practices followed by American presidents in settling claims between American citizens and hostile nations, despite the lack of explicit statutory authority.[68] The situation in *Dames & Moore*, however, involved more than legislative acquiescence inferred from inaction. There, Congress had repeatedly passed related legislation, "thus demonstrating Congress' continuing acceptance of the President's claim settlement authority."[69] In other words, the situation in *Dames & Moore* is more like that of those states that passed statutes facilitating *parens patriae* against the tobacco companies than like the situation of state legislatures that have merely failed to act.

It is widely accepted that the legislature's failure to act does not necessarily indicate its opposition to a proposed piece of legislation.[70] Hence, the legislature's failure to stop *parens patriae* litigation should not be construed as acquiescence in such litigation. It is far easier to kill a legislative proposal than it is to enact it. If the attorney general indicates his intention to file a regulatory action against product manufacturers, a legislature wanting to stop the litigation must undertake a difficult process that requires action by each house of the legislature, the signature of the governor, and the time and energy required to accomplish these steps during a crowed and busy legislative session. Adding to the legislature's challenge is the fact that the governor and the majority in each legislative chamber will often not be of the same political party.

Further, in some instances, any legislative attempt to prevent *parens patriae* litigation against product manufacturers by the attorney general may not be legally effective, because legislation is almost inevitably prospective in nature, while the common law generally operates retroactively. For example, in December 2006, the Ohio General Assembly passed legislation making public nuisance actions against product manufacturers subject to the requirements of the Ohio Product Liability Act[71]—in other words, eliminating public nuisance as a separate claim. Despite this enactment, in April

2007, the Ohio attorney general filed a *parens patriae* public nuisance action against manufacturers of lead-based paint and lead pigment, which later was dismissed.[72] If it had not been dismissed, the Ohio courts would have faced the issue of whether the legislation applied to harms caused by defendants' conduct occurring prior to the passage of the amendments to the Ohio Products Liability Act.[73] Regardless of how this retroactivity issue might have been resolved, the respective actions of the Ohio legislature and that state's attorney general illustrate the difficulties with assuming that a state legislature can always enact legislation to prevent the attorney general from filing a regulatory lawsuit establishing new state policy.

In summary, the goals of public products litigation represent a major shift in regulatory power from the legislative branch and administrative agencies to the state attorney general, in a manner unforeseeable even fifteen years ago. These assertions of power by the attorney general, when acting without explicit or implicit legislative approval, appear to be out of line with traditional notions of allocations of powers within our tripartite systems.

THE DISTORTION OF PUBLIC POLICY RESULTING FROM PUBLIC HIRING OF PRIVATE CONTINGENT FEE ATTORNEYS

One further factor suggests that the attorney general is less likely than the legislature and its authorized administrative agencies to pursue cost-effective solutions to public health problems and to fairly place fiscal responsibility for the costs of remediation on the appropriate parties. Public products litigation inherently seeks funds, whether characterized as damages or as "the costs of abatement," from defendants. In most, but not all, instances of *parens patriae* litigation against product manufacturers, the state attorneys general or municipal officials hire private attorneys, almost inevitably chosen from a small group of sophisticated plaintiffs' firms specializing in mass products litigation, to prosecute the litigation for them. For example, one of the firms that provided leadership for the tobacco litigation later litigated against lead pigment manufacturers in Rhode Island.

These arrangements between state attorneys general and plaintiffs' firms specializing in mass torts provide that outside counsel will be paid on a contingent fee basis. In other words, the retained attorneys receive a percentage of the state or local government's recovery as compensation for its services. Without the use of contingent fee arrangements, most states would not be able to match the quantity and quality of legal resources committed to mass products litigation by defendant-manufacturers. For example, the State of Rhode Island's brief in opposition to a motion challenging its employment of outside counsel paid on a contingent fee basis noted that the roster of the appearance of counsel in the lead pigment litigation included twenty-nine local counsel and ninety-two out-of-state counsel representing the defendant-manufacturers; at the time, the entire Government Litigation Unit of the Civil Division of the Rhode Island Attorney General's Office consisted of thirteen attorneys, of whom three were assigned to the case.[74]

The hiring of plaintiffs' firms by state and local governments on a contingent fee basis has been highly controversial during the past decade. Before the initiation of the tobacco litigation, states entered into fee agreements that resulted in the privately retained tobacco attorneys being entitled to fees that were estimated to exceed twenty-five billion dollars.[75] Not surprisingly, state governments and voters sometimes balked at the payment of these fees,[76] and the actual amount eventually paid was probably closer to fifteen billion dollars.[77]

In later cycles of litigation against other industries, defendant-manufacturers have challenged the legality of government officials retaining private attorneys on a contingent fee basis, with mixed results. The Rhode Island Supreme Court rejected such a challenge in the litigation against lead pigment manufacturers, concluding that "such contractual relationships may well, in some circumstances, lead to results that will be beneficial to society—results which otherwise might not have been attainable."[78] At the same time, the court indicated that the attorney general must retain "absolute and total control over all critical decision-making"[79] when private counsel are retained on a contingent fee basis.

The most dangerous aspect of the attorney general's use of contingent fee attorneys—an aspect not yet acknowledged by any court—flows naturally from the analysis, previously presented, of the allocation of powers among the coordinate branches of government. The government's decision regarding which course to select to solve highly complex public health problems probably is inherently influenced by the possible presence of a "deep-pocket" manufacturing defendant. Mass plaintiffs' firms routinely lobby state attorneys general, urging them to litigate against one industry or another. The evolving partnership between contingent fee counsel and state attorneys general thus determines which public health problems receive public attention. Allergies to mold, dust, and other substances, particularly among children from lower-income backgrounds, are far more pervasive problems than childhood lead poisoning,[80] yet these public health issues are unlikely to be addressed, because the funds for prevention must come from tax dollars appropriated by the legislature and not from "out-of-state" corporations with substantial resources.[81]

The selection of health risks to be addressed through *parens patriae* litigation is analogous to what Justice Stephen Breyer has characterized as the flawed process that administrative agencies use to select potentially toxic substances as the targets of regulation.[82] He asserts that regulatory priorities more closely track public perception, politics, or random chance than they do any actual ranking of levels of risk by experts. Justice Breyer goes on to say that the tort system is ill equipped to "serve as a substitute for government regulation." He acknowledges that the tort system is "much criticized for its random, lottery-like results and its high 'transaction costs' (i.e., legal fees) which eat up a large fraction of compensation awards."[83]

The possibility of identifying defendants who have significant resources or applicable insurance coverages probably also influences the decisions of state governments regarding how a particular public health problem should be addressed and who should be expected to pay for the remedial measures. For example, the existence of the well-heeled, out-of-state manufacturer of OxyContin probably distorted a more neutral public policy analysis, which might have shown that ending OxyContin abuse in West Virginia should focus on disciplining physicians who substantially overprescribed the painkiller. The perceived deep pockets of former lead pigment and lead paint manufacturers have led Rhode Island, Ohio, and a number of municipalities to conclude that manufacturers, not property owners, should bear responsibility for eliminating lead-based paint hazards. Despite prior state legislative[84] and federal administrative[85] determinations that the appropriate response to lead-based paint hazards is to implement cost-effective, so-called interim controls, the financial interests of contingent fee attorneys favor advocating abatement as the appropriate level of regulation, at much greater cost. The greater the recovery or settlement proceeds paid by defendant-manufacturers, the larger the fee will be for plaintiffs' counsel retained on a contingent fee basis.

Not surprisingly, plaintiffs' lawyers hired by state attorneys general to sue lead pigment manufacturers often are significant contributors to the political campaigns of the attorneys general.[86] Lobbying and campaign contributions, of course, are not unheard of in state legislatures. The legislature, however, is understood to be a political institution responding to clashing perceptions of the popular will and, probably to an unfortunate extent, campaign contributors. Influencing the state attorney general's decision whether to file litigation on behalf of the state through campaign contributions and persistent lobbying is arguably less consistent with traditional conceptions of his public office.

In any event, *parens patriae* litigation against product manufacturers shifts the locus of policy-making away from the popularly elected legislature and the administrative agencies it creates to implement its regulatory vision. The regulatory power inherent in sponsoring such litigation devolves to the attorney general, an official whose traditional regulatory powers were far more limited. More important, many attorneys general today exercise their increased policy discretion by focusing the blame for creating costly social problems on those manufacturers who possess considerable resources, in order to avoid politically costly tax increases. Plaintiffs' firms that promise millions or billions of dollars of new resources for the state without any financial obligation on its part augment this distortion of the attorney general's decision-making process. These firms, of course, stand to enrich themselves in the process.

CONCLUSION

To a large extent, the constitutional issues inherent in litigation driven by state attorneys general against manufacturers of products that cause latent diseases have escaped judicial scrutiny precisely because such claims have settled.[87] In the few instances when these claims have been judicially tested, notably the recent opinion in the Rhode Island litigation against lead pigment manufacturers, manufacturers prevailed more often than not, usually on the grounds that public nuisance and other vague torts cannot be expanded beyond traditional doctrinal boundaries. Lurking in the background of these opinions, there appears to be a more significant judicial concern—a nascent understanding that the attorney general's filing of *parens patriae* litigation against manufacturers of products already regulated through the legislative process distorts our constitutional structure.[88]

Attorneys general, many of whom became lawyers during an era that revered judicial activism, should not be faulted for seeking to use the full authority of their offices (and perhaps then some) to solve critical social problems. The notion of benefiting public health and public safety without spending tax revenues is a beguiling one, relentlessly espoused by a handful of national plaintiffs' firms that specialize in mass torts law and stand to profit handsomely. As imperfect as the functioning of Congress, federal administrative agencies, and state legislatures may be in reality, however, the attorney general's appropriate role within the constitutional framework is not to replace the legislatively approved provisions regulating products with a regulatory scheme, whether resulting from settlement or judicial decree, which implements his own vision of social engineering. Nor will public policy-making be improved by a process that prioritizes regulatory goals depending on whether corporations with perceived deep pockets can be blamed for causing a particular public health problem.

CONCLUSION

Nature abhors a vacuum. At the close of the twentieth century, tobacco-related diseases and childhood lead poisoning, along with diseases caused by asbestos exposure, constituted product-caused public health problems of an unprecedented scope. Public health advocates perceived that the ordinary regulatory processes of the legislative branch and the administrative agencies it creates had failed to prevent and to end these epidemics. In addition, victims of childhood lead poisoning and tobacco-related diseases were usually unable to sue and recover under the traditional rules of common-law torts. In short, when it came to preventing these illnesses and compensating the victims, the legislative and judicial branches had created a vacuum.

To their credit, the often-maligned plaintiffs' bar responded to this vacuum in a creative, albeit also self-enriching, manner. These attorneys argued that the ability of a state government to sue, as *parens patriae*, on behalf of its residents should be expanded and combined with novel interpretations of ancient, but vague, torts to enable billions of dollars of recovery for either damages resulting from product-caused disease or the costs of abatement designed to prevent them. Their proposed judicial solutions fit the spirit of the times. The public deeply mistrusted the political branches of government—Congress and administrative bureaucracies—but simultaneously expected protection from products that caused public health harms.

THE FAILURE OF GOVERNMENT PRODUCTS LITIGATION

Fifteen years after the filing of the first actions against tobacco manufacturers and a decade after the initiation of lead pigment litigation in Rhode Island, it is now possible, at least to a considerable extent, to assess the use of *parens patriae* litigation against product manufacturers as a means to solve product-caused public health problems. The Master Settlement Agreement (MSA), the negotiated resolution of the tobacco litigation, has widely been judged, even among the most ardent advocates of the litigation, to have been a disappointment and a lost opportunity. The agreement did provide modest regulation of cigarette advertising and promotion, but it failed to provide for FDA regulation of cigarettes. In significant ways, it protected the market positions of the largest tobacco companies and thus redounded to the benefit of those companies most responsible for millions of deaths. Additionally, very few of the funds paid by the tobacco companies under the agreement have in fact been used for smoking prevention and the treatment of tobacco-related diseases.

The results of the litigation against pigment manufacturers in Rhode Island are even more distressing to those who hoped courts would compensate for the failings of legislatures and regulatory agencies. By the time of that litigation, the tort of public nuisance had emerged as the strongest claim in public health litigation against product manufacturers. The recent decisions by the supreme courts of Rhode Island,[1] New Jersey,[2] and Missouri[3] likely portend, however, a more universal judicial rejection of the use of public nuisance to collectivize recovery for individual harms. State and municipal governments may have a better chance of prevailing on public nuisance claims in such states as California[4] and Wisconsin,[5] each of which have judicial precedents or legislative definitions of nuisance that differ from Rhode Island's. But the judicial trend now appears to be strongly opposed to liability.

For the decade following the states' tobacco litigation, public nuisance looked like the magic tort that collective entities, such as states and municipalities, could use to bypass defenses based on the individual victim's own conduct or the inability of any individual victim to identify the product manufacturer that harmed him. Some mass plaintiffs' attorneys argued—and some judges opined—that by viewing mass products torts as an environmental harm, public nuisance could substitute for traditional products liability theories that did not enable the victims to recover. To make this happen, however, it was necessary to stretch and distort, beyond recognition, the often unarticulated or vaguely stated boundaries of public nuisance.

The supreme courts of Rhode Island and New Jersey applied a principled common-law approach when they rejected the assertion that the tort of public nuisance was broad enough to encompass public health problems caused by mass products. These decisions were justified by the historical limits of the tort. Further, if the violation of a public right, which traditionally had been understood to be an integral element of any public nuisance claim, would have been redefined to include an individual's right not to be victimized by any public health problem caused by products, it is difficult to see how public nuisance would have been limited to claims against tobacco manufacturers and lead pigment manufacturers. Any mass product that causes harm will, by its very nature, cause repetitive harms to individuals. If such harms are to be viewed collectively as a public health problem and hence as a public nuisance, the resulting liability would extend to the manufacturers of any mass-produced widget that causes harm. To allow liability under such circumstances, even in the absence of proof of an individualized causation requirement between the manufacturer and the person directly harmed, would be a sharp departure from common-law principles governing the liability of a manufacturer.

In my view, tort law is not static and should develop and transform itself in response to new factual situations and changing prevailing public ideologies. The extension of public nuisance law to state actions seeking recovery for the public health problems caused by cigarettes and lead-based paint, however, is not justified by objectives of loss minimization or loss distribution, much less by more traditional corrective justice principles. Instead, the use of public nuisance law by a comparatively small number of judges seems to be nothing more than a well-intentioned, but nevertheless opportunistic, attempt to use the vaguely stated requirements of public nuisance as a means to address a public health problem when the legislature is not fulfilling its responsibilities.

Even if courts were able to identify a substantive common-law basis for holding manufacturers liable for the costs generated by product-caused public health problems, the Rhode Island litigation suggests that any victory for public health likely would be more illusory than real. Until he was rescued by the decision of the Rhode Island Supreme Court reversing his judgment ordering pigment manufacturers to abate the conditions that facilitated the widespread incidence of childhood lead poisoning in his state, a trial court judge had put himself in charge of remediating lead-based paint hazards in an estimated 240,000 individual housing units.[6] This task was probably more daunting than any ever undertaken by a court, even those involved in complex, but at least constitutionally mandated, cases of school desegregation and prison reform.

McCann and Haltom have identified at least one way in which the states' antitobacco litigation significantly advanced efforts at tobacco control. They found that the *parens patriae* litigation did help reframe the public perception of tobacco-related diseases as maladies attributable to corporate irresponsibility, in a manner that private litigation had not, thus generating public support for stronger regulation.[7] One must question, however, the legitimacy of a state attorney general, regarded as a "minister of justice" under prevailing professional ethics norms, suing corporations under legal claims of questionable validity in an effort to paint corporate misconduct as quasi-criminal and thereby reframe the debate within the political branches. Such a litigation strategy, while arguably appropriate for a public interest attorney or private plaintiffs' counsel, seems out of place given the state attorney general's unique status as the lawyer who speaks for the sovereign and yields enormous power to influence the remedial outcomes of either settlement or litigation.

The tobacco and lead pigment *parens patriae* litigation cycles not only largely failed to accomplish their public health objectives; they also inflicted intangible, but important, harm to the constitutional allocation of powers among the three coordinate branches of government. The assumption by the state attorneys general and the courts of regulatory powers that traditionally have been allocated to the legislature and the administrative agencies it creates represents one of the greatest attempts to change the allocations of powers among the coordinate branches of government since the New Deal. The filing of the *parens patriae* actions against product manufacturers by a state attorney general is something far different than the mere filing of claims by the state's chief government attorney on behalf of his client. In the recent waves of litigation, the attorney general intends to impose on product manufacturers a new regulatory regime that often is inconsistent with a product regulatory regime previously adopted by the legislature. His bargaining power in settlement talks with manufacturers is overwhelming, because he seeks damages resulting from the amalgamated harms initially and more directly experienced by millions of cancer victims or owners of older properties containing lead-based paint. The combination of these factors poses serious separation of powers concerns under our state and federal constitutions, because these transparently legislative actions of the attorney general are out of character with his traditional and constitutionally allocated role.

The Master Settlement Agreement, which ended the states' tobacco litigation, illustrates this phenomenon. The detailed regulatory regime imposed by the MSA to govern tobacco manufacturers resembles comprehensive congressional regulatory legislation or administrative regulations. Similarly, the Rhode Island attorney general's proposed plan for the remedial decree in the lead pigment legislation was one of the most detailed and comprehensive sets of rules ever promulgated by any branch of government for the remediation of lead hazards. It rivaled both HUD regulations governing lead-based paint in low-income housing and similarly comprehensive legislative and regulatory requirements enacted in such states as Massachusetts and Maryland. These *ex ante* macroeconomic regulatory functions typically are handled by legislatures and agencies that they create, not by attorneys general and courts. At the same time, the MSA specifically enabled tobacco manufacturers to raise the price of cigarettes in order to fund their obligations under the agreement, and in doing so, it blatantly protected them from price competition. These provisions functionally replicate the actions of a legislature in exercising its taxing authority. The attorney general takes on responsibilities constitutionally assigned to the legislative branch when he promotes a comprehensive regulatory regime that governs the marketing or distribution of a product, creates the functional equivalent of new taxes, and/or implements massive new programs of housing renovation in hundreds of thousands of residences in an attempt to solve public health problems. In a generation when both business interests and conservative commentators frequently assert that courts are exceeding their proper authority and entering into the legislative realm, an argument with which I often personally disagree, this does appear to be a bridge too far.

LEGISLATION TO ADDRESS PRODUCT-CAUSED PUBLIC HEALTH PROBLEMS

When plaintiffs' lawyers convinced state attorneys general and municipal solicitors to sue the manufacturers of tobacco products, handguns, and lead pigment, they justified their actions on the grounds that Congress and state legislatures had failed. Ultimately, this book is about the limits of the judicial process in solving product-caused public health problems. Nevertheless, it seems incumbent upon me to mention briefly the importance of reforming the legislative process through such measures as greater public transparency regarding the contacts between industry lobbyists and legislators, closing loopholes in campaign finance laws that enable lobbyists to bundle contributions made by others to public officials or otherwise to circumvent the regulation of such laws, and more vigorously and effectively enforcing existing laws.[8]

An industry group, such as tobacco manufacturers in Congress or rental property owners in state legislatures, typically represents a reasonably small number of firms, each with significant amounts of money at stake and hence intensely motivated to influence the regulatory process. Historically, victims of tobacco-related diseases and childhood lead poisoning lacked comparable resources and lobbying expertise. In recent decades, however, antismoking organizations have become increasingly effective in lobbying at the federal, state, and local levels of government, and advocates for prevention of childhood lead poisoning lobby effectively in many states, including New York and Maryland. The ability of President Barack Obama to use the internet during his presidential campaign to raise immense sums of political contributions from millions of small donors offers a precedent for public interest organizations acting on behalf of the victims of tobacco-related diseases and childhood lead poisoning to lessen the inherent advantages of large corporations in the campaign financing process.

Further, recent developments in Congress and in federal administrative agencies suggest the potential viability of legislative solutions to address tobacco-related illness. I address these debates and potential compensation systems for victims of tobacco-related illness in the next two sections. I then outline, in very broad strokes, what legislative solutions to childhood lead poisoning might look like.

Tobacco and the Ongoing Political Debate Regarding Further Regulation

A decade after the Master Settlement Agreement, the question is not only whether the MSA constituted an effective policy of tobacco control but also how effectively the political branches of government have been regulating tobacco products in the meantime. The wave of state and local ordinances that began even before the states' *parens patriae* litigation, particularly those laws protecting nonsmokers from secondhand smoke, has continued to swell. Within the past year, Congress has enacted and President Obama has signed into law, strong tobacco control legislation that gives the FDA authority to regulate cigarettes (including the authority to control and to reduce their nicotine content), prohibit additives designed to flavor cigarettes, more vigorously regulate advertising and package warning labels, require premarket approval of new cigarette brands and products, and regulate tobacco companies' claims that their products are less dangerous to health than other cigarettes.[9] The new legislation resembles the FDA proposal of the mid-1990s to regulate cigarettes, but it is even stronger.

A number of even stronger alternative approaches for regulating cigarettes have been proposed, and any one of them could be enacted if Congress chooses to do so. For example, the American Medical Association has endorsed a proposal that the FDA should use its regulatory authority to gradually reduce cigarettes' nicotine content and hence their addictive capacity.[10] W. Kip Viscusi has proposed a somewhat similar, two-part approach that would rely more heavily on the market economy and less on government regulation.[11] Viscusi recommends that the FDA develop an easily understood rating system for cigarettes that would communicate to consumers the comparative risk level of various brands. He further suggests that the FDA vigorously participate in the development of a safer cigarette. Viscusi's proposals have precipitated a backlash among some antitobacco activists who believe that prohibition is the only viable alternative, but Viscusi points out that prohibition of a product used by nearly a quarter of the American population probably is not realistic. So far, however, consumers have decisively rejected cigarettes and other tobacco products engineered to substantially reduce exposure to carcinogens.[12] The most extreme proposal to regulate cigarettes is that of former FDA commissioner David Kessler, who argues that only a congressionally chartered nonprofit monopoly should be allowed to sell tobacco products and then only in "brown paper wrappers."[13]

Public health officials and antitobacco activists continue to be understandably frustrated by the federal government's pace in responding to the public health crises caused by cigarettes. The lessons of the past decade, however, are clear. The regulation needed to substantially reduce the incidence of tobacco-related diseases lies within the province of popularly elected officials and those to whom they assign regulatory roles. The problem has been that some members of Congress and some appointees to federal regulatory authorities are less interested in stronger antitobacco responses than are others. In the political process, public health advocates will sometimes win and sometimes lose. None of these recent developments suggests that there is anything inherent in the legislative and administrative regulatory processes to prevent those interested in public health from enacting strong measures to regulate the distribution and promotion of a product that causes a public health problem. Recent successes within the political branches, when combined with the disappointing results of the state *parens patriae* tobacco litigation, suggest that legislation, not litigation, is the solution to the public health crises caused by tobacco.

Compensation for Victims of Tobacco-Related Diseases

Chapter 3 of this book described how traditional principles of the common law have made it virtually impossible for victims of tobacco-related diseases to recover from cigarette manufacturers, most often because of notions that the dangers of smoking were common knowledge or that the smoker assumed the risk. The reality, however, is that many victims of tobacco-related illness became addicted when they were young. Further, they often became addicted before the release of the 1964 surgeon general's report or at least before evidence emerged during the 1990s that representatives of tobacco companies had lied about both the dangers of smoking and the manipulation of nicotine levels in cigarettes. Finally, some victims of tobacco-related diseases may be able to prove that their illness resulted solely from environmental tobacco smoke (ETS).

In these circumstances, in addition to the goal of substantially reducing the incidence of tobacco-related illnesses in the future, Congress should establish an administrative compensation system, somewhat similar to workers' compensation, to provide at least partial compensation to victims of tobacco-related disease.[14] My purpose here is not to provide a comprehensive blueprint for such a system, as others have done.[15] Rather, I sketch out a few ideas to suggest, at least in a rough manner, the feasibility of such a system.

An administrative compensation system for victims of tobacco-related diseases could be financed most easily by an excise tax on cigarettes. The amounts paid by each manufacturer under this approach would not necessarily reflect its respective causal contribution over the course of decades to the current incidence of tobacco-related illnesses, because the respective market shares of tobacco manufacturers have changed through the years. Nevertheless, such a financing mechanism would provide a politically acceptable means of compensating victims of tobacco-related illness while maintaining the integrity of the common law. At the same time, it would assess the costs caused by tobacco products to the activity causing the harm—the manufacture and distribution of such products—albeit not necessarily in the proper proportions reflecting each manufacturer's contribution to the aggregate harm. Legislatures are unlike courts. The legislature is not required to prove that the proceeds of taxes imposed on a particular manufacturer is necessarily proportionate to the amount of harm to which its products contributed during earlier decades.

Compensation benefits under the system would be determined by an administrative compensation board. Those eligible to recover would include smokers currently diagnosed with "signature" diseases resulting from smoking, that is, diseases almost always caused by smoking, such as lung cancer, emphysema, or certain kinds of bronchitis. Victims of these diseases who could show substantial exposure to ETS, such as family members exposed to cigarette smoke in the home, also would be able to recover. More difficult are issues concerned with whether there should be compensation for smokers or those exposed to ETS who now suffer from a disease (e.g., heart disease) for which smoking demonstrably increases the risk but for which many other causes exist.

Finally, what damages would be recoverable? Workers' compensation systems typically award compensation only for medical and rehabilitation expenses, lost income not exceeding modest minimums, and so-called scheduled benefits in the case of various categories of disability. A similar approach appears appropriate for the tobacco compensation system. To be sure, victims of tobacco-related disease typically suffer horribly, and the economic losses recoverable under the compensation plan would not compensate them fully even for their economic losses, much less their intangible losses. Limiting recovery to specified economic losses, however, can be justified on at least two bases. First, such a limitation is a rough way of attributing to the smokers/victims themselves some of the fault and, accordingly, some of the financial responsibility. Second, a compensation system that does not compensate for noneconomic losses such as pain and suffering is easier to administer and far less expensive.

The Political Processes and the Prevention of Childhood Lead Poisoning

Eliminating childhood lead poisoning requires a three-pronged legislative approach, including (1) screening all children less than the age of six (and therefore at risk of childhood lead poisoning) who have lived in housing units constructed before 1978 or who otherwise have been exposed to lead-based paint, perhaps at a caretaker's or relative's home or in preschool, in order to determine whether they have elevated blood lead levels (EBLs); (2) educating parents, landlords, child care workers, physicians, and others about preventing childhood lead poisoning; and (3) proactively taking steps to eliminate the sources of childhood lead poisoning, including lead-based paint hazards in pre-1978 housing. The last prong is the most demanding. As I suggested in chapter 5, eliminating the sources of childhood lead poisoning in housing requires the federal or state government to resolve two separate issues:

(1) Should the government require full abatement or only the undertaking of interim controls in any residence where a child resides?
(2) Which party or parties—the property owner, past producers of lead pigment or lead paint, current producers of paint (which does not contain lead), and/or the government itself—should be required to share in the costs of such measures?

In my judgment (though some will disagree), the first question is the easier one to answer. Both federal standards for low-income housing and the statutory frameworks in those states that have been comparatively effective in substantially reducing the incidence of children with EBLs, notably Maryland and Massachusetts, suggest that only interim controls should be required. As described previously, the cost of interim controls typically is a modest fraction of the cost of full abatement, yet studies suggest that the reduction in the quantity of lead-bearing dust in the home is comparable. By way of review, a nonexclusive list of required interim controls would consist of the following measures: removing chipping, peeling, or flaking paint and repainting; replacing windows and window frames painted with lead-based paint or preventing them from rubbing against each other; and preventing hanging doors from rubbing against the door frame.

Existing state laws and federal regulations, where they are applicable, require compliance with interim control standards only in rental housing. Most childhood lead poisoning does in fact occur in low-income rental housing. Implicitly, at least, the assumption seems to be that property owners who own their own residences presumably already have sufficient incentives to maintain their properties, for two reasons. First, they do not want to risk harming their own children. Second, as they undertake the upkeep, renovations, and repairs that otherwise maintain the appearance, functionality, and market value of their own homes, they also tend to minimize lead-based paint hazards. Further, I have observed state legislators recoil at the suggestion that they should place on homeowners legal obligations to maintain their own homes in order to protect their own children. Such a reaction seems to reveal that legislators do not fully understand the severity of the risks posed by childhood lead poisoning. In a very real sense, for someone to recklessly subject his own child to lead-based paint hazards is every bit as much a form of child abuse as failing to provide the child with food or medicine. Ultimately, all pre-1978 residential properties where children reside, including owner-occupied properties, need to comply with interim controls designed to substantially reduce the risks of childhood lead poisoning.

These issues regarding what needs to be required in order to eliminate the incidence of childhood poisoning are easy ones compared with the question of who should pay for it. At the current time, most of these costs are borne by property owners, and there is an element of fairness in this allocation of financial responsibility. In recent decades, the fair market value of older, low-income housing has been reduced by the risks of lead-based paint.[16] Accordingly, at least in many markets, property owners in recent decades probably acquired older residential properties at a discount reflecting the presence of lead paint. The implementation of either interim controls or full abatement likely will increase their market prices. Further, both the risks to children and the costs of implementing interim controls are greatest in those units that property owners have most neglected.

At the same time, property owners understandably protest that they should not be required to bear the full brunt of the financial responsibility for eliminating lead-based paint hazards. They argue that the producers of lead pigment and the manufacturers of lead-based paint that distributed these products should bear all or at least some of the financial responsibility. It is unlikely, particularly following the decisions in Rhode Island and New Jersey, that the manufacturers who produced lead pigment or the paint containing it can be held liable under the common law. For all the reasons described in this book, these decisions appear to be well reasoned and legally well grounded. Altering common-law doctrines to hold pigment or paint manufacturers liable for the costs of abatement is ill advised.

An alternative to relying on the judicial system to force a financial contribution from manufacturers to help end lead-based paint hazards would be to impose a federal excise tax on the sale of paint. Of course, the paint sold today does not contain lead and poses no risk. Such a tax, therefore, would not be a proxy for legal liability, culpability, or even causation. Popularly elected legislatures, however, are supposed to tax and spend to solve social problems. Unlike courts, they need not find that the assessed party caused the harm or that its actions fell within the legal definitions of tortious conduct. An excise tax would simply be a means of financing a solution to a public health problem for which, in some colloquial, broadly interpreted sense, paint was a factual cause. It probably is not politically feasible for most individual states to impose such an excise tax, because, in most cases, it encourages paint consumers, particularly major contractors, to cross state borders to buy their paint elsewhere, disadvantaging local retailers and, in the process, reducing the government revenues the legislature intended to generate by imposing the tax. A federal excise tax imposed on all paint purchased in the United States, whether produced here or in other countries, appears to be both fairer and more politically palatable.

A 50-cents-per-gallon federal excise tax on paint, including paint manufactured outside the United States, would produce nearly four hundred million dollars annually.[17] Over a period of less than a decade, such a tax could finance a large share of the costs of interim controls for all pre-1978 housing in the United States.[18] The harm to paint manufacturers, which now sell only paint not containing lead, is likely to be minimal in terms of decreased sales revenues. Presumably, the demand for paint and other architectural coverings is relatively price-inelastic. Faced with modest industry-wide price increases, consumers are unlikely to stop painting houses. Relatively few alternatives to residential paint exist. It seems unlikely that most property owners and housing contractors would turn to alternatives, such as vinyl siding, shingles, or wallpaper, just because of a modest tax on paint.

The funds from the federal excise tax should be distributed to states and municipalities for purposes of childhood lead education, screening, and, primarily, the remediation of lead-based paint hazards. In short, property owners and paint companies would share the costs of remediating lead-based paint hazards and implementing interim controls. During an initial three-year funding cycle, the amount distributed to each state should depend on a formula factoring in the number of older housing units, the number of children with EBLs, and the percentage of children living below the poverty line (who face a greater risk of lead poisoning) in each state. Federal funding in subsequent grant cycles should depend on a state's progress in achieving certain benchmarks, including testing a higher percentage of children for EBLs, inspecting more pre-1978 housing for lead-based paint hazards, and bringing pre-1978 housing units into compliance with interim control standards.

In addition to funding from property owners and paint manufacturers, it does not seem unreasonable to expect that significant state and federal revenues should be employed to help finance the prevention of childhood lead poisoning—specifically the costs of screening children, public education, and inspection to assure that interim control standards are being implemented. As things stand now, the federal government and especially state governments pay a heavy price resulting from childhood lead poisoning through Medicaid expenditures, increased education and delinquency costs, and reduced tax revenues from those whose productivity as adults was diminished by lead exposure during childhood.

Federal and state governments also share the ethical responsibility for childhood lead poisoning, both because of their direct contributions and because they have ignored the problem for decades. For years, the federal government required that the paint it purchased for such projects as low-income housing include a very high concentration of lead.[19] Unlike the tobacco situation, there is no evidence that paint manufacturers or pigment manufacturers concealed information that was not available to the government and other sophisticated consumers regarding the dangers of lead poisoning.[20] If paint and pigment manufacturers are to be considered culpable for selling paint containing lead, so should be federal and state governments that had the same access to knowledge of the dangers caused by the presence of lead in paint. Finally, childhood lead poisoning is a public health problem, even if it is not a public nuisance. The federal government routinely funds massive programs to address such health problems as HIV, diabetes, and breast cancer. It has a responsibility to victims of childhood lead poisoning as well, especially because those victims are our youngest and often our poorest—and hence our most vulnerable—citizens.

In short, effective legislative proposals to substantially reduce the incidence of childhood lead poisoning are available, as are ones to reduce the number of tobacco-related deaths and diseases. Such proposals are more likely to be effective than litigation solutions, and they distort neither the constitutional allocation of powers nor fundamental principles of common-law development.

Compensation for Lead-Poisoned Children

The remaining issue is how the substantially fewer number of children that still develop childhood lead poisoning in the future should be compensated. As described in chapter 2, under the common law, an individual lead-poisoned child has no realistic chance of successfully suing the manufacturers of lead pigment or lead-based paint. More frequently, the child and its parents sue their landlord for negligence in failing to reasonably maintain the dwelling unit where they resided and, in the process, exposing the child to lead-based paint hazards. The lawsuit typically requires the plaintiffs' attorney to hire an expert on maintaining older residential units that contain lead, a medical expert to testify that exposure to lead caused the child's impairment, and an economist to estimate the child's loss of future income. Often, the plaintiff loses when the jury finds that the child was not impaired, that his impairment did not result from exposure to lead, that his impairment occurred not in the residence owned by the defendant but in another house where the child spent time, or that the defendant was not negligent. Even when the jury rules in favor of the plaintiff, his attorney may have difficulty collecting from the landlord.

Lead-poisoned children and their parents or guardians should be afforded two alternative sources of compensation for the harms suffered by the lead-poisoned child. First, the child should be able to sue the landlord for negligence or any other existing common-law or statutory claim if his parent or guardian believes that such an approach is viable. In those instances in which the property owner has failed to comply with a state's requirements of interim control standards of the type I recommended earlier, the violation of such a statute or local ordinance should constitute negligence per se.[21]

Second, as a mutually exclusive alternative to suing the landlord and regardless of whether or not the property owner has complied with state requirements mandating interim controls, the family of the lead-poisoned child should have the option of receiving compensation from a no-fault compensation system. The no-fault compensation system should be funded from two separate sources: (1) a portion of the federal excise tax on paint and (2) the proceeds of a substantial annual assessment on the owners of properties (both rental and owner-occupied) that are not in compliance with interim control standards. The amount of compensation should fully compensate the child and his parents for medical and rehabilitation expenses, as well as relocation expenses. If a child's future income-producing ability has been impaired by the exposure to lead, he also should be reasonably compensated for his loss of earning capacity. In short, realistic legislative solutions to end childhood lead poisoning and to compensate victims of childhood lead poisoning can be devised.

CONCLUSION

Fifteen years after the filing of the first state *parens patriae* litigation against manufacturers seeking damages or the costs of abatement resulting from product-caused diseases, the early returns are in. The most promising substantive claim in such litigation, public nuisance, appears to be dead in the water. Even if plaintiffs' attorneys and activist judges identify other viable claims that would enable states and municipalities to recover funds from manufacturers resulting from product-cased diseases, such litigation is unlikely to provide effective and feasible solutions to public health problems. At the same time, such lawsuits driven by attorneys general fundamentally distort the constitutionally provided structure of government. Finally, the important role played by a small handful of mass plaintiffs' law firms, paid on a contingent fee basis, alters both the determination of which public health problems become the focus of government attention and the choice of how they are resolved.

Congress, state legislatures, and federal and state administrative agencies can and must do better to prevent tobacco-related illnesses and childhood lead poisoning. The alluring mirage of litigation solutions has distracted public health advocates and other concerned citizens from more vigorously pursuing the more promising legislative path. Admittedly, the legislative and administrative processes for regulating tobacco products and ending childhood lead poisoning throughout the late twentieth century were deeply flawed, supporting Winston Churchill's observation that "no one pretends that democracy is perfect or all-wise." In a classic and oft-repeated quotation, however, Churchill continued, "Indeed, it has been said that democracy is the worst form of Government except all those other forms that have been tried from time to time."[22] The wave of recent attempts by state attorneys general and judges to replace the programs adopted by democratically elected legislatures with regimes established through *parens patriae* litigation, however well intended, is no exception.

NOTES

Introduction

1. Kevin Sack, "Tobacco Industry's Dogged Nemesis," *New York Times*, Apr. 6, 1997, sec. 1, 22.

2. Peter B. Lord, "3 Companies Found Liable in Lead-Paint Nuisance Suit," *Providence Journal*, Feb. 23, 2006, A1.

3. *State v. Lead Indus. Ass'n*, 951 A.2d 428, 458 (R.I. 2008).

4. John P. Coale, remarks under "Panel Three: Government-Sponsored Regulation—What's Next?" in Center for Legal Policy at the Manhattan Institute, *Regulation by Litigation: The New Wave of Government-Sponsored Litigation* (New York: Center for Legal Policy at the Manhattan Institute, 1999), 64, available at http://www.manhattan-institute.org/pdf/mics1.pdf.

5. Complaint at 14, ¶ 3, *State v. Lead Indus. Ass'n, Inc.*, C.A. No. 99-5226 (R.I. Super. Ct. Oct. 12, 1999) (Relief Requested), available at http://www.riag.ri.gov /documents/reports/lead/lead_complaint.pdf.

6. Robert A. Kagan, *Adversarial Legalism: The American Way of Life* (Cambridge, MA: Harvard University Press, 2001), 93.

7. Ibid., 40; see also 229.

8. *Alfred L. Snapp & Son, Inc. v. Puerto Rico*, 458 U.S. 592, 600 (1982); Lawrence B. Custer, "The Origins of the Doctrine of *Parens Patriae*," *Emory Law Journal* 27 (1978): 195, 196.

9. See, e.g., *Kansas v. Colorado*, 206 U.S. 46, 117 (1907); *Missouri v. Illinois*, 180 U.S. 208 (1901).

10. Carrick Mollenkamp et al., *The People vs. Big Tobacco: How the States Took on the Cigarette Giants* (Princeton: Bloomberg Press, 1998), 29 (quoting Don Barrett).

11. William L. Prosser, "Private Action for Public Nuisance," *Virginia Law Review* 52 (1966): 997, 999.

12. *Awad v. McColgan*, 98 N.W.2d 571, 573 (Mich. 1959).

13. Scott Harshbarger, remarks under "Panel One: State Attorneys General and the Power to Change Law," in Center for Legal Policy at the Manhattan Institute, *Regulation by Litigation*, 13–14.

14. *Protection of Lawful Commerce in Arms Act*, Public Law 109-92, *U.S. Statutes at Large* 119 (2005): 2095 (codified at 15 U.S.C. §§ 7901–7903 (2006)).

15. Harry Shulman, Fleming James Jr., Oscar S. Gray, and Donald G. Gifford, *Law of Torts: Cases and Materials*, 4th ed. (New York: Foundation Press, 2003), iv.

16. *County of Santa Clara v. Atl. Richfield Co.*, 40 Cal. Rptr. 3d 313, 348 (Ct. App. 2006).

17. Robert B. Reich, "Don't Democrats Believe in Democracy?" *Wall Street Journal*, Jan. 12, 2000, A22. Reich reserves his strongest criticism for actions brought by the federal government seeking damages against product manufacturers.

Chapter 1

1. *World's Columbian Exposition of 1893: The Dream City* (St. Louis: N. D. Thompson, 1893), unpaginated, available at http://columbus.iit.edu/dream city/00034032.html.

2. "Fair's New Robot is Schizophrenic," *New York Times*, Apr. 23, 1939, 36; Corie Lok, "Technology as Hope," *Technology Review*, Apr. 2005, 88.

3. *The Tax Burden on Tobacco: Historical Compilation*, vol. 41 (Arlington, VA: Orzechowski and Walker, 2006), 6. In 1890, per capita cigarette consumption was 35. In 1930, it was 977.

4. Robert Sobel, *They Satisfy: The Cigarette in American Life* (Garden City, NY: Anchor Books Press, 1978), 84.

5. Allan M. Brandt, *The Cigarette Century: The Rise, Fall, and Deadly Persistence of the Product That Defined America* (New York: Basic Books, 2007), 1.

6. Gary A. Giovino, Michael W. Schooley, Bao-Ping Zhu, Jeffrey H. Chrismon, Scott L. Tomar, John P. Peddicord, Robert K. Merritt, Corinne G. Huston, and Michael P. Eriksen, "Surveillance for Selected Tobacco-Use Behaviors, United States, 1900–1994," *Morbidity and Mortality Weekly Report Surveillance Summaries* 43, no. SS-3 (Nov. 18, 1994): 8, table 2.

7. Tara Parker-Pope, *Cigarettes: Anatomy of an Industry from Seed to Smoke* (New York: New Press, 2001), 6–13.

8. Richard Kluger, *Ashes to Ashes: America's Hundred-Year Cigarette War, the Public Health, and the Unabashed Triumph of Philip Morris* (New York: Vintage Books, 1997), 5–8.

9. Ibid., 8.

10. Brandt, *Cigarette Century*, 24.

11. Ibid., 27.

12. Ibid., 31–34, 54–56, 61–101; Kluger, *Ashes to Ashes*, 80–111.

13. "Prizes for Ill-Doing," *New York Times*, Dec. 25, 1888, 8. For examples, see Wisconsin Historical Society, "Wisconsin Historical Images," http://www.wisconsin-history.org/whi/, and search the online collection for the keywords "cigarette" and "trading cards."

14. Brandt, *Cigarette Century*, 97.

15. Kluger, *Ashes to Ashes*, 105.

16. Christian Warren, *Brush with Death: A Social History of Lead Poisoning* (Baltimore: Johns Hopkins University Press, 2000), 18–23.

17. Ibid., 117–27.

18. Ibid., 123–24, 130.

19. Ibid., 208. See generally ibid., 203–22.

20. U.S. Environmental Protection Agency, *Air Quality Criteria for Lead,* vol. 1 (Research Triangle Park, NC: Environmental Criteria and Assessment Office, Environmental Protection Agency, 1986), 90.

21. *Clean Air Act,* 42 U.S.C. § 7545(n) (2000).

22. Warren, *Brush with Death,* 63.

23. Ibid., 9.

24. Ibid., 25 (quoting National Lead Company, *Uncle Sam's Experience with Paints* (New York: National Lead Company, 1900)).

25. Warren, *Brush with Death,* 63.

26. *Adulterated or Mislabeled Paint, Turpentine, or Linseed Oil: Hearings before the S. Comm. on Manufacturers,* 61st Cong., 2d sess., 1910, Committee Print, 17–18, reprinted in *Painters Magazine* 37, Mar. 1910, 259.

27. Warren, *Brush with Death,* 49.

28. Gerald N. Grob, *The Deadly Truth: A History of Disease in America* (Cambridge, MA: Harvard University Press, 2002), 192.

29. Ibid., 200.

30. Rachel Carson, *Silent Spring* (Boston: Houghton Mifflin; Cambridge, MA: Riverside Press, 1962).

31. Linda Lear, introduction to *Silent Spring,* 40th anniversary ed. (Boston: Mariner Books, 2002), x–xi, xv–xix.

32. Robert V. Percival et al., *Environmental Regulation: Law, Science, and Policy,* 5th ed. (New York: Aspen, 2006), 90.

33. William Osler and Thomas McCrae, *The Principles and Practice of Medicine,* 8th ed. (New York: D. Appleton, 1915), 393–95, 402–07.

34. John Duffy, *The Sanitarians: A History of American Public Health* (Urbana: University of Illinois Press, 1992), 286–87.

35. Ibid., 286.

36. Christopher Sellers, "A Prejudice Which May Cloud the Mentality: The Making of Objectivity in Early Twentieth-Century Occupational Health," in *Silent Victories: The History and Practice of Public Health in Twentieth-Century America,* ed. John W. Ward and Christian Warren (Oxford: Oxford University Press, 2007), 240.

37. King James I, *A Counter-Blaste to Tobacco* (London: R.B., 1604), transcribed by Risa S. Bear (Salem, OR: Renascence, 2003), available at http://uoregon.edu/~rbear/james1.html.

38. Kluger, *Ashes to Ashes,* 38–39.

39. Brandt, *Cigarette Century,* 46–48.

40. Ibid., 47.

41. Allan L. Benson, "Smokes for Women," *Good Housekeeping,* Aug. 1929, 193 (quoting Cobb).

42. Albert G. Ingalls, "If You Smoke," *Scientific American* 154 (1936): 354–55.

43. William Weiss, "Cigarette Smoking and Lung Cancer Trends: A Light at the End of the Tunnel?" *Chest* 111 (1997): 1414–16.

44. Kluger, *Ashes to Ashes*, 107.

45. Franz Hermann Müller, "Abuse of Tobacco and Carcinoma of Lungs," Abstract, *Journal of the American Medical Association* 113 (1939): 1372.

46. Brandt, *Cigarette Century*, 105.

47. Kluger, *Ashes to Ashes*, 132–33.

48. Ernst L. Wynder and Evarts A. Graham, "Tobacco Smoking as a Possible Etiologic Factor in Bronchiogenic Carcinoma: A Study of Six Hundred and Eighty-Four Proved Cases," *Journal of the American Medical Association* 143 (1950): 336.

49. Roy Norr, "Cancer by the Carton," *Reader's Digest*, Dec. 1952, 7–8, available at http://legacy.library.ucsf.edu/tid/ncr66c00/pdf.

50. Brandt, *Cigarette Century*, 134–41.

51. Richard Doll and Austin Bradford Hill, "Smoking and Carcinoma of the Lung: Preliminary Report," *British Medical Journal* 2 (1950): 742–43, 747; Richard Doll, "Conversation with Sir Richard Doll," *British Journal of Addiction* 86 (1991): 367–68.

52. E. Cuyler Hammond and Daniel Horn, "The Relationship between Human Smoking Habits and Death Rates," *Journal of the American Medical Association* 155 (1954): 1316–28.

53. Kluger, *Ashes to Ashes*, 164.

54. Ibid., 151, 186.

55. Tobacco Industry Research Committee, "A Frank Statement to Cigarette Smokers," Jan. 4, 1954, available at http://tobaccodocuments.org/ness/10245.html?pattern=frank%5Ba-z%5D%2A%5CW%2Bstatem%5Ba-z%5D%2A&#p4.

56. Tobacco Industry Research Committee, "Confidential Report—Tobacco Industry Research Committee Meeting," Oct. 19, 1954, Bates Nos. CTRMN007 295–97, available at http://tobaccodocuments.org/ctr/CTRMN007295-7297.html.

57. Brandt, *Cigarette Century*, 183.

58. Allan Rodgman, "The Analysis of Cigarette Smoke Condensate. I. The Isolation and/or Identification of Polycyclic Aromatic Hydrocarbons in Camel Cigarette Smoke Condensate," Sept. 28, 1956, available at http://tobaccodocuments.org/rjr/501008241-8293.html.

59. Hekmut Wakeham, "Tobacco and Health—R&D Approach: Presentation to R&D Committee at Meeting Held in New York Office," Nov. 15, 1961, available at http://legacy.library.ucsf.edu/tid/uxc85e00/pdf.

60. L. E. Burney, "Smoking and Lung Cancer: A Statement of the Public Health Service," *Journal of the American Medical Association* 171 (1959): 1835–36.

61. John H. Talbott, "Smoking and Lung Cancer," *Journal of the American Medical Association* 171 (1959): 2104.

62. Kluger, *Ashes to Ashes*, 222.

63. Brandt, *Cigarette Century*, 211–39; Kluger, *Ashes to Ashes*, 242–62.

64. U.S. Department of Health, Education, and Welfare, *Smoking and Health: Report of the Advisory Committee to the Surgeon General of the Public Health Service* (Washington, DC: Government Printing Office, 1964), 37.

65. U.S. Department of Health and Human Services, *The Health Consequences*

of Smoking: Cardiovascular Diseases; A Report of the Surgeon General (Washington, DC: Government Printing Office, 1983), v, available at http://profiles.nlm.nih.gov/NN/B/B/T/D/_/nnbbtd.pdf.

66. U.S. Department of Health and Human Services, *The Health Consequences of Smoking: Chronic Obstructive Lung Disease; A Report of the Surgeon General* (Washington, DC: Government Printing Office, 1984), vii, available at http://profiles.nlm.nih.gov/NN/B/B/B/F/_/nnbbbf.pdf.

67. U.S. Department of Health and Human Services, *The Health Consequences of Smoking: Nicotine Addiction; A Report of the Surgeon General* (Washington, DC: Government Printing Office, 1988), i, available at http://profiles.nlm.nih.gov/NN/B/B/2/D/_/nnbbzd.pdf.

68. Addison Yeaman, "Implications of Battelle Hippo I & II and the Griffith Filter," July 17, 1963, Bates No. 2074459290-9294, available at http://legacy.library.ucsf.edu/tid/ari52c00/pdf.

69. Kluger, *Ashes to Ashes*, 744.

70. *Regulation of Tobacco Products (Part I): Hearings before the Subcommittee on Health and the Environment of the House Committee on Commerce and Energy*, 103d Cong., 2d sess., 1994, 620–21.

71. Ibid., 579, 585.

72. Takeshi Hirayama, "Non-Smoking Wives of Heavy Smokers Have a Higher Risk of Lung Cancer: A Study from Japan," *British Medical Journal* 282 (1981): 183–85; Dimitrios Trichopoulos, Anna Kalandidi, Louhas Sparros, and Brian MacMahon, "Lung Cancer and Passive Smoking," *International Journal of Cancer* 27 (1981): 3

73. J. L. Repace and A. H. Lowrey, "Indoor Air Pollution, Tobacco Smoke, and Public Health," *Science* 208 (1980): 464–72

74. Brandt, *Cigarette Century*, 299.

75. Ibid., 298 (quoting Merryman).

76. Philip Morris U.S.A., *Great American Smoker's Kit*, 1986, Bates No. 2024274465, available at http://legacy.library.ucsf.edu/tid/irho4e00/pdf.

77. U.S. Department of Health and Human Services, Centers for Disease Control and Prevention, "Tobacco Use among Adults—United States, 2005," *Morbidity and Mortality Weekly Report* 55, no. 42 (Oct. 27, 2005): 1145, available at http://www.cdc.gov/mmwr/pdf/wk/mm5542.pdf.

78. U.S. Department of Health and Human Services, Centers for Disease Control and Prevention, "Cigarette Use among High School Students—United States, 1991–2003," *Morbidity and Mortality Weekly Report* 53, no. 23 (June 18, 2004): 499–502, available at http://www.cdc.gov/mmwr/preview/mmwrhtml/mm5323a1.htm.

79. Jacqueline Karnell Corn, *Response to Occupational Health Hazards: A Historical Perspective* (New York: Van Nostrand Reinhold, 1992), 72.

80. Warren, *Brush with Death*, 3.

81. Ibid., 50.

82. Ibid., 68 (quoting Carroll Davidson Wright, *The Working Girls of Boston* (Boston: Wright & Potter, 1889), 73).

83. Gerald Markowitz and David Rosner, *Deceit and Denial: The Deadly Politics of Industrial Pollution* (Berkeley: University of California Press, 2002), 20–21.

84. Warren, *Brush with Death*, 114. Today, workplace exposure is governed by the regulations of the Occupational Safety and Health Administration. See 29 C.F.R. § 1910.1025 (2008).

85. Warren, *Brush with Death*, 116–34, 203–23.

86. Ibid., 33–38.

87. Peter C. English, *Old Paint: A Medical History of Childhood Lead-Paint Poisoning in the United States to 1980* (New Brunswick, NJ: Rutgers University Press, 2001), 61.

88. L. Emmett Holt, *The Diseases of Infancy and Childhood, for the Use of Students and Practitioners of Medicine*, 8th ed. (New York: D. Appleton, 1922), 693; J. P. Crozier Griffith, *The Diseases of Infants and Children* (Philadelphia: W. B. Saunders, 1919), 287.

89. Warren, *Brush with Death*, 182.

90. Markowitz and Rosner, *Deceit and Denial*, 43–44.

91. English, *Old Paint*, 46–48.

92. Ibid., 14–15.

93. Markowitz and Rosner, *Deceit and Denial*, 140–43; Warren, *Brush with Death*, 140–43, 49–51.

94. English, *Old Paint*, 83.

95. U.S. Department of Commerce, *United States Government Master Specification for Paint, White, and Tinted Paints Made on a White Base, Semipaste, and Ready Mixed, Fed. Specification Board, Standard Spec. No 10B*, 3rd ed., Circular of the Bureau of Standards, no. 89 (Washington, DC: Government Printing Office, 1927), 2.

96. Randolph K. Byers and Elizabeth E. Lord, "Late Effects of Lead Poisoning on Mental Development," *American Journal of Diseases of Children* 66 (1943): 471–94.

97. Markowitz and Rosner, *Deceit and Denial*, 56.

98. English, *Old Paint*, 92.

99. Ibid., 99–100.

100. Centers for Disease Control and Prevention, *Preventing Lead Poisoning in Young Children* (Atlanta: Department of Health and Human Services, 2005).

101. Markowitz and Rosner have testified as expert witnesses in litigation brought against the paint industry and are highly critical of the industry's response. See Markowitz and Rosner, *Deceit and Denial*, 39, 45–49, 65–66. English has testified as a witness on behalf of the industry in litigation and is more favorable. See English, *Old Paint*, 37–41, 69–70, 161. Warren's approach appears to be more balanced. See Warren, *Brush with Death*, 97–99, 149–50.

102. English, *Old Paint*, 38, 70.

103. Ibid., 38.

104. Abell Foundation, "Childhood Lead Poisoning in Baltimore: A Generation Imperiled as Laws Ignored," *Abell Report*, 15, no. 5 (2002): 2.

105. Markowitz and Rosner, *Deceit and Denial*, 102.

106. American Standards Association, *Standard No. Z66.1* (1955).

107. 16 C.F.R. § 1303.1 (2009) (originally promulgated at *Federal Register* 42 (Sept. 1, 1977): 44,149)).

108. Department of Health and Human Services, Agency for Toxic Substances and Disease Registry, *Case Studies in Environmental Medicine—Lead Toxicity*, Course SS3059, Continuing Education Program in Environmental Medicine and Health (Atlanta: ATSDR, 2000), 22–26, available at http://www.atsdr.cdc.gov/HEC/CSEM/lead/docs/lead.pdf.

109. Philip J. Landrigan, "Pediatric Lead Poisoning: Is There a Threshold?" *Public Health Reports* 115 (2000): 530–31.

110. Centers for Disease Control, *Preventing Lead Poisoning in Young Children* (Atlanta: Department of Health and Human Services, 1991).

111. E.g., Bruce P. Lanphear et al., "Cognitive Deficits Associated with Blood Lead Concentrations <10 µg/dL in US Children and Adolescents," *Public Health Reports* 115 (2000): 521–29.

112. William Kovarik, "Ethyl-Leaded Gasoline: How a Classic Occupational Disease Became an International Public Health Disaster," *International Journal of Occupational and Environmental Health* 11 (2005): 384, 394.

113. U.S. Department of Health and Human Services, Centers for Disease Control and Prevention, "Blood Lead Levels—United States, 1999–2002," *Morbidity and Mortality Weekly Report* 54, no. 20 (May 27, 2005): 513–16, available at http://www.cdc.gov/mmwr/PDF/wk/mm5420.pdf.

114. U.S. Environmental Protection Agency, *Air Quality Criteria for Lead (Final)* (Research Triangle Park, NC: Office of Research and Development, National Center for Environmental Assessment-RTP Division, Environmental Protection Agency, 2006), 4-22, tbl. 4-1, available at http://cfpub.epa.gov/ncea/cfm/recordisplay.cfm?deid=158823.

115. David E. Jacobs et al., "The Prevalence of Lead-Based Paint Hazards in US Housing," *Environmental Health Perspectives* 110 (2002): A599–A606.

116. Carson, *Silent Spring*, 2.

Chapter 2

1. Philip Soper, *A Theory of Law* (Cambridge, MA: Harvard University Press, 1984), 111. See generally Gregory C. Keating, "Fidelity to Pre-existing Law and the Legitimacy of Legal Decision," *Notre Dame Law Review* 69 (1993): 2.

2. Keating, "Fidelity to Pre-existing Law," 4.

3. Lawrence M. Friedman, *A History of American Law*, 3rd ed. (New York: Simon and Schuster, 2005), 350.

4. Oliver Wendell Holmes Jr., *The Common Law*, ed. Mark deWolfe Howe

(1881; reprint, Cambridge, MA: Harvard University Press, Belknap Press, 1963), 76.

5. Ibid., 77. Holmes defended a strict liability standard in limited circumstances where the defendant's activity both posed a high degree of risk and had great social utility. See Oliver Wendell Holmes Jr., "The Theory of Torts," *American Law Review* 7 (1873): 652, 653, 663.

6. *Losee v. Buchanan*, 51 N.Y. 476, 484 (1873).

7. *Ives v. S. Buffalo Ry. Co.*, 94 N.E. 431, 441 (N.Y. 1911).

8. *Escola v. Coca Cola Bottling Co.*, 150 P.2d 436, 462 (Cal. 1944) (Traynor, J., concurring).

9. E.g., *Winterbottom v. Wright*, 10 M. & W. 109, 114–15, 152 Eng. Rep. 402, 405 (Exch. 1842).

10. 111 N.E. 1050, 1053 (N.Y. 1916).

11. See, e.g., *Greenman v. Yuba Power Prods., Inc.*, 377 P.2d 897, 900 (Cal. 1963).

12. *Restatement (Second) of Torts* § 402A (1965).

13. Robert W. Miller, "Significant New Concepts of Tort Liability—Strict Liability," *Syracuse Law Review* 17 (1965): 25, 29 (quoting American Law Institute Meeting, *U.S. Law Week* 32 (1964): 2623, 2627).

14. *Greenman*, 377 P.2d at 900–901.

15. *Escola*, 150 P.2d at 441 (Traynor, J., concurring).

16. E.g., *Henningsen v. Bloomfield Motors, Inc.*, 161 A.2d 69, 84 (N.J. 1960).

17. Robert L. Rabin, "Essay: A Sociolegal History of the Tobacco Tort Litigation," *Stanford Law Review* 44 (1992): 853, 871. For examinations of the history of early tobacco litigation, see ibid.; Marcia L. Stein, "Cigarette Products Liability Law in Transition," *Tennessee Law Review* 54 (1987): 631.

18. A few older opinions had stated that sellers were not liable if they were not aware of possible harm to a few individuals with "peculiar idiosyncrasies" (e.g., *Flynn v. Bedell Co.*, 136 N.E. 252, 254 (Mass. 1922)).

19. 328 F.2d 3 (8th Cir. 1964).

20. Ibid., 9.

21. Ibid., 8.

22. Ibid., 9.

23. Ibid., 6.

24. 317 F.2d 19 (5th Cir. 1963).

25. Ibid., 36.

26. 154 So. 2d 169, 172 (5th Cir. 1963) (Fla. law).

27. *Pritchard v. Liggett & Myers Tobacco Co.*, 295 F.2d 292, 302 (3d Cir. 1961).

28. *Escola*, 150 P.2d at 440–41.

29. See, e.g., *Boyl v. Cal. Chem. Co.*, 221 F. Supp. 669, 674 (D. Or. 1963); *Babylon v. Scruton*, 138 A.2d 375, 378 (Md. 1958); *Braun v. Roux Distrib. Co.*, 312 S.W.2d 758, 763 (Mo. 1958); Fleming James Jr., "Qualities of the Reasonable Man in Negligence Cases," *Missouri Law Review* 16 (1951): 1, 13.

30. *Restatement (Third) of Torts: Products Liability* § 2 (1998).

31. *Restatement (Second) of Torts* § 402A cmt. i (1965). See also *Pritchard*, 295 F.2d at 302.

32. Both at 15 U.S.C. §§ 1331–40 (2006) (as amended).

33. 505 U.S. 504 (1992).

34. However, the Court also held that claims of tobacco-related disease victims alleging either express warranties or intentional fraud by misrepresentation or concealment of material facts were not preempted. Ibid., 525–29.

35. See Rabin, "Sociolegal History of the Tobacco Tort Litigation," 857–60, 867–68.

36. John Fabian Witt, "Toward a New History of American Accident Law: Classical Tort Law and the Cooperative First-Party Insurance Movement," *Harvard Law Review* 114 (2001): 690, 744.

37. See *Tiller v. Atl. Coast Line R.R.*, 318 U.S. 54, 68–69 (1943) (Frankfurter, J., concurring); *Blackburn v. Dorta*, 348 So. 2d 287, 292 (Fla. 1977).

38. 667 So. 2d 1289 (Miss. 1995).

39. Ibid., 1290.

40. Ibid., 1291.

41. E.g., *Payton v. Brown & Williamson Tobacco Co.*, 1995 U.S. Dist. LEXIS 22070, °12 (E.D. Tex. May 2, 1995); cf. *Paugh v. R. J. Reynolds Tobacco Co.*, 834 F. Supp. 228, 230 (N.D. Ohio 1993) ("ordinary knowledge").

42. *Restatement (Second) of Torts* § 402A cmt. i (1965).

43. See, e.g., *Cal. Civ. Code* § 1714.45(a)(1) (West 1998 and Supp. 2008)); *Louisiana Products Liability Act, La. Rev. Stat. Ann.* §§ 9:2800.57(B)(1) (West 1997); *N.J. Stat. Ann.* § 2A:58C-3(a)(2) (West 2000).

44. See, e.g., Holmes, *Common Law*, 64; see also *Claytor v. Owens-Corning Fiberglas Corp.*, 662 A.2d 1374, 1382 (D.C. App. 1995); Richard L. Abel, "A Critique of Torts," *UCLA Law Review* 37 (1990): 785, 811; Louis Kaplow and Steven Shavell, "Fairness versus Welfare," *Harvard Law Review* 114 (2001): 966, 1101.

45. William Prosser, *Handbook of the Law of Torts*, 4th ed. (St. Paul, MN: West, 1971), § 41, 237.

46. *Ingersoll v. Liberty Bank of Buffalo*, 14 N.E.2d 828, 829–30 (N.Y. 1938).

47. 437 N.E.2d 171 (Mass. 1982).

48. Ibid., 188.

49. David Rosenberg, "The Causal Connection in Mass Exposure Cases: A 'Public Law' Vision of the Tort System," *Harvard Law Review* 97 (1984): 851, 907.

50. Ibid., 924–25.

51. See, e.g., *Brenner v. Am. Cyanamid Co.*, 732 N.Y.S.2d 799, 800 (App. Div. 2001); *Skipworth v. Lead Indus. Ass'n, Inc.*, 690 A.2d 169, 171 (Pa. 1997).

52. *Skipworth*, 690 A.2d at 172–73.

53. See, e.g., Christy Plumer, "Setting Priorities for Prevention of Childhood Lead Poisoning in Providence" (master's thesis abstract, Brown University, 2000), http://envstudies.brown.edu/oldsite/dept/thesis/master9900/christy_plumer.htm.

54. See, e.g., Christian Warren, *Brush with Death: A Social History of Lead Poisoning* (Baltimore: Johns Hopkins University Press, 2000), 170–71; J. Julian Chisolm Jr., "Lead Poisoning," *Scientific American* 224 (Feb. 1971): 15, 21.

55. 823 N.E.2d 126 (Ill. App. Ct. 2005), *appeal denied*, 833 N.E.2d 1 (Ill. 2005).

56. *Chicago v. Am. Cyanamid Co.*, 823 N.E.2d at 139.

57. *In re Lead Paint Litig.*, 924 A.2d 484, 492, 500 (N.J. 2007).

58. James Brugger, "Lead-Paint Bill to Protect Children May Contain Flaws," *Louisville (KY) Courier-Journal*, Feb. 11, 2004, 1A (quoting Ralph Scott of the Alliance for Healthy Homes as saying that legislation sponsored by paint industry "would allow paint manufacturers to 'shift the blame' for dangerous lead-based paints they made decades ago to property owners").

59. 563 N.E.2d 684, 686 (Mass. 1990), *superseded by statute, Mass. Gen. Laws Ann.* ch. 231, § 85 (West 2000).

60. *Schmidt v. Merch. Despatch Trans. Co.*, 200 N.E. 824, 828 (N.Y. 1936); see also *Garrett v. Raytheon Co.*, 368 So. 2d 516, 521 (Ala. 1979).

61. *Schmidt*, 200 N.E. at 827–28.

62. 493 F.2d 1076 (5th Cir. 1973) (Judge Wisdom), *reh'g denied*, 493 F.2d 1103 (5th Cir. 1974), *cert. denied*, 419 U.S. 869 (1974).

63. The facts of Borel's exposure and disease, as well as the account of his trial, are taken both from the appellate court opinion (*Borel*, 493 F.2d 1076) and from the classic account of the litigation contained in Paul Brodeur, *Outrageous Misconduct: The Asbestos Industry on Trial* (New York: Pantheon, 1985), 3–70.

64. *Borel*, 493 F.2d at 1082.

65. Ibid., 1088.

66. Ibid. See also *Restatement (Second) of Torts* § 402A cmt. k (1965).

67. *Borel*, 493 F.2d at 1088.

68. Ibid., 1089.

69. Ibid., 1089–90.

70. Ibid., 1083.

71. Ibid., 1082 (quoting plaintiff's deposition).

72. Ibid., 1104.

73. Ibid., 1082.

74. Ibid., 1104.

75. See *Restatement (Second) of Torts* § 402A cmt. n (1965).

76. *Borel*, 493 F.2d at 1098.

77. See *Restatement (Third) of Torts: Products Liability* § 17 cmt. a (1998).

78. *Borel*, 493 F.2d at 1099.

79. See, e.g., *Corey v. Havener*, 65 N.E. 69, 69 (Mass. 1902); *Restatement (Second) of Torts* § 433A cmt. i and illus. 12–17 (1965).

80. *Borel*, 493 F.2d at 1094.

81. Ibid., 1091.

82. See Barry L. Castleman, *Asbestos: Medical and Legal Aspects*, 5th ed. (Englewood Cliffs, NJ: Aspen Law and Business, 2005), 437–43.

83. *MacPherson v. Buick Motor Co.*, 111 N.E. 1050.

84. E.g., *Elmore v. Am. Motors Co.*, 451 P.2d 84, 89 (Cal. 1969); *West v. Caterpillar Tractor Co.*, 336 So. 2d 80, 92 (Fla. 1976).

85. See *Borel*, 493 F.2d at 1086; *Falise v. Am. Tobacco Co.*, 94 F. Supp. 2d 316, 324–25 (E.D.N.Y. 2000).

86. 337 U.S. 163 (1949).

87. Ibid., 169.

88. Ibid., 170.

89. E.g., *Fernandi v. Strully*, 173 A.2d 277, 279 (N.J. 1961); *Flanagan v. Mt. Eden Gen. Hosp.*, 248 N.E.2d 871, 873–74 (N.Y. 1969).

90. *Borel*, 493 F.2d at 1102.

91. E.g., *Colo. Rev. Stat. Ann.* § 13-21-403(3) (1977) (amended 2003); *N.H. Rev. Stat. Ann.* § 507-D:2(II) (1979), *invalidated by Heath v. Sears, Roebuck & Co.*, 464 A.2d 288, 295 (N.H. 1983); *R.I. Gen. Laws* § 9-1-13(2)(b) (1981), *invalidated by Kennedy v. Cumberland Eng'g Co.*, 471 A.2d 195, 201 (R.I. 1984).

92. E.g., *Ind. Code Ann.* §§ 34-20-3-1 to 34-20-3-2 (1999); *Iowa Code Ann.* § 614.1(2A)(b) (West Supp. 2006); *Neb. Rev. Stat. Ann.* § 25-224 (2004).

93. E.g., *Gardner v. Asbestos Corp.*, 634 F. Supp. 609, 612 (W.D.N.C. 1986)

94. E.g., *Moran v. Johns-Manville Sales Corp.*, 691 F.2d 811, 816 (6th Cir. 1982); *Karjala v. Johns-Manville Prods. Corp.*, 523 F.2d 155 (8th Cir. 1975); *Hammond v. N. American Asbestos*, 435 N.E.2d 540, 547 (Ill. App. 1982).

95. *Henningsen*, 161 A.2d 69.

96. *Greenman*, 377 P.2d 897.

97. See *Restatement (Second) of Torts* § 402A cmt. l, illus. 1(1965).

Chapter 3

Portions of this chapter consist of revised excerpts from Donald G. Gifford, "The Challenge to the Individual Causation Requirement in Mass Products Torts," *Washington and Lee Law Review* 62 (2005): 873–935 (reprinted by permission).

1. See, e.g., *Borel v. Fibreboard Paper Products Corp.*, 493 F.2d 1076, 1094 (5th cir. 1973); *Rutherford v. Owens-Illinois, Inc.*, 941 P.2d 1203 (Cal. 1997).

2. David Rosenberg, "The Causal Connection in Mass Exposure Cases: A 'Public Law' Vision of the Tort System," *Harvard Law Review* 97 (1984): 906.

3. I do not consider here certain voluntary practices of plaintiffs' attorneys in aggregating similar cases for the purposes of preparation, discovery, and trial, including voluntary joinder, the creation of networks among plaintiffs' attorneys, or the combination of a test case with pattern settlements. See, e.g., Fed. R. Civ. P. 20 (joinder); Michael D. Green, "The Inability of Offensive Collateral Estoppel to Fulfill Its Promise: An Examination of Estoppel in Asbestos Litigation," *Iowa Law Review* 70 (1984): 141, 183–84 (joinder); American Law Institute, *Enterprise Responsibility for Personal Injury*, vol. 2, *Approaches to Legal and Institutional Change* (Philadelphia: American Law Institute, 1991), 40 (networks among plaintiffs' attorneys), 404–5 (joinder), 405–6 (test case and pattern settlements); Paul D.

Rheingold, "The MER/29 Story: An Instance of Successful Mass Disaster Litigation," *California Law Review* 56 (1968): 116, 122–23 (networks among plaintiffs' attorneys); Jack B. Weinstein, "Revision of Procedure: Some Problems in Class Actions," *Buffalo Law Review* 9 (1960): 433, 447–48 (test case and pattern settlements). By themselves, these procedural and tactical devices do not result in the imposition of collective liability in a way that substitutes for proof of the traditional causal link between a particular plaintiff and a particular defendant.

4. Guido Calabresi, "Concerning Cause and the Law of Torts: An Essay for Harry Kalven, Jr.," *University of Chicago Law Review* 43 (1975): 69, 85.

5. For welfare economics, Keith N. Hylton, "The Theory of Tort Doctrine and the *Restatement (Third) of Torts*," *Vanderbilt Law Review* 54 (2001): 1413, 1416–17. For the reformist zeal of the 1960s, see John C. P. Goldberg and Benjamin C. Zipursky, "Accidents of the Great Society," *Maryland Law Review* 64 (2005): 364, 370.

6. Guido Calabresi, *The Costs of Accidents: A Legal and Economic Analysis* (New Haven: Yale University Press, 1970), 68.

7. Ibid., 27–28. The aim of Calabresi's third subgoal, "tertiary cost reduction," is to reduce the costs of achieving the other two goals, primary and secondary cost avoidance.

8. Ibid., 40.

9. Ibid., 297.

10. Calabresi, "Concerning Cause and the Law of Torts," 85.

11. See Richard A. Posner, "Guido Calabresi's *The Costs of Accidents:* A Reassessment," *Maryland Law Review* 64 (2005): 12; see also Izhak Englard, *The Philosophy of Tort Law* (Brookfield, VT: Dartmouth, 1993), 31–42; Keith N. Hylton, "Calabresi and the Intellectual History of Law and Economics," *Maryland Law Review* 64 (2005): 85, 90.

12. William M. Landes and Richard A. Posner, "Causation in Tort Law: An Economic Approach," *Journal of Legal Studies* 12 (1983): 109, 131.

13. Landes and Posner, "Causation in Tort Law," 124–25; see also William M. Landes and Richard A. Posner, "Joint and Multiple Tortfeasors: An Economic Analysis," *Journal of Legal Studies* 9 (1980): 517, 540–41.

14. Judith Jarvis Thomson, "The Decline of Cause," *Georgetown Law Journal* 76 (1987): 137.

15. See generally Jules L. Coleman, *Risks and Wrongs* (New York: Cambridge University Press, 1992); Ernest J. Weinrib, *The Idea of Private Law* (Cambridge, MA: Harvard University Press, 1995).

16. Jules Coleman, among others, expounds a corrective justice explanation for tort law that is distinct from that of Weinrib. See Coleman, *Risks and Wrongs,* 197–431. Coleman claims that his description of the wrongful losses for which the injurer should be held liable, unlike Weinrib's, is derived from existing social practice and not from abstract principles. Ibid., 433.

17. Ernest J. Weinrib, "Corrective Justice," *Iowa Law Review* 77 (1992): 403, 409 (emphasis added); see also Weinrib, *Idea of Private Law*, 1, 142–44.

18. Weinrib, *Idea of Private Law*, 56.

19. Weinrib, "Corrective Justice," 409.

20. Weinrib, *Idea of Private Law*, 142.

21. Coleman, *Risks and Wrongs*, 405. See also Richard W. Wright, "Causation, Responsibility, Risk, Probability, Naked Statistics, and Proof: Pruning the Bramble Bush by Clarifying the Concepts," *Iowa Law Review* 73 (1988): 1001, 1073); Claire Finkelstein, "Is Risk A Harm?" *University of Pennsylvania Law Review* 151 (2003): 963, 967–90; Glen O. Robinson, "Multiple Causation in Tort Law: Reflections on the *DES* Cases," *Virginia Law Review* 68 (1982): 713, 739.

22. 493 F.2d 1076, 1094 (5th Cir. 1973).

23. 941 P.2d 1203 (Cal. 1997).

24. Ibid., 1206.

25. Ibid., 1218.

26. Ibid., 1219.

27. *Summers v. Tice*, 199 P.2d 1, 3 (Cal. 1948); see also *Restatement (Second) of Torts* §433B(3) (1965); *Restatement (Third) of Torts: Liability for Physical Harm* §28(b) (Proposed Final Draft No. 1, Apr. 6, 2005).

28. See, e.g., *Poole v. Alpha Therapeutic Corp.*, 696 F. Supp. 351, 356 (N.D. Ill. 1988); *Abel v. Eli Lilly & Co.*, 343 N.W.2d 164, 176–77 (Mich. 1984); *Ferrigno v. Eli Lilly & Co.*, 420 A.2d 1305, 1316 (N.J. Super. Ct. Law Div. 1980).

29. 861 F.2d 1453, 1474 (10th Cir. 1988).

30. Ibid., 1468.

31. Ibid., 1469.

32. *Abel*, 343 N.W.2d at 172.

33. Ibid., 172–73.

34. Rosenberg, "Causal Connection," 883; see also *In re "Agent Orange" Prod. Liab. Litig.*, 597 F. Supp. 740, 823 (E.D.N.Y. 1984).

35. See, e.g., *Smith v. Cutter Biological, Inc.*, 823 P.2d 717, 725 (Haw. 1991); *Goldman v. Johns-Manville Sales Corp.*, 514 N.E.2d 691, 697 (Ohio 1987).

36. *Abel*, 343 N.W.2d at 173; see also *Menne v. Celotex Corp.*, 861 F.2d at 1466.

37. See, e.g., *In re Methyl Tertiary Butyl Ether ("MTBE") Prods. Liab. Litig.*, 175 F. Supp. 2d 593, 622 n. 42 (S.D.N.Y. 2001); *Sindell v. Abbott Labs.*, 607 P.2d 924, 930–31 (Cal. 1980); *Hymowitz v. Eli Lilly & Co.*, 539 N.E.2d 1069, 1074 (N.Y. 1989).

38. See, e.g., Fowler V. Harper, Fleming James Jr., and Oscar S. Gray, *Harper, James, and Gray on Torts*, 3rd ed. (New York: Aspen, 2007), §20.2; Finkelstein, "Is Risk a Harm?" 980–81; Robinson, "Multiple Causation in Tort Law," 739–40.

39. See *Sindell*, 607 P.2d at 937.

40. Ibid.

41. Ibid., 936.

42. Ibid.

43. *Hymowitz*, 539 N.E.2d 1069.

44. See Coleman, *Risks and Wrongs*, 399–400 (noting that the court assigned "the defendant liability reflecting his share of the national market"); see also *Collins v. Eli Lilly Co.*, 342 N.W.2d 37, 50–51 (Wis. 1984) (allowing the plaintiff to proceed against one or more manufacturers of DES on the theory that each defendant contributed to the "risk of injury"); Robinson, "Multiple Causation in Tort Law," 739 (justifying market share liability on the basis that fairness requires only that the particular defendant held liable be one that created a risk of injury to the particular plaintiff, not the injury itself).

45. *Hymowitz*, 539 N.E.2d at 1078.

46. Ibid., 1078 n. 3.

47. E.g., *Goldman v. Johns-Manville Sales Corp.*, 514 N.E.2d 691, 702 (Ohio 1987); *Shackil v. Lederle Labs.*, 561 A.2d 511, 529 (N.J. 1989).

48. *Brenner v. Am. Cyanamid Co.*, 699 N.Y.S.2d 848, 853 (App. Div. 1999).

49. *Skipworth v. Lead Indus. Ass'n*, 690 A.2d 169, 173 (Pa. 1997); accord *Brenner*, 699 N.Y.S.2d at 852.

50. *Skipworth*, 690 A.2d at 173.

51. See ibid.; accord *Brenner*, 699 N.Y.S.2d at 852–53.

52. 701 N.W.2d 523, 567 (Wis. 2005). Although I was not involved in the litigation of *Thomas v. Mallett*, I testified in favor of a bill that would have undone some of the consequences of the opinion. I testified on behalf of the Wisconsin Coalition for Civil Justice before a joint hearing of the Wisconsin legislature's judiciary committees.

53. Ibid., 557–58.

54. Ibid., 562.

55. Ibid., 594 (Prosser, J., dissenting).

56. See ibid., 552–54 (majority opinion) (acknowledging the liability of landlords for failure to maintain lead-based paint, but noting the absence of an effective remedy in litigation against landlords because of insurance policy exclusions and state legislation granting immunity).

57. *Collins*, 342 N.W.2d at 53.

58. *City of St. Louis v. Benjamin Moore & Co.*, 226 S.W.3d 110, 116–17 (Mo. 2007).

59. Marie Rohde, "Paint Makers Win Verdict: Milwaukee Boy Ingested Lead, but Other Factors Harmed Him, Jury Finds," *Milwaukee Journal Sentinel*, Nov. 6, 2007, A1.

60. 345 F. Supp. 353, 386 (E.D.N.Y. 1972).

61. See, e.g., *Sindell*, 607 P.2d at 928.

62. 514 F. Supp. 1004, 1017 (D.S.C. 1981).

63. Ibid., 1017.

64. *Lillge v. Johns-Manville Corp.*, 602 F. Supp. 855, 856 (E.D. Wis. 1985); see also *210 E. 86th St. Corp. v. Combustion Eng'g, Inc.*, 821 F. Supp. 125, 148

(S.D.N.Y. 1993); *Vigiolto v. Johns-Manville Corp.*, 643 F. Supp. 1454, 1459 (W.D. Pa. 1986).

65. *Blackston v. Shook & Fletcher Insulation Co.*, 764 F.2d 1480, 1483 (11th Cir. 1985).

66. *City of Philadelphia v. Lead Indus. Ass'n*, 994 F.2d 112, 129 (3d Cir. 1993); *Santiago v. Sherwin-Williams Co.*, 794 F. Supp. 29, 33–34 (D. Mass. 1992); *Swartzbauer v. Lead Indus. Ass'n, Inc.*, 794 F. Supp. 142, 145–46 (E.D. Pa. 1992); *Thomas*, 701 N.W.2d at 567.

67. See, e.g., *Aetna Cas. Sur. Co. v. P&B Autobody*, 43 F.3d 1546, 1564 (1st Cir. 1994); *In re Methyl Tertiary Butyl Ether ("MTBE") Prods.*, 175 F. Supp. 2d at 622–23; *Boyle v. Anderson Fire Fighters Ass'n Local 1262*, 497 N.E.2d 1073, 1079 (Ind. Ct. App. 1986).

68. See, e.g., *Doe v. Baxter Healthcare Corp.*, 178 F. Supp. 2d 1003, 1020 (S.D. Iowa 2001).

69. See, e.g., *In re Related Asbestos Cases*, 543 F. Supp. 1152, 1159 (N.D. Cal. 1982); *Abel*, 343 N.W.2d at 176.

70. 525 A.2d 146, 148 (Del. 1987).

71. Ibid., 147.

72. *Bichler v. Eli Lilly & Co.*, 436 N.E.2d 182, 187 (N.Y. 1982).

73. See ibid., 188.

74. E.g., *In re High Fructose Corn Syrup Antitrust Litig.*, 295 F.3d 651, 654–55 (7th Cir. 2002); *Sindell*, 607 P.2d at 933; *Hymowitz*, 539 N.E.2d at 1074–75; *Martin v. Abbott Labs.*, 689 P.2d 368, 379 (Wash. 1984).

75. 794 F. Supp. 29, 31 (D. Mass. 1992).

76. *In re Sch. Asbestos Litig.*, 789 F.2d 996, 1001, 1010 (3d Cir. 1986).

77. See generally *Cimino v. Raymark Indus., Inc.*, 751 F. Supp. 649 (E.D. Tex. 1990), *rev'd in part*, 151 F.3d 297 (5th Cir. 1998).

78. See generally Laurens Walker and John Monahan, "Sampling Liability," *Virginia Law Review* 85 (1999): 329.

79. Stephen J. Carroll et al., *Asbestos Litigation Costs and Compensation: An Interim Report*, Doc. No. DB-397-ICS (Santa Monica, CA: RAND Institute for Civil Justice, 2002), 61, 69, http://www.rand.org/publications/DB/DB397/DB397.pdf.

80. See 751 F. Supp. at 652, 659–65.

81. See *Cimino*, 151 F.3d at 301.

82. See *Cimino*, 751 F. Supp. at 664–66.

83. 151 F.3d at 315–21.

84. See ibid., 319 (quoting *In re Fibreboard Corp.*, 893 F.2d 706, 711 (5th Cir. 1990)).

85. See, e.g., *In re Copley Pharm., Inc.*, 158 F.R.D. 485, 488–93 (D. Wyo. 1994); *In re W. Va. Rezulin Litig. v. Hutchinson*, 585 S.E.2d 52, 62–76 (W. Va. 2003).

86. E.g., *In re St. Jude Med., Inc.*, MDL No. 01-1316, 2004 U.S. Dist. LEXIS 149, *39 (D. Minn. Jan. 5, 2004) (certifying class for purposes of a medical monitoring claim but denying class certification for injury claims); *In re Simon II Litig.*,

211 F.R.D. 86, 108, 190 (E.D.N.Y. 2002) (certifying class for punitive damages only), rev'd, 407 F.3d 125, 137–38 (2d Cir. 2005).

87. See, e.g., *Valentino v. Carter-Wallace, Inc.*, 97 F.3d 1227, 1235 (9th Cir. 1996); *In re Rhone-Poulenc Rorer, Inc.*, 51 F.3d 1293, 1304 (7th Cir. 1995).

88. Fed. R. Civ. P. 23(a).

89. Fed. R. Civ. P. 23(b).

90. E.g., *Mahoney v. R. J. Reynolds Tobacco Co.*, 204 F.R.D. 150, 154–56 (D. Iowa 2001).

91. See, e.g., *Perez v. Metabolife Int'l Inc.*, 218 F.R.D. 262, 266 (S.D. Fla. 2003).

92. *In re Rhone-Poulenc Rorer*, 51 F.3d at 1300.

93. 160 F.R.D. 544 (E.D. La. 1995), rev'd, 84 F.3d 734, 752 (5th Cir. 1996).

94. *Castano*, 160 F.R.D. at 559.

95. *Castano*, 84 F.3d at 738.

96. Ibid., 752.

97. Ibid., 743.

98. Laura Parker and Deborah Sharp, "Sentiment on Tobacco Shifts; Jurors in Fla. Smokers' Case Show Disdain for the Industry," *USA Today*, July 17, 2000, 3A.

99. *Liggett Group Inc. v. Engle*, 853 So. 2d 434, 449 (Fla. Dist. Ct. App. 2003).

100. 521 U.S. 591, 597 (1997), rev'g *Georgine v. Amchem Prods., Inc.*, 83 F.3d 610 (3d Cir. 1996).

101. 527 U.S. 815, 821 (1999).

102. *In re Asbestos Prods. Liab. Litig. (No. VI)*, 771 F. Supp. 415, 416, 421–22 (J.P.M.L. 1991).

103. *Amchem*, 521 U.S. at 632–33 (Breyer, J., concurring in part and dissenting in part).

104. Ibid., 601–3.

105. Ibid., 597.

106. Ibid., 624 (quoting the lower court in *Georgine*, 83 F.3d at 626).

107. Ibid., 626.

108. 527 U.S. at 864–65.

109. Fed. R. Civ. P. 23(b)(1)(B).

110. *Ortiz*, 527 U.S. at 828.

111. Ibid., 843.

112. Ibid., 838.

113. Ibid., 839.

114. Ibid., 862, 864.

115. *In re Simon II*, 211 F.R.D. at 184–86.

116. *In re Simon II*, 407 F.3d at 137–38.

117. *Class Action Fairness Act of 2005*, Public Law No. 109-2, *U.S. Statutes at Large* 119 (2005): 4 (codified as amended in scattered sections of 28 U.S.C.).

118. Richard A. Nagareda, *Mass Torts in a World of Settlement* (Chicago: University of Chicago Press, 2007), 76–77; see also Donald G. Gifford, "The Death of

Causation: Mass Products Torts' Incomplete Incorporation of Social Welfare Principles," *Wake Forest Law Review* 41 (2006): 943, 963–77.

119. Nagareda, *Mass Torts*, 77.

120. *Ortiz*, 527 U.S. at 821.

121. *Amchem*, 521 U.S. at 628.

122. Ibid., 598; *Ortiz*, 527 U.S. at 821 n. 1.

123. See Fed. R. Civ. P. 42(a) (allowing the court in actions involving a common question of law or fact to either consolidate the actions or join any matters at issue for hearing or trial purposes); *AC&S, Inc. v. Godwin*, 667 A.2d 116, 119–23 (Md. 1995).

124. *In re E. & S. Dists. Asbestos Litig.*, 772 F. Supp. 1380, 1387–88 (E.D. & S.D.N.Y. 1991), *aff'd in part, rev'd in part on other grounds sub nom. In re Brooklyn Navy Yard Asbestos Litig.*, 971 F.2d 831 (2d Cir. 1992).

125. See *In re Brooklyn Navy Yard*, 971 F.2d at 836.

126. Ibid., 837.

127. But see *Leverence v. PFS Corp.*, 532 N.W.2d 735, 741 (Wis. 1995) (reversing the trial court's judgment in consolidated cases on the grounds that the aggregative process adopted by the trial court, where all plaintiffs were awarded judgments calculated on the basis of the average jury awards in a few test cases, was inconsistent with the defendant's due process right to a jury trial on the issues of causation, contributory negligence, and damages).

128. See, e.g., *Malcolm v. Nat'l Gypsum Co.*, 995 F.2d 346, 353–54 (2d Cir. 1993).

129. See *In re New York Asbestos Litig.*, 145 F.R.D. 644, 653–56 (S.D.N.Y. 1993), *later proceedings at* 149 F.R.D. 490 (S.D.N.Y. 1993).

130. See *In re New York Asbestos Litig.*, 145 F.R.D. at 656.

Chapter 4

Portions of the section "Back to the Future" in this chapter consist of revised excerpts from Donald G. Gifford, "Public Nuisance as a Mass Products Liability Tort," *University of Cincinnati Law Review* 71 (2003): 741–837 (reprinted by permission).

1. Public Law 91-190, *U.S. Statutes at Large* 83 (1970): 852 (codified at 42 U.S.C. §§ 4332–4347 (2006)).

2. Public Law 91-604, *U.S. Statutes at Large* 84 (1970): 1676 (codified as amended at 42 U.S.C. §§ 7401–7642 (2006)).

3. Public Law 92-500, *U.S. Statutes at Large* 86 (1972): 816 (codified as amended at 33 U.S.C. §§ 1251–1376 (2006)).

4. Public Law 93-523, *U.S. Statutes at Large* 88 (1974): 1660 (codified as amended at 42 U.S.C. §§ 300f–300j (2006)).

5. Public Law 94-469, *U.S. Statutes at Large* 90 (1976): 2003 (codified at 15 U.S.C. §§ 2601–92 (2006)).

6. Public Law 94-580, *U.S. Statutes at Large* 90 (1976): 2795 (codified as amended at 42 U.S.C. §§ 6901–87 (2006)).

7. 42 U.S.C. §§ 9601–75 (2006).

8. Reorganization Plan No. 3 of 1970, 3 C.F.R. § 199 (Comp. 1970), reprinted in 5 U.S.C. app. (2000).

9. Public Law 102-550, *U.S. Statutes at Large* 106 (1992): 3897 (codified at 42 U.S.C. §§ 4851–56 (2006)).

10. *Safe Drinking Water Act Amendments of 1986*, Public Law 99-339, sec. 109, § 1417, *U.S. Statutes at Large* 100 (1986): 642, 651–53 (codified at 42 U.S.C. § 300g–6 (2006)); *Federal Register* 56 (June 7, 1991): 26,460 (codified at 40 C.F.R. pts. 141, 142) (2008).

11. *Federal Cigarette Labeling and Advertising Act*, Public Law 89-92, sec. 2, *U.S. Statutes at Large* 79 (1965): 282 (codified as amended at 15 U.S.C. §§ 1331–40 (2006)); *Public Health Cigarette Smoking Act of 1969*, Public Law 91-222, §§ 6, 7(b), *U.S. Statutes at Large* 84 (1970): 87, 89 (codified as amended at 15 U.S.C. §§ 1331–40 (2006)).

12. 42 U.S.C. § 9607.

13. David Sive, "The Litigation Process in the Development of Environmental Law," *Pace Environmental Law Review* 13 (1995): 1, 19.

14. *Massachusetts v. EPA*, 127 S. Ct. 1438, 1455 (2007) (holding that Massachusetts had standing to challenge the EPA's denial to regulate "greenhouse gases").

15. E.g., *Swain v. Tennessee Copper Co.*, 78 S.W. 93, 95 (Tenn. 1903); see also generally Fowler V. Harper, Fleming James Jr., and Oscar S. Gray, *Harper, James, and Gray on Torts*, 3rd ed. (New York: Aspen, 2007), § 20.3.

16. E.g., *Tucker Oil Co. v. Matthews*, 119 S.W.2d 606, 607–8 (Tex. Civ. App. 1938).

17. E.g., *D & W Jones v. Collier, Inc.*, 372 So. 2d 288, 294 (Miss. 1979); *Velsicol Chem. Corp. v. Rowe*, 543 S.W.2d 337, 343 (Tenn. 1976).

18. 495 F.2d 213, 218–19 (6th Cir. 1974) (Mich. law).

19. E.g., *United States v. Burlington N. & Santa Fe Ry.*, 502 F.3d 781, 793 (9th Cir. 2007).

20. *Flo-Sun, Inc. v. Kirk*, 783 So. 2d 1029, 1036 (Fla. 2001) (omitting internal quotations and citation).

21. *City of Chicago v. Festival Theatre Corp.*, 438 N.E.2d 159, 164 (Ill. 1982).

22. Donald G. Gifford, "Public Nuisance as a Mass Products Liability Tort," *University of Cincinnati Law Review* 71 (2003): 741, 790–800.

23. *London Assize of Nuisance, 1301–1431: A Calendar*, ed. Helena M. Chew and William Kellaway (Chatham, Kent: London Record Society, 1973), xii and, e.g., cases 142, 449, 454, 456, 457 (1973).

24. E.g., ibid., cases 140, 453, 459.

25. E.g., *People v. Gold Run Ditch & Mining Co.*, 4 P. 1152, 1153 (Cal. 1884); *Chenowith v. Hicks*, 5 Ind. 224, 224 (1854); *Luning v. State*, 2 Pin. 215, 218 (Wis. 1849).

26. E.g., *Smiths v. McConathy*, 11 Mo. 517, 519 (1848); *Price v. Grantz*, 11 A. 794, 795 (Pa. 1888).

27. 4 P. at 1155–56 (citations omitted).

28. E.g., *Brown v. E. & Midlands Ry. Co.*, 22 Q.B.D. 391, 391 (C.A. 1889) (U.K.); *Wesson v. Washburn Iron Co.*, 95 Mass. (13 Allen) 95, 95 (1866); *Francis v. Schoellkopf*, 53 N.Y. 152, 153 (1873).

29. *Luning*, 2 Pin. at 219.

30. Ibid., 221; see also *Chenowith*, 5 Ind. at 224.

31. 200 U.S. 496, 497 (1906). See generally Robert V. Percival, "The Clean Water Act and the Demise of the Federal Common Law of Interstate Nuisance," *Alabama Law Review* 55 (2004): 717.

32. 206 U.S. 230, 236 (1907).

33. Ibid., 237.

34. *New York v. New Jersey*, 256 U.S. 296, 298 (1921); *New Jersey v. City of New York*, 283 U.S. 473, 476–77 (1931).

35. 451 U.S. 304, 317 (1981).

36. Ibid.

37. Ibid., 325. A few years later, the Supreme Court held that the Clean Water Act also preempted state nuisance actions seeking, under state common law, to abate discharges covered by the regulatory framework created by the act, at least in cases that seek to apply the common law of the "receiving state" where the pollution caused harm instead of the law of the state where the source of the pollution is located. *Int'l Paper Co. v. Ouellette*, 479 U.S. 481, 487–91 (1987).

38. No. C06-05755 MJJ, 2007 U.S. Dist. LEXIS 68547, °2 (N.D. Cal., Sept. 17, 2007).

39. Ibid., °29.

40. Ibid., °42.

41. See John W. Wade, "Environmental Protection, the Common Law of Nuisance and the Restatement of Torts," *Forum* 8 (1972): 165, 168.

42. *Restatement (Second) of Torts* 6, 16–44 (Tentative Draft No. 15, 1969).

43. Presentation of *Restatement of Law, Second, Torts*, Tentative Draft No. 16, *American Law Institute Proceedings* 47 (1970): 287, 291 (remarks of John P. Frank); see generally Denise E. Antolini, "Modernizing Public Nuisance: Solving the Paradox of the Special Injury Rule," *Ecology Law Quarterly* 28 (2001): 755, 819–51.

44. John E. Bryson and Angus Macbeth, "Public Nuisance, the Restatement (Second) of Torts, and Environmental Law," *Ecology Law Quarterly* 2 (1972): 241, 276.

45. *Restatement (Second) of Torts* § 821B (1979).

46. *United States v. Hooker Chems. & Plastics Corp.*, 722 F. Supp. 960, 961–70 (W.D.N.Y. 1989).

47. Ibid., 962 (quoting *United States v. Hooker Chems. & Plastics Corp.*, 680 F. Supp. 546, 549 (W.D.N.Y. 1988) (alterations in original)).

48. Barry L. Johnson, *Environmental Policy and Public Health* (Boca Raton, FL: CRC Press, 2007).

49. *Hooker,* 680 F. Supp. at 559.

50. 42 U.S.C. § 9607(a).

51. *Hooker,* 680 F. Supp. at 556.

52. *Hooker,* 722 F. Supp. at 967.

53. 459 N.Y.S.2d 971, 976–77 (Sup. Ct. 1983) (citation omitted) (alteration in original).

54. Ibid., 977.

55. 722 F. Supp. at 964 (citing *Webster's Third New International Dictionary*).

56. *Restatement (Second) of Torts* § 834 (1979) provides, "One is subject to liability for a nuisance caused by an activity, not only when he carries on the activity but also when he participates to a substantial extent in carrying it on."

57. E.g., *In re StarLink Corn Prods. Liab. Litig.,* 212 F. Supp. 2d 828, 833 (N.D. Ill. 2002); *City of Cincinnati v. Beretta U.S.A. Corp.,* 768 N.E.2d 1136, 1148–49 (Ohio 2002).

58. Susan P. Baker and William Haddon Jr., "Reducing Injuries and Their Results: The Scientific Approach," *Milbank Memorial Fund Quarterly: Health and Society* 52 (1974): 377, 377–80; John E. Gordon, "The Epidemiology of Accidents," *American Journal of Public Health* 39 (1949): 504, 504–5.

59. Ralph Nader, *Unsafe at Any Speed: The Designed-In Dangers of the American Automobile* (New York: Grossman, 1972), 338 (quoting Kennedy).

60. William Randolph Hearst Jr., "The Traffic Accident Problem and the U.S. President's Committee for Traffic Safety," *Journal of Criminal Law, Criminology, and Police Science* 51 (1960): 90.

61. David Klein and Julian A. Waller, *Causation, Culpability, and Deterrence in Highway Crashes, Department of Transportation Automobile Insurance and Compensation Study* (Washington, DC: U.S. Department of Transportation, 1970), 62–74, 209–18.

62. William Haddon Jr., "The Changing Approach to the Epidemiology, Prevention, and Amelioration of Trauma: The Transition to Approaches Etiologically Rather Than Descriptively Based," *American Journal of Public Health* 58 (1968): 1431, 1434–35.

63. Daniel P. Moynihan, "Public Health and Traffic Safety," *Journal of Criminal Law, Criminology, and Police Science* 51 (1960): 93, 97.

64. Nader, *Unsafe at Any Speed,* 5.

65. Public Law 89-563, *U.S. Statutes at Large* 80 (1966): 718 (codified at 15 U.S.C. § 1381 et seq.) (current version at 49 U.S.C. § 30101 et seq. (2006)).

66. See *Federal Register* 35 (Oct. 29, 1970): 16,927; "Motor Vehicle Safety Standard No. 208, Occupant Crash Protection," 49 C.F.R. § 571.208, S.4.1.1 (2008).

67. *Congressional Record* 120 (1974): 30,837.

68. *Congressional Record* 120 (1974): 30,847.

69. Jerry L. Mashaw and David L. Harfst, *The Struggle for Auto Safety* (Cambridge, MA: Harvard University Press, 1990), 139.

70. See Mashaw and Harfst, *Struggle for Auto Safety,* 84–146.

71. David G. Owen, *Products Liability Law*, 2nd ed. (St. Paul, MN: West, 2008), § 17.2, 1128–30.

72. E.g. *Larsen v. Gen. Motors Corp.*, 391 F.2d 495, 496 (8th Cir. 1968); *Cardullo v. Gen. Motors Corp.*, 378 F. Supp. 890, 891 (E.D. Pa. 1974); *Gen. Motors Corp. v. Bryant*, 582 S.W.2d 521, 523 (Tex. Civ. App. 1979).

73. E.g., *Clay v. Ford Motor Co.*, 215 F.3d 663, 671 (6th Cir. 2000) (Ford Bronco II); *McCathern v. Toyota Motor Corp.*, 23 P.3d 320, 333 (Or. 2001) (Toyota 4Runner).

74. E.g., *Kallio v. Ford Motor Co.*, 407 N.W.2d 92, 94 (Minn. 1987); *Preissman v. Ford Motor Co.*, 82 Cal. Rptr. 108, 109–10 (Ct. App. 1969).

75. Owen, *Products Liability Law*, § 17.3, 1132.

76. 359 F.2d 822 (7th Cir. 1966) (Ind. law).

77. Ibid., 824.

78. 391 F.2d 495 (8th Cir. 1968) (Mich. law).

79. Ibid., 502.

80. Ibid., 503.

81. 529 U.S. 861, 864 (2000).

82. See "Federal Motor Vehicle Safety Standards: Occupant Crash Protection," *Federal Register* 58 (Sept. 2, 1993): 46,551 (codified at 49 C.F.R. § 571.208 (2008)).

83. See Timothy D. Lytton, "Using Litigation to Make Public Health Policy: Theoretical and Empirical Challenges in Assessing Product Liability, Tobacco, and Gun Litigation," *Journal of Law, Medicine, and Ethics* 32 (2004): 556, 556.

84. Jon S. Vernick et al., "Role of Litigation in Preventing Product-Related Injuries," *Epidemiologic Reviews* 25 (2003): 90, 96.

85. United States Department of Commerce, *Interagency Task Force on Product Liability: Final Report* (Washington, D.C.: Department of Commerce, 1978), VI 49 to VI 50.

86. Robert A. Kagan, *Adversarial Legalism: The American Way of Life* (Cambridge, MA: Harvard University Press, 2001), 144; see also 141–44.

87. Mashaw and Harfst, *Struggle for Auto Safety*, 238–41.

88. *Grimshaw v. Ford Motor Co.*, 174 Cal. Rptr. 348, 391 (Ct. App. 1981).

89. Stephen J. Carroll et al., *Asbestos Litigation Costs and Compensation: An Interim Report*, Doc. No. DB 397-ICS (Santa Monica, CA: RAND Institute for Civil Justice, 2002), 40, 53, 71, 78, 80, http://www.rand.org/pubs/documented_briefings/2005/DB397.pdf. Many claimants, although exposed to asbestos, currently have little or no functional impairment. Ibid., 45.

90. Stephen J. Carroll et al., *Asbestos Litigation*, Doc. No. MG-162-ICJ (Santa Monica, CA: RAND Institute for Civil Justice, 2005), 23–24, http://www.rand.org/pubs/monographs/2005/RAND_MG162.pdf.

91. Herbert M. Kritzer, "From Litigators of Ordinary Cases to Litigators of Extraordinary Cases: Stratification of the Plaintiffs' Bar in the Twenty-First Century," *DePaul Law Review* 51 (2002): 219, 231.

92. Saundra Torry, "Lead Paint Could Be Next Big Legal Target," *Washington Post*, June 10, 1999, A1.

93. Stephen P. Teret, "Litigating for the Public's Health," *American Journal of Public Health* 76 (1986): 1027, 1027.

94. J. S. Todd et al., "The Brown and Williamson Documents: Where Do We Go from Here?" *Journal of the American Medical Association* 274 (1995) 256, 258.

95. American Public Health Association, "Resolution No. 9704: Responsibilities of the Lead Pigment Industry and Others to Support Efforts to Address the National Child Lead Poisoning Problem," *American Journal of Public Health* 88 (1998): 498, 499.

Chapter 5

1. See Peter J. Boyer, "Big Guns: The Lawyers Who Brought Down the Tobacco Industry Are Taking on the Gunmakers and the NRA," *New Yorker*, May 17, 1999, 54, 61. The efforts of John Coale and others to regulate the firearms industry through litigation were cut short by Congress's enactment of the *Protection of Lawful Commerce in Arms Act*, Public Law 109-92, *U.S. Statutes at Large* 119 (2005): 2095 (codified at 15 U.S.C. §§ 7901–7903 (2006)).

2. U.S. Department of Health, Education, and Welfare, *Smoking and Health: Report of the Advisory Committee to the Surgeon General of the Public Health Service* (Washington, DC: Government Printing Office, 1964).

3. Lars Noah and Barbara A. Noah, "Nicotine Withdrawal: Assessing the FDA's Effort to Regulate Tobacco Products," *Alabama Law Review* 48 (1996): 1, 19.

4. *Federal Register* 29 (July 2, 1964): 8325.

5. Public Law 89-92, *U.S. Statutes at Large* 79 (1965): 282 (codified as amended at 15 U.S.C. §§ 1331–1341 (2006).

6. Allan M. Brandt, *The Cigarette Century: The Rise, Fall, and Deadly Persistence of the Product that Defined America* (New York: Basic Books, 2007), 254.

7. 15 U.S.C. § 1334 (2006).

8. *Station WCBS-TV*, 8 F.C.C.2d 381, 382 (1967), *aff'd sub nom. Banzhaf v. FCC*, 405 F.2d 1082 (D.C. Cir. 1968).

9. Brandt, *Cigarette Century*, 268.

10. Public Law 91-222, §§ 6, 7(b), *U.S. Statutes at Large* 84 (1970): 87, 89 (codified as amended at 15 U.S.C. §§ 1331–1341 (2006)).

11. *Comprehensive Smoking Education Act*, Public Law 98-474, *U.S. Statutes at Large* 98 (1984): 2201 (codified as amended at 15 U.S.C. §§1331, 1333, 1335a–1341 (2006)).

12. See, e.g., *Fair Packaging and Labeling Act of 1966*, Public Law 89-755, § 10(a)(1), *U.S. Statutes at Large* 80 (1966): 1296, 1301 (codified as amended at 15 U.S.C. §§ 1451–1461); *Comprehensive Drug Abuse Prevention and Control Act of 1970*, Public Law 91-513, § 102(6), *U.S. Statutes at Large* 84 (1970): 1236, 1243 (codified as amended at 26 U.S.C. § 801 et seq.); *Consumer Product Safety Commission Improvements Act of 1976*, Public Law 94-284, sec. 3(c), *U.S. Statutes at Large* 90 (1976): 503. 503 (amending *Federal Hazardous Substances Act* § 2(f)(2); codified as amended at 15 U.S.C. § 1261); *Toxic Substances Control Act*, Public

Law 94-469, § 3(2)(B)(iii), *U.S. Statutes at Large* 90 (1976): 2003, 2004 (codified as amended at 15 U.S.C. § 2601 et seq.).

13. Public Law 101-164, sec. 335, *U.S. Statutes at Large* 103 (1989): 1069, 1098–99 (originally codified at 49 U.S.C. app. § 1374(d); currently codified as amended at 49 U.S.C. § 41706) (interstate flights)).

14. See, e.g., Takeshi Hirayama, "Non-Smoking Wives of Heavy Smokers Have a Higher Risk of Lung Cancer: A Study from Japan," *British Medical Journal* 282 (1981): 183; J. L. Repace and A. H. Lowrey, "Indoor Air Pollution, Tobacco Smoke, and Public Health," *Science* 208 (1980): 464; John D. Spengler and Ken Sexton, "Indoor Air Pollution: A Public Health Perspective," *Science* 221 (1983): 9; Dimitrios Trichopoulos et al., "Lung Cancer and Passive Smoking," *International Journal of Cancer* 27 (1981): 1. See generally National Research Council, Committee on Passive Smoking, *Environmental Tobacco Smoke: Measuring Exposures and Assessing Health Effects* (Washington, DC: National Academy Press, 1986); U.S. Department of Health and Human Services, *The Health Consequences of Involuntary Smoking: A Report of the Surgeon General* (Washington, DC: Government Printing Office, 1986).

15. Dana M. Shelton, Marianne Haenlein Alciati, Michele M. Chang, Julie M. Fishman, Liza A. Fues, Jennifer Michaels, Ronald J. Bazile, et al., "State Laws on Tobacco Control—United States, 1995," *Morbidity and Mortality Weekly Report Surveillance Summaries* 44, no. SS 6 (Nov. 3, 1995), 6, available at http://www.cdc.gov/mmwr/PDF/ss/ss4406.pdf.

16. *Departments of Commerce, Justice, and State, the Judiciary, and Related Agencies Appropriation Act,* Public Law 100-202, § 328, *U.S. Statutes at Large* 101 (1988): 1329-1, 1329-382 (originally codified at 49 U.S.C. app. § 1374; currently codified as amended at 49 U.S.C. § 41706) (prohibiting smoking on domestic flights of two hours or less); *Department of Transportation and Related Agencies Appropriations Act,* Public Law 101-164, *U.S. Statutes at Large* 103 (1990): 1069 (originally codified at 49 U.S.C. app. § 1374; currently codified as amended at 49 U.S.C. § 41706) (prohibiting smoking on domestic flights of six hours or less); *Wendell H. Ford Aviation Investment and Reform Act for the 21st Century,* Public Law 106-181, § 708, *U.S. Statutes at Large* 114 (2000) 61, 159 (codified as amended at 49 U.S.C. § 41706) (extending prohibition to foreign air carriers).

17. *Federal Register* 53 (Aug. 30, 1988): 33,122, 33,124.

18. Public Law 102-321, sec. 202, § 1926(b), *U.S. Statutes at Large* 106 (1992): 323, 394–95 (codified at 42 U.S.C. § 300x-26 (2006)); 45 C.F.R. § 96.130(c), (d) (2008).

19. Joseph R. DiFranza, "State and Federal Compliance with the Synar Amendment: Federal Fiscal Year 1998," *Archives of Pediatric and Adolescent Medicine* 155 (2001): 572, 572.

20. Robert A. Kagan and William P. Nelson, "The Politics of Tobacco Regulation in the United States," in *Regulating Tobacco,* ed. Robert L. Rabin and Stephen D. Sugarman (New York: Oxford University Press, 2001), 27, table 2-2.

21. Carrick Mollenkamp et al., *The People vs. Big Tobacco: How the States Took on the Cigarette Giants* (Princeton, NJ: Bloomberg Press, 1998), 39–48.

22. Ibid., 40 (quoting Addison Yeaman).

23. Ibid., 111–13; David Kessler, *A Question of Intent: A Great American Battle with a Deadly Industry* (New York: Public Affairs, 2001), 183–97.

24. *Regulation of Tobacco Products (Part I): Hearings before the Subcommittee on Health and the Environment of the House Committee on Energy and Commerce*, 103d Cong., 2d sess., 1994, 628.

25. Martha A. Derthick, *Up in Smoke: From Legislation to Litigation in Tobacco Politics*, 2nd ed. (Washington, DC: CQ Press, 2004), 50–67; Kessler, *Question of Intent*, 63, 87, 270–72, 291–304, 325–36.

26. 655 F.2d 236, 236 (D.C. Cir. 1980).

27. Ibid., 240 (quoting 21 U.S.C. § 321(g)(1)(C) (1976)).

28. *Regulation of Tobacco Products*, 25–27; David A. Kessler, letter to Scott D. Ballin, Chairman, Coalition on Smoking and Health, Feb. 25, 1994, available at http://tobaccodocuments.org/batco/500821335-1337.html.

29. Kessler, *Question of Intent*, 328–33.

30. *Federal Register* 61 (Aug. 28, 1996): 44,396.

31. Kessler, *Question of Intent*, 336–37.

32. *Coyne Beahm, Inc. v. U.S. Food & Drug Admin.*, 966 F. Supp. 1374, 1400 (M.D.N.C. 1997).

33. *Brown & Williamson Tobacco Corp. v. FDA*, 153 F.3d 155, 184 (4th Cir. 1998).

34. *FDA v. Brown & Williamson Tobacco Corp.*, 529 U.S. 120, 161 (2000).

35. Ibid., 143.

36. Ibid., 139.

37. Stephen Moore et al., *Contributing to Death: The Influence of Tobacco Money on the U.S. Congress* (Washington, DC: Public Citizen's Health Research Group, 1993), 20.

38. Taft Wireback, "Cigarettes Unregulated—For Now," *Greensboro News & Record*, Sept. 28, 1992, A5 (quoting Representative Durbin).

39. Center for Responsive Politics, "Tobacco: Long-Term Contribution Trends," http://www.opensecrets.org/industries/indus.asp?Ind=A02.

40. Stephen Moore et al., "Epidemiology of Failed Tobacco Control Legislation," *Journal of the American Medical Association* 272 (1994): 1171, 1171. See also Stanton A. Glantz and Michael E. Begay, "Tobacco Industry Campaign Contributions Are Affecting Tobacco Control Policymaking in California," *Journal of the American Medical Association* 272 (1994): 1176, 1176.

41. John Wright, "Campaign Contributions and Congressional Voting on Tobacco Policy, 1980–2000," *Business and Politics* 6, no. 3 (2004): 1, 3.

42. Morris P. Fiorina and Paul E. Peterson, *The New American Democracy*, alternate 6th ed. (New York: Longman, 2009), 191; John R. Wright, *Interest Groups*

and Congress: Lobbying, Contributions and Influence (Boston: Allyn and Bacon, 1995), 137.

43. Douglas D. Roscoe and Shannon Jenkins, "A Meta-Analysis of Campaign Contributions' Impact on Roll Call Voting," *Social Science Quarterly* 86 (2005): 52, 63.

44. Thomas W. Merrill, "Agency Capture Theory and the Courts: 1967–1983," *Chicago-Kent Law Review* 72 (1997): 1039, 1053.

45. See 16 C.F.R. §§ 1303.1(a)–(b), 1303.4(a) (2008) (banning "lead-containing paint" from consumer use as a hazardous product) (originally promulgated at *Federal Register* 42 (Sept. 1, 1977): 44,199). See also *Lead-Based Paint Poisoning Prevention Act*, Public Law 91-695, § 401, *U.S. Statutes at Large* 84 (1971): 2078, 2079 (codified as amended at 42 U.S.C. § 4831 (2006)) (banning lead paint in federally funded programs).

46. David E. Jacobs et al., "The Prevalence of Lead-Based Paint Hazards in U.S. Housing," *Environmental Health Perspectives* 110 (2002): A599, A599.

47. See, e.g., *Md. Code Ann., Envir.* §§ 6-811 to 6-824 (West 2002 & Supp. 2007); *N.Y. Pub. Health Law* §§ 1370-a to 1370-e (McKinney 2002 & Supp. 2008); *R.I. Gen. Laws* §§ 23-24.6-1 to 23-24.6-27 (1996 & Supp. 2006).

48. 42 U.S.C. §§ 4851–4856 (2006); 24 C.F.R. §§ 35.100–35.175 (2008); *Federal Register* 64 (Sept. 15, 1999): 50,140, 50,219.

40. 15 U.S.C. §§ 2681–2692 (2006).

50. 24 C.F.R. § 35.110 (2008).

51. 24 C.F.R. § 35.1330 (2008); *Md. Code Ann., Envir.* §§ 6-815.

52. President's Task Force on Environmental Health Risks and Safety Risks to Children, *Eliminating Childhood Lead Poisoning: A Federal Strategy Targeting Lead Paint Hazards* (2000), 5, table 2, http://www.cdc.gov/nceh/lead/about/fedstrategy2000.pdf.

53. Jonathan Wilson et al., "Evaluation of HUD-funded Lead Hazard Control Treatments at 6 Years Post-Intervention," *Environmental Research* 102 (2006): 237, 244.

54. 1994 *Md. Laws* ch. 114, 1282 (codified at *Md. Code Ann., Envir.* §§ 6-801 to 6-852). I chaired the Maryland Lead Paint Poisoning Commission that made the recommendations leading to the enactment of this statutory framework.

55. 1993 *Mass. Acts* 1422 (codified at *Mass. Gen. Laws Ann.* ch. 111, §§ 189A to 199B (West 2003)).

56. *Mass. Gen. Laws Ann.* ch. 111, § 199.

57. 1994 *Md. Laws* ch. 114, § 1, 1332 (codified at *Md. Code Ann., Envir.* § 6-836).

58. *R.I. Gen. Laws* §§ 23-24.6-1 to 23-24.6-27.

59. E.g., *Ill. Comp. Stat. Ann.* 45/8 (West 2005 and Supp. 2008); *Mo. Rev. Stat. Ann.* § 701.304 (West 2006).

60. James D. Sargent et al., "The Association between State Housing Policy and

Lead Poisoning in Children," *American Journal of Public Health* 89 (1999): 1690, 1690.

61. Maryland Department of the Environment, Lead Poisoning Prevention Program, *Childhood Blood Lead Surveillance in Maryland, 2004 Annual Report* (Baltimore: Maryland Dept. of Environment, 2005), 13, table 5.

62. *Cal. Health & Safety Code* § 105310 (West 2006); *Cal. Rev. & Tax. Code* §§ 43057, 43152.14 & 43554 (West 2004); *Me. Rev. Stat. Ann.* tit. 22, § 1322-E (Supp. 2007).

63. *Conn. Gen. Stat. Ann.* § 8-219e (West Supp. 2008) (providing for grants and loans); *Mass. Gen. Laws Ann.* ch. 111, § 197E (loans); *R.I. Gen. Laws* § 42-55-27 (2006).

64. National Institute on Money in State Politics, http://www.followthemoney.org/Institute/index.phtml.

65. Common Cause/NY, *Lead Poisoning Legislation and the Political Power of Real Estate in New York City* (New York: Common Cause/NY, 2003), http://www.nmic.org/nyccelp/Documents/articles/common-cause-report.pdf.

66. Ibid., 6.

Chapter 6

1. See Frank J. Vandall, "The Legal Theory and the Visionaries that Led to the Proposed $368.5 Billion Tobacco Settlement," *Southwestern University Law Review* 27 (1998): 473, 478–82. See also Raymond E. Gangarosa, Frank J. Vandall, and Brian M. Willis, "Suits by Public Hospitals to Recover Expenditures for the Treatment of Disease, Injury, and Disability Caused by Tobacco and Alcohol," *Fordham Urban Law Journal* 22 (1994): 81.

2. Carrick Mollenkamp et al., *The People vs. Big Tobacco: How the States Took On the Cigarette Giants* (Princeton, NJ: Bloomberg Press, 1998), 29 (quoting Don Barrett).

3. A different and more colorful account of how the idea of a direct *parens patriae* against product manufacturers first came to the attention of Mississippi attorney general Michael Moore is offered by Mollenkamp. According to Mollenkamp, in May 1993, Mike Lewis, a small-town Mississippi lawyer, visited a friend, a smoker who now was dying of cancer, in the hospital. Ibid., 23–30. It occurred to him that the state of Mississippi spent millions of dollars treating people suffering and dying from tobacco-related illnesses. Lewis also realized that Mississippi's action would not be blocked by the conduct of smokers who had become victims of tobacco-related diseases.

4. Complaint, *Moore ex rel. State v. Am. Tobacco Co.*, No. 94-1429 (Miss. Ch. Ct. May 23, 1994), available at http://www.library.ucsf.edu/sites/all/files/ucsf_assets/ms_complaint.pdf.

5. Copies of complaints filed by the states are available at the Galen Digital Library, University of California, San Francisco, http://www.library.ucsf.edu/tobacco/litigation/states.html.

6. E.g., Complaint, *County of Los Angeles v. R. J. Reynolds Tobacco Co.*, No. 707651 (Super. Ct. Aug. 5, 1996), available at http://www.library.ucsf.edu/tobacco/litigation/other/lacomplaint.html.

7. E.g., *Blue Cross & Blue Shield of New Jersey, Inc. v. Philip Morris, Inc.*, 113 F. Supp. 2d 345 (E.D.N.Y. 2000).

8. E.g., *Steamfitters Local Union No. 420 Welfare Fund. v. Philip Morris, Inc.*, 171 F.3d 912 (3d Cir. 1999); *Iron Workers Local Union No. 17 Ins. Fund v. Philip Morris, Inc.*, 23 F. Supp. 2d 771 (N.D. Ohio 1998).

9. See http://www.library.ucsf.edu/tobacco/litigation/states.html for complaints filed by Arizona, Connecticut, Iowa, Kansas, Massachusetts, Michigan, New Jersey, Oklahoma, Texas, Utah, Washington, and West Virginia.

10. Ibid., complaints filed by Kansas, Michigan, Massachusetts, Oklahoma, Utah, and West Virginia.

11. Ibid., complaints filed by Kansas, Michigan, Minnesota, Texas, Washington, and West Virginia.

12. Ibid., complaints filed by New Jersey, Texas, and Utah.

13. *Medicaid Third-Party Liability Act*, 1994 *Fla. Laws* ch. 251, § 4 (codified at *Fla. Stat.* § 409.910 (9)(a) (repealed 1998)). See also *Vt. Stat. Ann.* tit. 33, § 1911 (2001); see generally Christa Sarafa, "Making Tobacco Companies Pay: The Florida Medicaid Third-Party Liability Act," *DePaul Journal of Health Care Law* 2 (1997): 123, 124; Elizabeth A. Frohlich, "Statutes Aiding States' Recovery of Medicaid Costs from Tobacco Companies: A Better Strategy for Redressing an Identifiable Harm?" *American Journal of Law and Medicine* 21 (1995): 445.

14. *Agency for Health Care Admin. v. Associated Indus. of Fla.*, 678 So. 2d 1239, 1253–55 (Fla. 1996).

15. Marc Lacey, "Tobacco Industry Accused of Fraud in Lawsuit by U.S.," *New York Times*, Sept. 23, 1999, A1 (quoting Attorney General Janet Reno).

16. *United States v. Philip Morris USA*, 396 F.3d 1190, 1199 (D.C. Cir. 2005), *cert. denied*, 546 U.S. 960 (2005).

17. See Richard P. Ieyoub and Theodore Eisenberg, "State Attorney General Actions, the Tobacco Litigation, and the Doctrine of *Parens Patriae*," *Tulane Law Review* 74 (2000): 1859, 1865–66.

18. See Donald G. Gifford, "The Challenge to the Individual Causation Requirement in Mass Products Torts," *Washington and Lee Law Review* 62 (2005): 873, 915–19.

19. 42 U.S.C. § 1396a (a)(25)(B) (2006).

20. E.g., *Tex. Hum. Res. Code Ann.* § 32.033 (Vernon 2001); *Md. Code Ann., Health-Gen.* §§ 15-121.1 to 15-121.2 (West 2002).

21. See, e.g., *E. H. Ashley & Co. v. Wells Fargo Alarm Servs.*, 907 F.2d 1274, 1277 (1st Cir. 1990).

22. See *Alfred L. Snapp & Son, Inc. v. Puerto Rico*, 458 U.S. 592, 609 (1982).

23. See Ieyoub and Eisenberg, "State Attorney General Actions," 1864.

24. *Snapp*, 458 U.S. 592, 602.

25. Ibid., 607.

26. See *Massachusetts v. EPA*, 549 U.S. 497, 537–38 (2007) (Roberts, C.J., dissenting).

27. Ibid.

28. E.g., *Ganim v. Smith & Wesson Corp.*, 780 A.2d 98, 108 (Conn. 2001); *State v. Philip Morris, Inc.*, 577 N.W.2d 401, 406 (Iowa 1998). Often, a functionally identical result is reached, but under a doctrinal pigeonhole other than standing. E.g., *District of Columbia v. Beretta, U.S.A., Corp.*, 872 A.2d 633, 650 (D.C. 2005) (dismissing claim because of tenuous causal chain); *City of San Francisco v. Philip Morris, Inc.*, 957 F. Supp. 1130, 1141 (N.D. Cal. 1997) (finding claims to be too remote); *People v. Sturm, Ruger & Co.*, 761 N.Y.S.2d 192, 201 (App. Div. 2003) (finding a lack of both duty and proximate causation); *City of Chicago v. Am. Cyanamid Co.*, 823 N.E.2d 126, 133 (Ill. App. Ct. 2005) (rejecting claim because of a lack of proximate causation).

29. *Ganim*, 780 A.2d 98.

30. Ibid., 123–24.

31. See, e.g., *Kansas v. Colorado*, 206 U.S. 46, 117 (1907); *Missouri v. Illinois*, 180 U.S. 208 (1901).

32. 206 U.S. 230 (1907).

33. E.g., *Hawaii v. Standard Oil Co. of Cal.*, 405 U.S. 251, 258–59 (1972); *Pennsylvania v. West Virginia*, 262 U.S. 553, 591 (1923).

34. *Snapp*, 458 U.S. at 607–8.

35. *Massachusetts v. EPA*, 549 U.S. 497.

36. Ibid., 519 (quoting *Snapp*, 458 U.S. at 607).

37. Ibid.

38. See Donald G. Gifford, "Public Nuisance as a Mass Products Liability Tort," *University of Cincinnati Law Review* 71 (2003): 741, 755–56.

39. See *City of Milwaukee v. NL Indus., Inc.*, 691 N.W.2d 888, 893 (Wis. App. Ct. 2004); but see *City of St. Louis v. Benjamin Moore & Co.*, 226 S.W.3d 110, 116–17 (Mo. 2007).

40. *City of Chicago v. Am. Cyanamid*, 823 N.E.2d at 131.

41. See Gifford, "Public Nuisance," 814–19.

42. *State v. Lead Indus. Ass'n*, C.A. No. 99-5226, 2001 R.I. Super. LEXIS 37, °50–51 (R.I. Super. Ct. Apr. 2, 2001); see also *NL Indus., Inc.*, 691 N.W.2d at 896–97.

43. *State v. Lead Indus. Ass'n*, C.A. No. 99-5226, 2007 R.I. Super. LEXIS 32, °6 (R.I. Super. Ct. Feb. 26, 2007).

44. See *Merch. Mut. Ins. Co. v. Newport Hosp.*, 272 A.2d 329, 332 (R.I. 1971).

45. See *Lead Indus. Ass'n, Inc.*, 2001 R.I. Super. LEXIS 37, °49 (quoting *R&B Elec. Co. v. Amco Constr. Co.*, 471 A.2d 1351, 1355–56 (R.I. 1984)).

46. See ibid., °50–51.

47. See, e.g., *Perry v. Am. Tobacco Co.*, 324 F.3d 845, 851 (6th Cir. 2003); *Or. Laborers-Employers Health & Welfare Trust Fund v. Philip Morris, Inc.*, 185 F.3d

957, 968–69 (9th Cir. 1999); *Steamfitters Local Union No. 420 Welfare Fund v. Philip Morris, Inc.*, 171 F.3d 912, 937 (3d Cir. 1999).

48. *Perry*, 324 F.3d at 851.

49. See *Restatement (Third) of Restitution and Unjust Enrichment* § 1 cmt. a (discussion draft Mar. 31, 2000).

50. See *State v. Lead Indus. Ass'n, Inc.*, C.A. No. 99-3226, 2001 R.I. Super. LEXIS 37, °53 (R.I. Super. Ct. Apr. 2, 2001), *rev'd*, 951 A.2d 428 (R.I. 2008)

51. Ibid., °51 (quoting *McCrory v. Spigel*, 740 A.2d 1274, 1276 (R.I. 1999)).

52. Ibid., °51 (quoting *Muldowney v. Weatherking Prods., Inc.*, 509 A.2d 441, 443 (R.I. 1986)).

53. Ibid., °53.

54. E.g., *Allegheny Gen. Hosp. v. Philip Morris, Inc.*, 116 F. Supp. 2d 610, 622 (W.D. Pa. 1999); *Serv. Employees Int'l Union Health & Welfare Fund v. Philip Morris, Inc.*, 83 F. Supp. 2d 70, 93 (D.D.C. 1999).

55. 116 F. Supp. 2d 610.

56. See ibid., 621–22 (quoting *Builders Supply Co. v. McCabe*, 77 A.2d 368 (Pa. 1951)).

57. See ibid., 622 (same).

58. E.g., ibid.; *Serv. Employees Int'l Union Health & Welfare Fund*, 83 F. Supp. 2d at 93.

59. Peter Pringle, *Cornered: Big Tobacco at the Bar of Justice* (New York: Henry Holt, 1998), 230.

60. "Carter Awarded $750,000 in Tobacco Case Against Brown & Williamson Corp.," *Mealey's Emerging Toxic Torts* 5, no. 10 (Aug. 23, 1996): 26.

61. Pringle, *Cornered*, 226–44; Alix M. Freedman, Suein L. Hwang, Steven Lipin and Milo Geyelin, "Breaking Away: Liggett Group Offers First-Ever Settlement of Cigarette Lawsuits," *Wall Street Journal*, Mar. 13, 1996, A1.

62. Mollenkamp et al., *People vs. Big Tobacco*, 120.

63. Ibid., 68–70.

64. Perhaps the best account of these negotiations is provided in ibid., 67–234.

65. Timothy D. Lytton, "Tort Claims against Gun Manufacturers for Crime-Related Injuries: Defining a Suitable Role for the Tort System in Regulating the Firearms Industry," *Missouri Law Review* 65 (2000): 1, 3.

66. See, e.g., Julie Samia Mair, Stephen Teret, and Shannon Frattaroli, "A Public Health Perspective on Gun Violence Prevention," in *Suing the Gun Industry: A Battle at the Crossroads of Gun Control and Mass Torts*, ed. Timothy D. Lytton (Ann Arbor: University of Michigan Press, 2005), 39.

67. Public Law 109-92, *U.S. Statutes at Large* 119 (2005): 2095 (codified at 15 U.S.C. §§ 7901–7903) (2006).

68. David Kairys, "The Origin and Development of the Governmental Handgun Cases," *Connecticut Law Review* 32 (2000): 1163, 1172. Professor Kairys served as counsel for the appellant in *Camden County Bd. of Chosen Freeholders v. Beretta, U.S.A. Corp.*, 273 F.3d 536 (3d Cir. 2001), and as a member of the City of Philadel-

phia's Handgun Violence Reduction Taskforce that promoted the litigation by the City of Philadelphia. See *City of Philadelphia v. Beretta U.S.A. Corp.*, 277 F.3d 415 (3d Cir. 2002).

69. *Morial v. Smith & Wesson Corp.*, 785 So. 2d 1, 6 (La. 2001) (quoting the city's petition).

70. Complaint, *Morial v. Smith & Wesson Corp.*, No. 98-18578 (Civ. Dist. Ct. Oct. 30, 1998), available at http://www.gunlawsuits.com/pdf/docket/neworleans .pdf .

71. Howard M. Erichson, "Private Lawyers, Public Lawsuits: Plaintiffs' Attorneys in Municipal Gun Litigation," in Lytton, *Suing the Gun Industry*, 129–51.

72. Complaint, *City of Chicago v. Beretta U.S.A. Corp.*, No. 98-CH15596 (Ill. Cir. Ct. Nov. 12, 1998), available at http://www.gunlawsuits.com/downloads/ chicago.pdf.

73. Erichson, "Private Lawyers, Public Lawsuits," 134, 138.

74. Tom Diaz, "The American Gun Industry: Designing and Marketing Increasingly Lethal Weapons," in Lytton, *Suing the Gun Industry*, 100–102.

75. Erichson, "Private Lawyers, Public Lawsuits," 137.

76. Timothy D. Lytton, "The NRA, the Brady Campaign, and the Politics of Gun Litigation," in Lytton, *Suing the Gun Industry*, 152.

77. Elisa Barnes, quoted in Walter K. Olsen, *The Rule of Lawyers: How the New Litigation Elite Threatens America's Rule of Law* (New York: St. Martin's, 2003), 101.

78. Douglas McCollam, "Long Shot," *American Lawyer*, June 1999, 86.

79. 273 F.3d 536 (N.J. law).

80. Second Amended Complaint and Jury Demand ¶ 15, *Camden County Bd. of Chosen Freeholders v. Beretta, U.S.A. Corp.*, No. 99CV2518 (D.N.J. Jan. 6, 2000), quoted in David Kairys, "The Governmental Handgun Cases and the Elements and Underlying Policies of Public Nuisance Law," *Connecticut Law Review* 32 (2000): 1175, 1179.

81. E.g., *City of Cincinnati v. Beretta U.S.A. Corp.*, 768 N.E.2d 1136, 1145 (Ohio 2002).

82. See Richard C. Ausness, "Tort Liability for the Sale of Non-defective Products: An Analysis and Critique of the Concept of Negligent Marketing," *South Carolina Law Review* 53 (2002): 907, 908–9. See, e.g., *Ileto v. Glock, Inc.*, 349 F.3d 1191, 1197, 1204 (9th Cir. 2003). Despite a number of judicial opinions made several years ago upholding negligent marketing and distribution claims—e.g., *Hamilton v. Accu-Tek*, 62 F. Supp. 2d 802, 829 (E.D.N.Y. 1999); *Merrill v. Navegar, Inc.*, 89 Cal. Rptr. 2d 146, 189 (Ct. App. 1999); *City of Cincinnati v. Beretta U.S.A. Corp.*, 768 N.E.2d 1136, 1145—more recent decisions tend to reject such a theory of recovery. E.g., *Merrill v. Navegar, Inc.*, 28 P.3d 116, 133 (Cal. 2001), *rev'g* 89 Cal. Rptr. 2d 146; *Hamilton v. Beretta U.S.A. Corp.*, 750 N.E.2d 1055, 1063 (N.Y. 2001).

83. *Camden County Bd. of Chosen Freeholders*, 273 F.3d at, 541 (N.J. law); *Ganim*, 780 A.2d at 118–28.

84. E.g., *Camden County Bd. of Chosen Freeholders*, 273 F.3d at 541.

85. *Ganim*, 780 A.2d at 118–28.

86. Ibid., 120.

87. Ibid., 123.

88. E.g., *City of Philadelphia v. Beretta U.S.A. Corp.*, 277 F.3d 415 (Pennsylvania law); *Camden County Bd. of Chosen Freeholders*, 273 F.3d 536 (N.J. law); *Ganim*, 780 A.2d 98.

89. *City of Cincinnati v. Beretta U.S.A. Corp.*, 768 N.E.2d 1136; *City of Boston v. Smith & Wesson Corp.*, No. 1999-02590, 2000 Mass. Super. LEXIS 352, °63–64 (Mass. Super. Ct. July 13, 2000).

90. Steven Harras, "Cincinnati Drops Suit against Gun Industry," *BNA Product Liability Daily*, May 22, 2003; Erichson, "Private Lawyers, Public Lawsuits," 140; Raja Mishra, "Boston Drops Lawsuit on Guns: Growing Cost Cited in Case vs. 31 Firms," *Boston Globe*, Mar. 28, 2002, A1.

91. E.g., *Fla. Stat. Ann.* § 790.331(2) (West 2007); *Idaho Code Ann.* § 5-247 (2004); *Tex. Civ. Prac. & Rem. Code Ann.* § 128.001(b) (Vernon 2005); Lytton, "The NRA, the Brady Campaign, and the Politics of Gun Litigation," 152.

92. Public Law 109-92, *U.S. Statutes at Large* 119 (2005): 2095 (codified at 15 U.S.C. §§ 7901–7903 (2006)).

93. See *District of Columbia v. Heller*, 128 S. Ct. 2783 (2008) (holding a municipal handgun ban and trigger-lock requirement to be unconstitutional under the Second Amendment and declaring that the Second Amendment "protects an individual right to possess a firearm unconnected with service in a militia").

Chapter 7

1. "Interest Intensifies over Suing Lead Pigment Industry," *Mealey's Litigation Report: Lead* 8, no. 18 (June 25, 1999), 2 (quoting Ronald L. Motley's interview with *CBS Evening News*, June 13, 1999).

2. Mark Curriden, "Tobacco Fees Give Plaintiffs' Lawyers New Muscle for Other Litigation," *Dallas Morning News*, Oct. 31, 1999, H1.

3. Complaint, *State v. Lead Indus. Ass'n*, C.A. No. 99-5226 (R.I. Super. Ct. Oct. 12, 1999), http://www.riag.ri.gov/documents/reports/lead/lead_complaint.pdf, and reprinted in *Mealey's Litigation Report: Lead* 9, no. 1 (Oct. 15, 1999), app. D; see also Peter B. Lord, "R.I. Is First State to Sue Paint Makers over Lead Poisoning," *Providence Journal*, Oct. 14, 1999, 1A.

4. *Wood v. Picillo*, 443 A.2d 1244, 1247 (R.I. 1982).

5. David Herzog, "In War on Lead Poisoning, R.I. Is Inadequately Armed," *Providence Journal*, Feb. 21, 1999, 1A.

6. Peter B. Lord, "Another Generation Caught in a Sad Cycle—Public Health Crisis—Poisoned," *Providence Journal*, May 13, 2001, 1E.

7. Herzog, "In War on Lead Poisoning" (quoting Christine Brackett).

8. Peter B. Lord, "Defense: State Did Not Prove Lead Case," *Providence Journal*, Feb. 9, 2006, B1.

9. Lord, "Another Generation" (quoting Liana Cassar, director of the HELP Lead Safe Center).

10. *State v. Lead Indus. Ass'n*, C.A. No. 99-5226, 2001 R.I. Super. LEXIS 37, °28 (R.I. Super. Ct. Apr. 2, 2001).

11. Ibid., °10 (quoting *Estados Unidos Mexicanos v. DeCoster*, 229 F.3d 332, 335 (1st Cir. 2000), quoting *Alfred L. Snapp & Son, Inc. v. Puerto Rico*, 458 U.S. 592, 602 (1982)).

12. Ibid., °44; see also, e.g., *State v. Philip Morris, Inc.*, 577 N.W.2d 401, 406–7 (Iowa 1998) (holding that state could not sue because damages were too remote and derivative).

13. *Lead Indus. Ass'n*, 2001 R.I. Super. LEXIS 37, °10.

14. E.g., *O'Connor v. Boeing N. Am., Inc.*, 197 F.R.D. 404, 417 (C.D. Cal. 2000); *Wade v. Campbell*, 19 Cal. Rptr. 173, 177 (Ct. App. 1962) (quoting *Turlock v. Bristow*, 284 P. 962, 965 (Cal. Ct. App. 1930)); *State ex rel. Smith v. Kermit Lumber & Pressure Treating Co.*, 488 S.E.2d 901, 925 (W. Va. 1997).

15. *Lead Indus. Ass'n*, 2001 R.I. Super. LEXIS 37, °29–34.

16. *State v. Lead Indus. Ass'n*, 951 A.2d 428, 440–41 (R.I. 2008).

17. *State v. Lead Indus. Ass'n*, C.A. No. 99-5226, 2007 R.I. Super. LEXIS 32, °6 (R.I. Super. Ct. Feb. 26, 2007).

18. *State v. Lead Indus. Ass'n*, 951 A.2d at 440–41.

19. Karl Llewellyn, *The Case Law System in America*, ed. Paul Gewirtz, trans. Michael Ansaldi (Chicago: University of Chicago Press, 1989), 1.

20. Ibid., 80.

21. Ibid., 77.

22. Compare John Rawls, *A Theory of Justice* (Cambridge, MA: Harvard University Press, Belknap Press, 1971), 235; Antonin Scalia, "The Rule of Law as a Law of Rules," *University of Chicago Law Review* 56 (1989): 1175, 1187; with Mark Tushnet, "The Dilemmas of Liberal Constitutionalism," *Ohio State Law Journal* 42 (1981): 411, 424 (the proper judicial role "is to make an explicitly political judgment: which result is, in the circumstances now existing, likely to advance the cause of socialism?").

23. See Henry M. Hart Jr. and Albert M. Sacks, *The Legal Process: Basic Problems in the Making and Application of Law*, ed. William N. Eskridge Jr. and Philip P. Frickey (Westbury, NY: Foundation Press, 1994), particularly Eskridge and Frickey, "An Historical and Critical Introduction to *The Legal Process*," li, lx–cxxxix.

24. Ibid., 568–69.

25. Supplemental Brief of Appellee on Procedural and Factual Background at 48, *State v. Lead Indus. Ass'n*, No. 07-121A (R.I. Mar. 17, 2008), available as attachment to electronic version of "Rhode Island Says Nuisance Is a Valid Theory," *Mealey's Litigation Report: Lead* 17, no. 6 (Apr. 2008), 4.

26. *Whitehouse v. Lead Indus. Ass'n*, C.A. No. 99-5226, 2002 R.I. Super. LEXIS 43, °1 (R.I. Super. Ct. Mar. 15, 2002).

27. Peter B. Lord, "Deliberations Begin in Lead-Paint Trial," *Providence Journal*, Oct. 25, 2002, B3.

28. Peter B. Lord, "Jury in Lead-Paint Lawsuit Asks Judge for Clarifications," *Providence Journal*, Oct. 26, 2002, A6.

29. "Mistrial Declared in Rhode Island Lead Paint Case; Motions for Judgment to be Filed, Retrial Likely," *Mealey's Litigation Report: Lead* (Oct. 30, 2002).

30. 2007 R.I. Super. LEXIS 32, °4.

31. Peter B. Lord, "No Witnesses Put Forth by 4 Paint Companies," *Providence Journal*, Jan. 27, 2006, B1.

32. Peter B. Lord, "Closing Arguments in Lead-Paint Case Begin Tomorrow," *Providence Journal*, Feb. 7, 2006, B2.

33. Peter B. Lord, "No Punitive Damages for Lead-Paint Companies," *Providence Journal*, Mar. 1, 2006, A1.

34. Peter B. Lord, "3 Companies Found Liable in Lead-Paint Nuisance Suit," *Providence Journal*, Feb. 23, 2006, A1.

35. Sacha Pfeiffer, "Group Asks States to Sue for Cleanup of Lead Paint," *Boston Globe*, June 29, 2006, D1.

36. *Lead Indus. Ass'n*, 2007 R.I. Super. LEXIS 32, °101; Lord, "No Punitive Damages for Lead-Paint Companies."

37. *Lead Indus. Ass'n*, 951 A.2d at 443.

38. 2001 R.I. Super. LEXIS 37, °19 (quoting Amended Complaint at 44 46, *State v. Lead Indus. Ass'n*, C.A. No. 99-5226 (R.I. Super. Ct. Oct. 14, 1999)).

39. Defendants' Reply Memorandum in Support of Their Motion to Dismiss at 46, No. 99-5226 (R.I. Super. Ct. Aug. 15, 2000).

40. 2001 R.I. Super. LEXIS 37, °27.

41. 951 A.2d at 447 (quoting 58 Am. Jur. 2d "Nuisances" § 39 (2002)).

42. Ibid., 448 (including emphasis not in the Restatement) (quoting *Restatement (Second) of Torts* § 821B cmt. g (1979)).

43. Ibid. (hyphens omitted) (quoting *City of Chicago v. Am. Cyanamid Co.*, 823 N.E.2d 126, 131 (Ill. App. Ct. 2005)).

44. Ibid. (quoting Donald G. Gifford, "Public Nuisance as a Mass Products Liability Tort," *University of Cincinnati Law Review* 71 (2003): 741, 817).

45. *Ganim v. Smith & Wesson Corp.*, 780 A.2d 98, 131–32 (Conn. 2001) (citations omitted) (alterations in original).

46. See *Young v. Bryco Arms*, 765 N.E.2d 1, 10 (Ill. App. Ct. 2001), *rev'd on other grounds*, 821 N.E.2d 1078, 1091 (Ill. 2004); accord *Ganim*, 780 A.2d at 113–15. Even in *Young*, the Illinois Supreme Court reversed the lower court's opinion on other grounds and, in doing so, expressed skepticism concerning "the existence of a public right to use the public space without undue risk of injury." *Young*, 821 N.E.2d at 1084. Other courts generally have held that the concept of unreasonable interference with public rights does not include the sale and distribution of handguns that pose a threat to public safety. E.g., *City of Philadelphia v. Beretta*

U.S.A. Corp., 277 F.3d 415, 421–22 (3d Cir. 2002); *Camden County Bd. of Chosen Freeholders v. Beretta, U.S.A. Corp.*, 273 F.3d 536, 541–42 (3d Cir. 2001).

47. *N.Y. Penal Law* § 400.05 (McKinney 2008).

48. *Am. Cyanamid Co.*, 823 N.E.2d 126 at 132–33 (quoting Gifford, "Public Nuisance," 818).

49. 2007 R.I. Super LEXIS 32, °56.

50. Ibid., °57 (quoting Dr. Philip Landrigan).

51. Brief of Appellant NL Industries on Public Nuisance at 23, *State v. Lead Indus. Ass'n*, No. SU-07-121-A (R.I. Jan. 31, 2008), available as attachment to electronic version of "Sherwin-Williams, NL: Rhode Island Lawsuit Barred; Errors Require New Trial," *Mealey's Litigation Report: Lead* 17, no. 4 (Feb. 2008): 1.

52. 2007 R.I. Super LEXIS 32, °147–48 (quoting Jury Instructions 14).

53. 951 A.2d at 449.

54. David Ibbetson, *A Historical Introduction to the Law of Obligations* (New York: Oxford University Press, 1999), 100.

55. William L. Prosser, "Nuisance without Fault," *Texas Law Review* 20 (1942): 399, 418.

56. 637 F. Supp. 646, 656 (D.R.I. 1986) (citations omitted).

57. E.g., *James v. Arms Tech., Inc.*, 820 A.2d 27, 52–53 (N.J. Super. Ct. App. Div. 2003); *State v. Schenectady Chems., Inc.*, 459 N.Y.S.2d 971, 976 (Sup. Ct. 1983); *City of Cincinnati v. Beretta U.S.A. Corp.*, 768 N.E.2d 1136, 1143 (Ohio 2002). Cf. *City of Boston v. Smith & Wesson Corp.*, No. 1999-02590, 2000 Mass. Super. LEXIS 352, °62 (Mass. Super. Ct. July 13, 2000) (circumventing the control requirement by interpreting the instrumentality to be "the creation and supply of [the] secondary market" of firearms for juveniles, criminals, and other unauthorized gun users).

58. *United States v. Hooker Chems. & Plastics Corp.*, 722 F. Supp. 960, 967 (W.D.N.Y. 1989) (emphasis added).

59. *Restatement (Second) of Torts* § 834 cmt. g (1979). See also ibid., ch. 40, introductory note (stating that the restatement's rules are applicable to both private and public nuisance unless otherwise stated).

60. *Comprehensive Environmental Response, Compensation, and Liability Act, U.S. Code* 42 (2006) §§ 9601–75 (2006).

61. Brief of Appellant NL Industries on Public Nuisance at 18.

62. Ibid., 17.

63. Ibid., 22.

64. 951 A.2d at 452.

65. 821 N.E. 2d 1099, 1117 (Ill. 2004).

66. 924 A.2d 484, 495 (N.J. 2007).

67. E.g., *City of Cincinnati v. Beretta U.S.A. Corp.*, 768 N.E.2d at 1142.

68. *Lead Indus. Ass'n*, 2007 R.I. Super LEXIS 32, °54.

69. 951 A.2d at 451 (quoting W. Page Keeton, Dan B. Dobbs, Robert E. Keeton,

and David G. Owen, *Prosser and Keeton on the Law of Torts*, 5th ed. (St. Paul, MN: West, 1984), § 41, 264).

70. Ibid. (quoting *City of Chicago v. Beretta U.S.A. Corp.*, 821 N.E.2d at 1127).

71. Contra *Am. Cyanamid Co.*, 823 N.E.2d 126, 139.

72. *R.I. Gen. Laws* § 23-24.6-3(1) (Supp. 2006).

73. *R.I. Gen. Laws* § 42-128.1-1 to 42-128.1-13 (2006). See also Scott MacKay, "At the Assembly—House Gives Lead-Paint Bill the Finishing Touch," *Providence Journal*, June 5, 2002, B1. I testified in favor of this legislation and also advised the bill sponsor, Senator Tom Izzo, and his staff. In this role, I was compensated by DuPont.

74. See generally Rhode Island Department of Health, *Rules and Regulations for Lead Poisoning Prevention*, R23-24.6-PB (Aug. 2007), available at http://www2 .sec.state.ri.us/dar/regdocs/released/pdf/DOH/4806.pdf; Rhode Island Department of Administration, Housing Resources Commission, *Rules and Regulations Governing Lead Hazard Mitigation* (Jan. 2006), available at http://www.hrc .ri.gov/documents/LHMR%2003.23.07%20Final.pdf.

75. Peter B. Lord, "Fewer Children Poisoned by Lead," *Providence Journal*, Aug. 3, 2006, B1.

76. 951 A.2d at 457 (emphasis added).

77. Ibid., 456.

78. *In re Lead Paint Litig.*, 924 A.2d at 501.

79. Ibid., 503.

80. "Lead Industry Files Third-Party Complaint, Says State, Property Owners Responsible," *Mealey's Litigation Report: Lead* 10, no. 19 (July 3, 2001): 4; Peter B. Lord, "State's Lead-Paint Suit to Proceed, Court Rules," *Providence Journal*, Feb. 6, 2002, A1.

81. Third Party Complaint at 3, *State v. Lead Indus. Ass'n*, C.A. No. 99-5226 (June 25, 2001), reprinted in *Mealey's Litigation Report: Lead* 10, no. 19 (July 3, 2001), app. D.

82. Bruce Landis, "Paint Companies Accused of Trying to Stall Lead Suit," *Providence Journal*, July 21, 2001, 1A.

83. *State v. Lead Indus. Ass'n*, C.A. No. 99-5226, 2005 R.I. Super. LEXIS 79, °1 (R.I. Super. Ct. May 18, 2005).

84. *Whitehouse v. Lead Indus. Ass'n*, C.A. No. 99-5226, 2002 R.I. Super. LEXIS 90, °1 (R.I. Super. Ct. July 3, 2002).

85. "Rhode Island Judge Allows Web Site Inquiry, Denies Property-Specific Evidence," *Mealey's Litigation Report: Lead* 14, no. 2 (Dec. 2004): 5.

86. Peter B. Lord, "Jurors in Lead-Paint Trial Say They're Proud of Verdict," *Providence Journal*, Mar. 12, 2006, B1 (quoting Lenau).

87. Ibid. (quoting Destefanis).

88. Brief of Appellee on Public Nuisance Claim at 4, *State v. Lead Indus. Ass'n*, No. SU-07-121-A (R.I. Mar. 17, 2008), available as attachment to electronic version

of "Rhode Island Says Nuisance Is a Valid Theory," *Mealey's Litigation Report: Lead*, 4.

89. Supplemental Brief of Appellee on Procedural and Factual Background at 24.

90. *Restatement (Second) of Torts* ch. 40, introductory note and § 822 (1979).

91. Henry de Bracton, *Bracton on the Laws and Customs of England*, ed. Samuel E. Thorne, vol. 3 (Cambridge, MA: Harvard University Press, Belknap Press, 1977), 100, 191, f. 232b.

92. William L. Prosser, "Private Action for Public Nuisance," *Virginia Law Review* 52 (1966): 997, 997.

93. *Restatement (Second) of Torts* § 821B cmt. e (1979); see also, e.g., *Keeney v. Town of Old Saybrook*, 676 A.2d 795, 810 (Conn. 1996); *Physicians Plus Ins. Corp. v. Midwest Mut. Ins. Co.*, 646 N.W.2d 777, 792 (Wis. 2002).

94. *Restatement (Second) of Torts* § 825 (1979).

95. 768 N.E.2d 1136.

96. Ibid., 1143 (quoting plaintiff's complaint).

97. 951 A.2d at 446.

98. *Wood*, 443 A.2d at 1247.

99. 951 A.2d at 447.

100. The same issue, but in the context of private nuisance, was debated at the time of the adoption of the *Restatement (Second) of Torts.* See Harry Shulman, Fleming James Jr., Oscar S. Gray, and Donald G. Gifford, *Cases and Materials on the Law of Torts*, 4th ed. (New York: Foundation Press, 2003), 91–94.

101. 951 A.2d at 447 (citations omitted).

102. Supplemental Brief of Appellee on Procedural and Factual Background at 22 (quoting trial testimony of plaintiff's expert Dr. Michael Shannon).

103. *Whitehouse*, 2002 R.I. Super. LEXIS 90, °7 n. 1 (quoting Rhode Island assistant attorney general Linn Freedman).

104. Brief of Coalition for Public Nuisance Fairness and Property Casualty Insurers Association of America as Amici Curiae in Support of Appellants at 2, *State v. Lead Indus. Ass'n*, No. 07-121A (R.I. Sup. Ct. Jan. 31, 2008), available as attachment to electronic version of "Sherwin-Williams, NL: Rhode Island Lawsuit Barred; Errors Require New Trial," *Mealey's Litigation Report: Lead* 17, no. 4 (Feb. 2008): 1.

105. Brief of Appellee on Public Nuisance Claim at 71.

106. *Lead Indus. Ass'n*, 2005 R.I. Super. LEXIS 95, °3.

107. *Lead Indus. Ass'n*, 2007 R.I. Super. LEXIS 32, °1.

108. *In re Lead Paint Litig.*, 924 A.2d 484, 505.

109. 951 A.2d at 456.

110. *Camden County Bd. of Chosen Freeholders*, 273 F.3d at 540.

111. *Tioga Pub. Sch. Dist. No. 15 v. U.S. Gypsum Co.*, 984 F.2d 915, 921 (8th Cir. 1993).

112. See *Restatement (Third) of Torts: Products Liability* § 2 (1998).

113. W. L. F. Felstiner and Peter Siegelman, "Neoclassical Difficulties: Tort Deterrence for Latent Injuries," *Law and Policy* 11 (1989): 309, 309.

114. Steven S. Cherensky, "Shareholders, Managers, and Corporate R&D Spending: An Agency Cost Model," *Santa Clara Computer and High Technology Law Journal* 10 (1994): 299, 328 (footnotes omitted). See also James A. Henderson Jr., "Product Liability and the Passage of Time: The Imprisonment of Corporate Rationality," *New York University Law Review* 58 (1983): 765, 775; Thomas Lee Hazen, "The Short-Term/Long-Term Dichotomy and Investment Theory: Implications for Securities Market Regulation and for Corporate Law," *North Carolina Law Review* 70 (1991): 137, 140.

115. See Ulf Lundberg, Johan M. von Wright, Marianne Frankenhaeuser, and Ulf-Johan Olson, "Involvement in Four Future Events as a Function of Temporal Distance," *Scandinavian Journal of Psychology* 16 (1975): 2.

116. Clayton P. Gillette and James E. Krier, "Risks, Courts, and Agencies," *University of Pennsylvania Law Review* 138 (1990): 1027, 1041.

117. E.g., *O'Connor v. Boeing N. Am.*, 197 F.R.D. 404, 416 (C.D. Cal. 2000); *Wade v. Campbell*, 19 Cal. Rptr. 173, 177 (Ct. App. 1962); *State ex rel. Smith v. Kermit Lumber & Pressure Treating Co.*, 488 S.E.2d 901, 925 (W. Va. 1997); cf. *Jamail v. Stoneledge Condo. Owners Ass'n*, 970 S.W.2d 673, 676 (Tex. App. 1998) ("Limitations is not a defense to an action to abate a continuing nuisance." (citation omitted)).

118. Guido Calabresi, *The Costs of Accidents: A Legal and Economic Analysis* (New Haven, CT: Yale University Press, 1970), 135–73.

119. E.g., *Fischer v. Johns-Manville Corp.*, 512 A.2d 466, 473 (N.J. 1986).

120. David Kairys, "The Origin and Development of the Governmental Handgun Cases," *Connecticut Law Review* 32 (2000): 1163, 1170.

121. David Kairys, "The Governmental Handgun Cases and the Elements and Underlying Policies of Public Nuisance Law," *Connecticut Law Review* 32 (2000): 1175, 1179 n. 16 (quoting Second Amended Complaint and Jury Demand ¶15(c), *Camden County Bd. of Chosen Freeholders v. Beretta U.S.A. Corp.*, No. 99CV2518 (D.N.J. filed Jan. 6, 2000)).

122. *Connecticut v. Am. Elec. Power Co.*, 406 F. Supp. 2d 265, 268, 274 (S.D.N.Y. 2005), *argued* No. 05-5104-cv (2d Cir. June 7, 2006); *California v. Gen. Motors Corp.*, No. C06-05755 MJJ, 2007 U.S. Dist. LEXIS 68547, °48 (N.D. Cal. Sept. 17, 2007).

123. Calabresi, *Costs of Accidents*, 39–42.

124. Stephen J. Carroll et al., *Asbestos Litigation*, Doc. No. MG-162-ICJ (Santa Monica, CA: RAND Institute for Civil Justice, 2005), chap. 5, http://www.rand.org/pubs/monographs/2005/RAND_MG162.pdf.

125. Lester Brickman, "Effective Hourly Rates of Contingency-Fee Lawyers: Competing Data and Non-Competitive Fees," *Washington University Law Quarterly* 81 (2003): 653, 700, 721.

126. See, e.g., *Brown & Williamson Tobacco Corp. v. Chesley*, 749 N.Y.S.2d 842,

849 (Sup. Ct. 2002); "Minn. High Court Asked to Rule on Tobacco Fee," *Mealey's Litigation Report: Tobacco* 13, no. 19 (Feb. 3, 2000): 5.

127. See Environmental Working Group, "Tobacco Subsidies in United States," *Farm Subsidy Database,* http://www.ewg.org/farm/progdetail.php?fips=00000& progcode =tobacco; see also Lucien J. Dhooge, "Smoke across the Waters: Tobacco Production and Exportation as International Human Rights Violations," *Fordham International Law Journal* 22 (1998): 355, 359.

128. Benjamin N. Cardozo, *The Nature of the Judicial Process* (New Haven, CT: Yale University Press, 1921), 141.

129. Melvin Aron Eisenberg, *The Nature of the Common Law* (Cambridge, MA: Harvard University Press, 1988), 150.

Chapter 8

1. Dan Zegart, *Civil Warriors: The Legal Siege on the Tobacco Industry* (New York: Bantam, 2000), 141 (quoting Motley).

2. Graham E. Kelder Jr. and Richard A. Daynard, "The Role of Litigation in the Effective Control of the Sale and Use of Tobacco," *Stanford Law and Policy Review* 8 (1997): 63, 63–64.

3. Frank J. Vandall, "The Legal Theory and the Visionaries that Led to the Proposed $368.5 Billion Tobacco Settlement," *Southwestern University Law Review* 27 (1998): 478–80.

4. Complaint, *State v. Lead Indus. Ass'n,* C.A. No. 99-5226 (R.I. Super. Ct. Oct. 12, 1999), reprinted in *Mealey's Litigation Report: Lead* 9, no. 1 (Oct. 15, 1999): app. D; see also Peter B. Lord, "Rhode Island Becomes First State to Sue Lead Pigment Industry," *Providence Journal,* Oct. 14, 1999, 1A.

5. Complaint, *State v. Lead Indus. Ass'n,* Relief Requested, ¶¶ 3, 4.

6. State of Rhode Island and Providence Plantations, Department of Attorney General, *Rhode Island Lead Nuisance Abatement Plan* (Sept. 14, 2007), available at http://www.riag.state.ri.us/documents/RILeadNuisanceAbatementPlan9-14-07.pdf and as attachment to LexisNexis electronic version of "Rhode Island Defendants to Respond to $2.4B Abatement Plan by Nov. 15," *Mealey's Litigation Report: Lead* 16, no. 12 (Oct. 2007): 7, 9.

7. *Rhode Island Lead Nuisance Abatement Plan,* 6.

8. The settlement agreement is reprinted in its entirety in Carrick Mollenkamp et al., *The People vs. Big Tobacco: How the States Took on the Cigarette Giants* (Princeton, NJ: Bloomberg Press, 1998), app., 265–317.

9. Ibid., 277.

10. David A. Kessler, *A Question of Intent: A Great American Battle with a Deadly Industry* (New York: PublicAffairs, 2001), 360.

11. Zegart, *Civil Warriors,* 263.

12. Martha A. Derthick, *Up in Smoke: From Legislation to Litigation in Tobacco Politics,* 2nd ed. (Washington, DC: CQ Press, 2004), 119–22.

13. *Universal Tobacco Settlement Act,* S. 1415, 105th Cong., 1st sess. (Nov. 7, 1997), 175–77, 15–40, 7; C. Stephen Redhead, *Tobacco Master Settlement Agreement (1998): Overview, Implementation by States, and Congressional Issues* (Washington, DC: Congressional Research Service, 1999), 13.

14. David E. Rosenbaum, "Cigarette Makers Quit Negotiations on Tobacco Bill," *New York Times,* Apr. 9, 1998, A1.

15. Barry Meier, "Court Rules F.D.A. Lacks Authority to Limit Tobacco," *New York Times,* Aug. 15, 1998, A1 (quoting Gregoire).

16. Derthick, *Up in Smoke,* 166–69.

17. Zegart, *Civil Warriors,* 329–31.

18. 153 F.3d 155, 184 (4th Cir. 1998), *aff'd, FDA v. Brown & Williamson Tobacco Corp.,* 529 U.S. 120 (2000).

19. *Master Settlement Agreement* (1998), 32–47, available at http://ag.ca.gov/tobacco/pdf/1msa.pdf.

20. The account of the settlement negotiations is taken from Mollenkamp et al., *People vs. Big Tobacco,* 67–72.

21. Ibid., 171–72.

22. See generally Jeremy Bulow and Paul Klemperer, "The Tobacco Deal," in *Brookings Papers on Economic Activity,* ed. Marin Neil Baily, Peter C. Reiss, and Clifford Winston (Washington, DC: Brookings Institution, 1998), 323.

23. V. J. Rock, A. Malarcher, J. W. Kahende, K. Asman, C. Husten, and R. Caraballo, Office on Smoking and Health, National Center for Chronic Disease Prevention and Health Promotion, Centers for Disease Control and Prevention, "Cigarette Smoking among Adults—United States, 2006," *Morbidity and Mortality Weekly Report* 56, no. 44 (Nov. 9, 2007), 1157, 1160, http://www.cdc.gov/mmwr/PDF/wk/mm5644.pdf.

24. Roni Caryn Rabin, "Teen Smoking Rates Decline," *New York Times,* Dec. 15, 2008, http://www.nytimes.com/2008/12/16/health/research/16smoking.html#.

25. Rock et al., "Cigarette Smoking among Adults," 1160.

26. Rick Hampson, "States Squander Chance to Fight Smoking," *USA Today,* Mar. 11, 2003, 1B (quoting Gregoire).

27. Allan M. Brandt, *The Cigarette Century: The Rise, Fall, and Deadly Persistence of the Product That Defined America* (New York: Basic Books, 2007), 435 (quoting Daynard).

28. Brandt, *Cigarette Century,* 435; Derthick, *Up In Smoke,* 188.

29. Steven A. Schroeder, "Tobacco Control in the Wake of the 1998 Master Settlement Agreement," *New England Journal of Medicine* 350 (2004): 293, 296.

30. Mark Curriden, "Up in Smoke: How Greed, Hubris, and High-Stakes Lobbying Laid Waste to the $246 Billion Tobacco Settlement," *American Bar Association Journal,* 93, no. 3 (Mar. 2007): 27, 28.

31. F. A. Sloan, C. A. Mathews, and J. G. Trogdon, "Impacts of the Master Settlement Agreement on the Tobacco Industry," *Tobacco Control* 13 (2004): 357–58.

32. Brandt, *Cigarette Century*, 449–91.

33. Richard A. Nagareda, *Mass Torts in a World of Settlement* (Chicago: University of Chicago Press, 2007), 201.

34. *Master Settlement Agreement*, part IX; Nagareda, *Mass Torts in a World of Settlement*, 199–207.

35. Nagareda, *Mass Torts in a World of Settlement*, 204; Roger Parloff, "Is the $200 Billion Tobacco Deal Going Up in Smoke?" *Fortune*, Mar. 7, 2005, 126.

36. *Master Settlement Agreement*, part IX.

37. Vanessa O'Connell, "Big Tobacco Seeks $1.2 Billion Cut in Payments to States," *Wall Street Journal*, Mar. 8, 2006, B1; Todd Sullivan, "Behind the Tobacco Payments," *ValuePlays*, Apr. 16, 2008, http://valueplays.blogspot.com/2008/04/behind-tobacco-payments-morai.html.

38. Derthick, *Up in Smoke*, 186–88.

39. E.g., *Sanders v. Brown*, 504 F.3d 903, 906 (9th Cir. 2007); *S&M Brands, Inc. v. Summers*, 228 Fed. Appx. 560 (6th Cir. 2007); *Mariana v. Fisher*, 338 F.3d 189, 206 (3d Cir. 2003); *Star Scientific, Inc. v. Beales*, 278 F.3d 339, 362 (4th Cir. 2002).

40. 357 F.3d 205, 226–33 (2d Cir. 2004).

41. Ibid., 230 (quoting statement of Pitofsky on Hearing on Tobacco Settlement before the Senate Committee on Commerce, Science, and Transportation, 105th Cong. (1998)).

42. Ibid., 230 n. 25.

43. Complaint, *Moore ex rel. State v. Am. Tobacco Co.*, No. 94-1429 (Miss. Ch. Ct. May 23, 1994), ¶ 2, http://www.library.ucsf.edu/tobacco/litigation/ms/2moore.html.

44. W. Kip Viscusi, *Smoke-Filled Rooms: A Postmortem on the Tobacco Deal* (Chicago: University of Chicago Press, 2002), 68–72, 98.

45. Ibid., 98.

46. Brandt, *Cigarette Century*, 435; Myron Levin, "States' Tobacco Settlement Has Failed to Clear the Air," *Los Angeles Times*, Nov. 9, 2003, C1. The CDC's guidelines for funding tobacco prevention programs were included in U.S. Department of Health and Human Services, Centers for Disease Control and Prevention, *Best Practices for Comprehensive Tobacco Control Programs* (Atlanta: Centers for Disease Control and Prevention, Aug. 1999), available at http://www.quitwith-yale.org/policy/action/downloads/CDC%201999%20bestprac.pdf (later updated by the CDC's *Best Practices for Tobacco Control Programs* (Atlanta: Centers for Disease Control and Prevention, 2007), available at http://www.cdc.gov/tobacco/tobacco_control_programs/stateandcommunity/best_practices/_pdfs/2007/BestPractices_Complete.pdf).

47. Levin, "States' Tobacco Settlement Has Failed to Clear the Air."

48. Curriden, "Up in Smoke," 30.

49. Viscusi, *Smoke-Filled Rooms*, 64.

50. E.g., *Floyd v. Thompson*, 227 F.3d 1029, 1037 (7th Cir. 2000); *Watson v.*

Texas, 261 F.3d 436, 444 (5th Cir. 2001); *Harris v. Owens*, 264 F.3d 1282, 1297 (10th Cir. 2001).

51. Curriden, "Up In Smoke," 29; Derthick, *Up In Smoke*, 193–94.

52. Viscusi, *Smoke-Filled Rooms*, 53.

53. See, e.g., *Brown & Williamson Tobacco Corp. v. Chesley*, 749 N.Y.S.2d 842, 850 (Sup. Ct. 2002), rev'd, 777 N.Y.S.2d 82 (App. Div. 2004); "Minn. High Court Asked to Rule on Tobacco Fee," *Mealey's Litigation Report: Tobacco* 13, no. 19 (Feb. 3, 2000): 5.

54. *Rhode Island Lead Nuisance Abatement Plan*, 82.

55. Ibid., 8.

56. Ibid., 122.

57. Ibid., 107, 115.

58. Ibid., 98–101.

59. Ibid., 7, 24.

60. Ibid., 7

61. Ibid (emphasis added).

62. Ibid., 91–93.

63. Ibid., 92.

64. Ibid., 52–53.

65. Ibid., 15, 83.

66. In an analogous situation, the scope of issues contained in judicial decrees in public interest litigation often expands to encompass the comprehensive management of public institutions. See Ross Sandler and David Schoenbrod, *Democracy by Decree: What Happens When Courts Run Government* (New Haven, CT: Yale University Press, 2003), 11–12, 55–56, 109–12, 199–203.

67. Lynn Mather, "Theorizing about Trial Courts: Lawyers, Policymaking, and Tobacco Litigation," *Law and Social Inquiry* 23 (1998): 908. See also Peter D. Jacobson and Kenneth E. Warner, "Litigation and Public Health Policy Marking: The Case of Tobacco Control," *Journal of Health Politics, Policy and Law* 24 (1999): 769.

68. William Haltom and Michael McCann, *Distorting the Law: Politics, Media, and the Litigation Crisis* (Chicago: University of Chicago Press, 2004), 295.

69. Timothy D. Lytton, "Using Tort Litigation to Enhance Regulatory Policy Making: Evaluating Climate-Change Litigation in Light of Lessons from Gun-Industry and Clergy-Sexual-Abuse Lawsuits," *Texas Law Review* 86 (2008): 1837, 1842. See also Timothy D. Lytton, "Using Litigation to Make Public Health Policy: Theoretical and Empirical Challenges in Assessing Product Liability, Tobacco, and Gun Litigation," *Journal of Law, Medicine and Ethics* 32 (2004): 556.

70. Michael McCann and William Haltom, "Seeing through the Smoke: Adversarial Legalism and U.S. Tobacco Politics" (unpublished conference paper made available by the authors), 17.

71. Michael McCann, telephone conversation with the author, Jan. 12, 2008.

72. Ibid.

73. Gerald N. Rosenberg, *The Hollow Hope: Can Courts Bring About Social Change?* 2nd ed. (Chicago: University of Chicago Press, 2008), 427.

74. Wendy E. Wagner, "When All Else Fails: Regulating Risky Products through Tort Litigation," *Georgetown Law Journal* 95 (2007): 695. See also ibid., 697–710; Wendy E. Wagner, "Rough Justice and the Attorney General Litigation," *Georgia Law Review* 33 (1999): 935.

75. Wagner, "When All Else Fails," 711–13, 717–19, 725–26.

76. 2002 *R.I. Pub. Laws* chap. 187.

77. Peter B. Lord, "Lead Paint Battle Spreads," *Providence Journal*, Apr. 8, 2007, B1.

Chapter 9

Portions of this chapter consist of revised excerpts from Donald G. Gifford, "Impersonating the Legislature: State Attorneys General and *Parens Patriae* Products Litigation," *Boston College Law Review* 49 (2008): 913 (reprinted by permission).

1. See Robert L. Rabin, "The Third Wave of Tobacco Tort Litigation," in *Regulating Tobacco*, ed. Robert L. Rabin and Stephen D. Sugarman (New York: Oxford University Press, 2001), 192; Robert L. Rabin, "The Tobacco Litigation: A Tentative Assessment," *DePaul Law Review* 51 (2001): 338 n. 34, 340.

2. See, e.g., "Illinois Jury Sides with Merck in a 10th Trial over Painkiller," *New York Times,* Mar. 28, 2007, C7.

3. See *In re Rhone-Poulenc Rorer Inc.*, 51 F.3d 1293, 1298 (7th Cir. 1995) (Posner, C.J.).

4. *Lead Poisoning Prevention Act*, R.I. Gen. Laws §§ 23-24.6-1 to 23-24.6-27 (1996 & Supp. 2006).

5. Ibid., § 23-24.6-23(d).

6. *State v. Lead Indus. Ass'n*, C.A. No. PC 99-5226, 2007 R.I. Super LEXIS 32, °307 (R.I. Super Ct. Feb. 26, 2007).

7. 924 A.2d 484, 501 (N.J. 2007).

8. Ibid., 494.

9. Ibid., 492 (quoting *N.J. Stat. Ann.* § 24:14A-5 (West 1997)).

10. Ibid., 505.

11. *State v. Lead Indus. Ass'n, Inc.*, 951 A.2d 428, 457 (R.I. 2008).

12. Gary T. Schwartz, "Considering the Proper Federal Role in American Tort Law," *Arizona Law Review* 38 (1996): 917, 936–37. See also Neil K. Komesar, *Imperfect Alternatives. Choosing Institutions in Law, Economics, and Public Policy* (Chicago: University of Chicago Press, 1994), 192–95.

13. *Master Settlement Agreement* (1998), 10–20, available at http://ag.ca.gov/tobacco/pdf/1msa.pdf.

14. State of Rhode Island and Providence Plantations, Department of Attorney General, *Rhode Island Lead Nuisance Abatement Plan* (Sept. 14, 2007), 68, available at http://www.riag.state.ri.us/documents/RILeadNuisanceAbatementPlan9-

14-07.pdf and as attachment to LexisNexis electronic version of "Rhode Island Defendants to Respond to $2.4B Abatement Plan by Nov. 15," *Mealey's Litigation Report: Lead* 16, no. 12 (Oct. 2007).

15. Ibid., 15, 83.

16. Ibid., 85.

17. Ibid., 17, 87.

18. Ibid., 12–13.

19. Ibid., 12.

20. *Lead Indus. Ass'n,* 2007 R.I. Super LEXIS 32, at °277, 316.

21. *Model Rules of Professional Conduct,* Rule 3.8 cmt. 1 (American Bar Association, 2002).

22. *Restatement (Third) of the Law Governing Lawyers* § 97 cmt. h (2000); *Freeport-McMoRan Oil & Gas Co. v. FERC,* 962 F.2d 45, 47 (D.C. Cir. 1992) (Mikva, J.).

23. *Berger v. United States,* 295 U.S. 78, 88 (1935).

24. Eliot Spitzer, remarks under "Panel One: State Attorneys General and the Power to Change Law," in Center for Legal Policy at the Manhattan Institute, *Regulation by Litigation: The New Wave of Government-Sponsored Litigation* (New York: Center for Legal Policy at the Manhattan Institute, 1999), 1, 21, available at http://www.manhattan-institute.org/pdf/mics1.pdf.

25. See *In re Rhone-Poulenc Rorer Inc.,* 51 F.3d 1293, 1298–1300 (7th Cir. 1995).

26. *Blankenship v. Gen. Motors Corp.,* 406 S.E.2d 781, 782–83 (W. Va. 1991) (Neely, J.).

27. Peter H. Schuck, *The Limits of Law: Essays on Democratic Governance* (Boulder: Westview Press, 2000), 419, 427–32.

28. Ibid., 432.

29. E.g., Gerald N. Rosenberg, *The Hollow Hope: Can Courts Bring About Social Change?* 2nd ed. (Chicago: University of Chicago Press, 2008), 422; Colin S. Diver, "The Judge as Political Powerbroker: Superintending Structural Change in Public Institutions," *Virginia Law Review* 65 (1979): 89; Paul J. Mishkin, "Federal Courts as State Reformers," *Washington and Lee Law Review* 35 (1978): 965–66, 970–71.

30. *Lead Indus. Ass'n,* 2007 R.I. Super. LEXIS 32, at °302 .

31. Schuck, *Limits of Law,* 432.

32. James A. Henderson Jr., "The Lawlessness of Aggregative Torts," *Hofstra Law Review* 34 (2005): 338.

33. In *Baker v. Carr,* the U.S. Supreme Court stated, "The nonjusticiability of a political question is primarily a function of the separation of powers." 369 U.S. 186, 210 (1962).

34. Ibid., 217. See also *Vieth v. Jubelirer,* 541 U.S. 267, 277–78 (2004).

35. *Baker,* 369 U.S. at 217.

36. *Vieth,* 541 U.S. at 277.

37. 406 F. Supp. 2d 265, 268, 274 (S.D.N.Y. 2005), *argued* No. 05-5104-cv (2d Cir. June 7, 2006).

38. Ibid., 272 (quoting *Vieth*, 541 U.S. at 278).

39. *California v. Gen. Motors Corp.*, No. C06-05755 MJJ, 2007 U.S. Dist. LEXIS 68547, °48 (N.D. Cal. Sept. 17, 2007). But see *In re Methyl Tertiary Butyl Ether ("MTBE") Prods. Liab. Litig.*, 438 F. Supp. 2d 291, 304 (S.D.N.Y. 2006); *In re "Agent Orange" Prod. Liab. Litig.*, 373 F. Supp. 2d 7, 78 (E.D.N.Y. 2005).

40. *California v. Gen. Motors Corp.*, 2007 U.S. Dist. LEXIS 68547, at °22–23.

41. Ibid., °48.

42. Ibid., °47–48.

43. See *Lorillard Tobacco Co. v. Reilly*, 533 U.S. 525, 550–51 (2001).

44. But see *McGraw ex rel. State v. Am. Tobacco Co.*, Civ. A. No. 94-C-1707, 1995 WL 569618, °2 (W. Va. Cir. Ct. June 6, 1995) (holding that "the Attorney General . . . possesses no common law authority or power" and thus lacks standing to institute or prosecute claims based on unjust enrichment, public nuisance, fraud, conspiracy, and other common-law claims).

45. See M. Elizabeth Magill, "The Real Separation in Separation of Powers Law," *Virginia Law Review* 86 (2000): 1149 (quoting *Misretta v. United States*, 488 U.S. 361, 381 (1989)).

46. See, e.g., *Fla. Const.* art. II, § 3; *Ill. Const.* art. II, § 1; *Mass. Const.* pt. 1, art. XXX; *Mich. Const.* art. III, § 2; *N.J. Const.* art. III, § 1; *Va. Const.* art. I, § 5.

47. See, e.g., James A. Gardner, *Interpreting State Constitutions* (Chicago: University of Chicago Press, 2005), 161; Jim Rossi, "Institutional Design and the Lingering Legacy of Antifederal Separation of Powers Ideals in the States," *Vanderbilt Law Review* 52 (1999): 1191.

48. See, e.g., Laurence Tribe, *American Constitutional Law*, 3rd ed., vol. 1 (New York: Foundation Press, 2000), 133–34. See also Michael C. Dorf, "The Relevance of Federal Norms for State Separation of Powers," *Roger Williams University Law Review* 4 (1998): 54–56, 59. But see *Sweezy v. New Hampshire*, 354 U.S. 234, 255 (1957) (holding that "the concept of separation of powers embodied in the United States Constitution is not mandatory in state governments"); *Dreyer v. Illinois*, 187 U.S. 71, 84 (1902) ("Whether the legislative, executive and judicial powers of a State shall be kept altogether distinct and separate . . . is for the determination of the State.").

49. *U.S. Const.* art. IV, § 4; see Tribe, *American Constitutional Law*, 133. In addition, the Seventeenth Amendment of the U.S. Constitution refers to both state legislatures and executives.

50. G. Alan Tarr, *Understanding State Constitutions* (Princeton: Princeton University Press, 1998), 16.

51. See, e.g., *City of Pawtucket v. Sundlun*, 662 A.2d 40, 44 (R.I. 1995); Harold H. Bruff, "Separation of Powers under the Texas Constitution," *Texas Law Review* 68 (1990): 1348.

52. See Tribe, *American Constitutional Law*, 134.

53. 343 U.S. 579 (1952).

54. Compare Sanford Levinson, "The Rhetoric of the Judicial Opinion," in *Law's Stories: Narrative and Rhetoric in the Law*, ed. Peter Brooks and Paul Gewirtz (New Haven: Yale University Press, 1996), 202 (characterizing Justice Jackson's opinion in *Youngstown* as "the most truly intellectually satisfying . . . opinion in our two-hundred-year constitutional history"), with Jesse H. Choper, *Judicial Review and the National Political Process* (Chicago: University of Chicago Press, 1980), 273–75 (arguing that the nonjudicial branches of government should be left alone to work out their own differences); Patricia L. Bellia, "Executive Power in *Youngstown*'s Shadows," *Constitutional Commentary* 19 (2002): 91 (asserting that "the lessons that the case . . . offers . . . are less clear and less helpful than is often believed").

55. See Rebecca L. Brown, "Separated Powers and Ordered Liberty," *University of Pennsylvania Law Review* 139 (1991): 1523. Scholarly analyses of separation of powers typically characterized as "formalist" include Martin H. Redish, *The Constitution as Political Structure* (New York: Oxford University Press, 1995); Steven G. Calabresi and Kevin H. Rhodes, "The Structural Constitution: Unitary Executive, Plural Judiciary," *Harvard Law Review* 105 (1992): 1153; Stephen L. Carter, "Constitutional Improprieties: Reflections on *Mistretta*, *Morrison*, and Administrative Government," *University of Chicago Law Review* 57 (1990): 364–76.

56. *Youngstown*, 343 U.S. at 582, 588–89.

57. Ibid., 585.

58. Ibid., 586.

59. See Brown, "Separated Powers and Ordered Liberty," 1527–28; Magill, "Real Separation," 1142–43. For a functionalist scholarly perspective, see, e.g., Harold H. Bruff, "Presidential Power and Administrative Rulemaking," *Yale Law Journal* 88 (1979): 451.

60. *Youngstown*, 343 U.S. at 634–55 (Jackson, J., concurring).

61. Ibid., 637.

62. Ibid.

63. Ibid.

64. See, e.g., Fla. Stat. Ann. § 409.910 (West 2005 & Supp. 2009); Md. Code Ann., Health-Gen. § 15 120(a) (West 2009); Vt. Stat. Ann. tit. 33, § 1911 (2001).

65. New York Penal Law § 400.05 (McKinney 2008).

66. E.g., Fla. Stat. Ann. § 790.331(2) (West 2007); Idaho Code Ann. § 5-247 (2004); Tex. Civ. Prac. & Rem. Code Ann. § 128.001(b) (Vernon 2005).

67. 453 U.S. 654 (1981).

68. Ibid., 679–82.

69. Ibid., 681.

70. See, e.g., Henry M. Hart Jr. and Albert M. Sacks, *The Legal Process: Basic Problems in the Making and Application of Law*, ed. William N. Eskridge Jr. and Philip P. Frickey (Westbury, NY: Foundation Press, 1994), 1358–60; Tribe, *American Constitutional Law*, 205 ("When the array of powers held by the executive, the

judiciary, or the states with respect to a given matter can be transformed only by congressional approval or disapproval, then it is essential that such approval or disapproval take the form of legislation made through [the formal constitutional procedures for passing laws].").

71. Amended Substitute S.B. 117, 126th Gen. Assem. (Ohio 2006) (codified at *Ohio Rev. Code Ann.* §§ 2307.71–2307.80 (LexisNexis 2005 & Supp. 2008)).

72. Complaint, *State of Ohio v. Sherwin-Williams Co.*, CA 07-CVC-044587 (Ohio Ct. Com. Pl., Apr. 2, 2007). The state voluntarily dismissed the lawsuit on February 9, 2009. See press release, "Cordray Dismisses Lead Paint Lawsuit," Office of Richard Cordray, Attorney General, Feb. 9, 2009, available at http://www.ag.state.oh.us/press/09/02/pr090206b.asp.

73. There are two strong arguments suggesting that the public nuisance legislation does apply to the action filed by the attorney general. First, any action seeking abatement of a public nuisance or recovery of costs of the abatement of public nuisance is, by its nature, an action for prospective injunctive relief. See *Landgraf v. USI Film Prods.*, 511 U.S. 244, 273 (1994); *Am. Steel Foundries v. Tri-City Cent. Trades Council*, 257 U.S. 184, 201 (1921). Second, because the Ohio legislation amended the existing products liability statutory framework, it arguably merely clarified existing legislation, and clarifying statutory amendments do apply retrospectively. See *Wilson v. AC&S, Inc.*, 864 N.E.2d 682, 702 (Ohio Ct. App. 2006).

74. Brief of the State of Rhode Island at 3 and n.5, *State v. Lead Indus. Ass'n, Inc.*, 951 A.2d 428 (R.I. 2008) (No. 2004-63-M.P.), 2005 R.I. S. Ct. Briefs LEXIS 12.

75. See Lester Brickman, "Effective Hourly Rates of Contingency-Fee Lawyers: Competing Data and Non-Competitive Fees," *Washington University Law Quarterly* 81 (2003): 721.

76. See, e.g., *Brown & Williamson Tobacco Corp. v. Chesley*, 749 N.Y.S.2d 842, 850 (Sup. Ct. 2002); "Minn. High Court Asked to Rule on Tobacco Fee," *Mealey's Litigation Report: Tobacco*, 13, no. 19 (Feb. 3, 2000): 5.

77. Mark Curriden, "Up in Smoke: How Greed, Hubris and High-Stakes Lobbying Laid Waste to the $246 Billion Tobacco Settlement," *American Bar Association Journal* 93, no. 3 (2007): 27.

78. See *Lead Indus. Ass'n*, 951 A.2d at 475; accord *County of Santa Clara v. Super. Ct. of Santa Clara County*, 74 Cal. Rptr. 3d 842, 853 (Ct. App. 2008).

79. *Lead Indus. Ass'n*, 951 A.2d at 475 (emphasis omitted).

80. According to the Centers for Disease Control and Prevention, by 1996, asthma affected 6.2 percent of all American children. It resulted in 14 million lost days of school and 266 deaths in 1996 and was the third leading cause of hospitalization among children. See Department of Health and Human Services, Centers for Disease Control and Prevention, *Asthma's Impact on Children and Adolescents*, web.archive.org/web/20080307050326/http://www.cdc.gov/asthma/children.htm. In comparison, the CDC estimates that during the period 1999–2000, 2.2 percent of American children ages 1–5 had elevated blood lead levels deemed to be of concern to the CDC. See Pamela A. Meyer, Timothy Pivetz, Timothy A. Dignam,

David M. Homa, Jaime Schoonover, and Debra Brody, et al., "Surveillance for Elevated Blood Lead Levels among Children—United States, 1997–2001," *Morbidity and Mortality Weekly Report Surveillance Summaries* 52, no. SS-10 (Sept. 12, 2003), available at http://www.cdc.gov/mmwR/PDF/ss/ss5210.pdf.

81. See Department of Health and Human Services, Centers for Disease Control and Prevention, *Healthy Homes Initiative*, http://www.cdc.gov/healthyplaces/healthyhomes.htm.

82. Stephen Breyer, *Breaking the Vicious Circle: Toward Effective Risk Regulation* (Cambridge, MA: Harvard University Press, 1993), 19–20.

83. Ibid., 59.

84. See, e.g., *Md. Code Ann., Envir.* § 6-815; *R.I. Gen. Laws* § 23-24.6-4(15) (West 2002 and Supp. 2008).

85. See, e.g., *U.S. Code* 42 (2006): §§ 4851–4856; 24 *Code of Federal Regulations* §§ 35.100–35.175 (2008); "Requirements for Notification, Evaluation, and Reduction of Lead-Based Paint Hazards in Federally Owned Residential Property and Housing Receiving Federal Assistance," *Federal Register* 64 (Sept. 15, 1999): 50,140, 50,142 (codified in scattered sections of 24 C.F.R.).

86. See, e.g., Rhode Island Board of Elections Contribution Reporting, http://www.ricampaignfinance.com/RIPublic/Contributions.aspx (enter "John McConnell" under Donor Name; last visited Aug. 18, 2008) (showing contributions of John J. McConnell Jr., one of the state's privately retained trial counsel in the Rhode Island litigation against pigment manufacturers, to "Friends of Patrick Lynch" (Lynch is the current Rhode Island attorney general)).

87. In the case of claims against firearm manufacturers, Congress, for all intents and purposes, ended the litigation cycle by enacting the *Protection of Lawful Commerce in Arms Act*, 15 U.S.C. §§ 7901–7903 (2006).

88. *In re Lead Paint Litig.*, 924 A.2d 484, 502 n.8, 505 (N.J. 2007).

Conclusion

1. *State v. Lead Indus. Ass'n*, 951 A.2d 428, 480 (R.I. 2008).

2. *In re Lead Paint Litig.*, 924 A.2d 484, 501 (N.J. 2007).

3. *City of St. Louis v. Benjamin Moore & Co.*, 226 S.W.3d 110, 116–17 (Mo. 2007) (holding that the city must prove an individualized causal connection between a specific manufacturer and the alleged harm, even under a public nuisance claim brought by the city).

4. *County of Santa Clara v. Atl. Richfield Co.*, 40 Cal. Rptr. 3d 313, 330 (Ct. App. 2006) (reversing the trial court's dismissal of a public nuisance cause of action). California statutes define public nuisance in a comparatively vague manner, without explicitly including the traditional common-law requirements; see *Cal. Civ. Code* §§ 3470–80 (West 2009). See also *Cal. Civ. Code* § 3491 (West 2009) (providing abatement and civil actions as remedies); *Cal. Gov't Code* § 26528 (West 2004) (allowing the state to bring a civil action).

5. *City of Milwaukee v. NL Indus., Inc.*, 691 N.W.2d 888, 894 (Wis. Ct. App.

2004) (allowing public nuisance claims to proceed against manufacturers of lead pigment or lead-based paint).

6. State of Rhode Island and Providence Plantations, Department of Attorney General, *Rhode Island Lead Nuisance Abatement Plan* (Sept. 14, 2007), 126, available at http://www.riag.state.ri.gov/documents/RILeadNuisanceAbatementPlan9-14-07.pdf and as attachment to "Rhode Island Defendants to Respond to $2.4B Abatement Plan by Nov. 15," *Mealey's Litigation Report: Lead* 16, no. 12 (Oct. 2007): 7.

7. Michael McCann and William Haltom, "Seeing through the Smoke: Adversarial Legalism and U.S. Tobacco Politics" (unpublished conference paper made available by the authors), 17; Michael McCann, telephone conversation with the author, Jan. 12, 2008.

8. See generally Trevor D. Dryer, "Gaining Access: A State Lobbying Case Study," *Journal of Law and Politics* 23 (2007): 283; Brian Griffith, Comment, "Lobbying Reform: House-Cleaning or Window Dressing?" *University of Cincinnati Law Review* 75 (2006): 863.

9. See, e.g., *Family Smoking Prevention and Tobacco Control Act*, S. 625, 110th Cong. (2007); H.R. 1108, 110th Cong. (2007).

10. Tammy O. Tengs, Sajjad Ahmad, Jennifer M. Savage, Rebecca Moore, and Eric Gage, "The AMA Proposal to Mandate Nicotine Reduction in Cigarettes: A Simulation of the Population Health Impacts," *Preventive Medicine* 40 (2005): 170, 170–71.

11. W. Kip Viscusi, *Smoke-Filled Rooms: A Postmortem on the Tobacco Deal* (Chicago: University of Chicago Press, 2002), 194–95, 213–14.

12. Vanessa O'Connell, "Altria Drops New Filter Cigarettes, in Strategy Setback," *Wall Street Journal*, June 23, 2008, B1.

13. David Kessler, *A Question of Intent: A Great American Battle with a Deadly Industry* (New York: PublicAffairs, 2001), 392.

14. See Robert L. Rabin, "Some Thoughts on the Efficacy of a Mass Toxics Administrative Compensation Scheme," *Maryland Law Review* 52 (1993): 951. Peter H. Schuck offers a more sobering assessment of administrative compensation schemes enacted by legislatures, in *The Limits of Law: Essays on Democratic Governance* (Boulder: Westview Press, 2000), 361, 370–72.

15. Jon D. Hanson, Kyle D. Logue, and Michael S. Zamore, "Smokers' Compensation: Toward a Blueprint for Federal Regulation of Cigarette Manufacturers," *Southern Illinois University Law Journal* 22 (1998): 519, 552–97; Richard C. Ausness, "Compensation for Smoking-Related Injuries: An Alternative to Strict Liability in Tort," *Wayne Law Review* 36 (1990): 1085, 1124–33. See generally Rabin, "Efficacy of a Mass Toxics Administrative Compensation Scheme," 964–78.

16. Deborah Ann Ford and Michele Gilligan, "The Effect of Lead Paint Abatement Laws on Rental Property Values," *American Real Estate and Urban Economics Association Journal* 16 (1988): 84, 92; see also President's Task Force on Environmental Health Risks and Safety Risks to Children, *Eliminating Childhood Lead*

Poisoning: A Federal Strategy Targeting Lead Paint Hazards (Washington, DC: Government Printing Office, 2000), app. A-27, table 25.

17. According to the U.S. Census Bureau, the paint industry annually produces a total of 794.5 million gallons of architectural coatings. See U.S. Census Bureau, "Current Industrial Reports, Paints and Allied Products: 2005" (June 2006), 2, table 1, http://www.census.gov/industry/1/ma325f05.pdf.

18. A Clinton administration task force in 2000 estimated that $1.84 billion would be required to implement interim controls on a nationwide basis. See President's Task Force on Environmental Health Risks and Safety Risks to Children, *Eliminating Childhood Lead Poisoning*, 5, table 2.

19. U.S. Department of Commerce, *United States Government Master Specification for Paint, White, and Tinted Paints Made on a White Base, Semipaste, and Ready Mixed, Federal Specifications Board Stand. and Spec. No. 10B*, 3rd ed., Circular of the Bureau of Standards, no. 89 (Washington, DC: Government Printing Office, 1927), 2 (providing that pigment in white base semipaste paint to be purchased by the federal government shall contain not more than 70 percent and not less than 45 percent white lead).

20. *Wright v. Lead Indus. Ass'n*, No. 94363042/CL190487, slip op. at 5 (Md. Cir. Ct. June 20, 1996). Similarly, a New York trial court rejected claims of fraud against the lead pigment industry, stating, "[T]he dangers associated with white lead pigment were well established decades prior to the injuries complained of, and in fact, decades prior to the birth of the infant plaintiffs. . . . As a result, plaintiffs, herein, cannot claim ignorance. Rather, the widely available information about the dangers of white lead pigment forecloses any presumption of reliance" *Sabater v. Lead Indus. Ass'n*, 704 N.Y.S.2d 800, 807 (Sup. Ct. 2000).

21. See, e.g., *Md. Code Ann., Envir.* § 6-030 (West 2008 and Supp. 2004) (violation of statute requiring interim controls creates a presumption of negligence). See also *Brooks v. Lewin Realty III, Inc.*, 835 A.2d 616, 627 (Md. 2003) (holding that violation of city housing code provisions designed to prevent childhood lead poisoning constituted a prima facie case of negligence).

22. Winston Churchill, speech, House of Commons, Nov. 11, 1947, included in *Winston S Churchill, His Complete Speeches, 1897–1963*, ed. Robert Rhodes James, vol. 7 (1974), 7566.

SELECTED BIBLIOGRAPHY

Selected Books, Articles, Reports, and Government Publications

Abell Foundation. "Childhood Lead Poisoning in Baltimore: A Generation Imperiled as Laws Ignored." *Abell Report* 15, no. 5 (2002): 1–8, available at http://www.abell.org/pubsitems/arn1002.pdf.

American Cancer Society. "Tobacco Use in the US, 1900–2004." *Cancer Statistics 2008*, PowerPoint no. 396, slide 27. http://www.cancer.org/downloads/STT/Cancer_Statistics_2008.ppt.

American Law Institute. *Enterprise Responsibility for Personal Injury.* Vol. 2, *Approaches to Legal and Institutional Change.* Philadelphia: American Law Institute, 1991.

American Public Health Association. "Resolution No. 9704: Responsibilities of the Lead Pigment Industry and Others to Support Efforts to Address the National Child Lead Poisoning Problem." *American Journal of Public Health* 88 (1998): 498–500.

Ausness, Richard C. "Compensation for Smoking-Related Injuries: An Alternative to Strict Liability in Tort." *Wayne Law Review* 36 (1990): 1085–1148.

Ausness, Richard C. "Tort Liability for the Sale of Non-defective Products: An Analysis and Critique of the Concept of Negligent Marketing." *South Carolina Law Review* 53 (2002): 907–65.

Baker, Susan P., and William Haddon Jr. "Reducing Injuries and Their Results: The Scientific Approach." *Milbank Memorial Fund Quarterly: Health and Society* 52 (1974): 377–89.

Bellia, Patricia L. "Executive Power in Youngstown's Shadows." *Constitutional Commentary* 19 (2002): 87–154.

Boyer, Peter J. "Big Guns." *New Yorker,* May 17, 1999, 54.

Bracton, Henry de. *Bracton on the Laws and Customs of England.* Ed. Samuel E. Thorne. Vol. 3. Cambridge, MA: Harvard University Press, Belknap Press, 1977.

Brandt, Allan M. *The Cigarette Century: The Rise, Fall, and Deadly Persistence of the Product That Defined America.* New York: Basic Books, 2007.

Breyer, Stephen. *Breaking the Vicious Circle: Toward Effective Risk Regulation.* Cambridge, MA: Harvard University Press, 1993.

Brickman, Lester. "Effective Hourly Rates of Contingency-Fee Lawyers: Competing Data and Non-Competitive Fees." *Washington University Law Quarterly* 81 (2003): 653–736.

Brodeur, Paul. *Outrageous Misconduct: The Asbestos Industry on Trial.* New York: Pantheon, 1985.

Brown, Rebecca L. "Separated Powers and Ordered Liberty." *University of Pennsylvania Law Review* 139 (1991): 1513–66.

Bruff, Harold H. "Presidential Power and Administrative Rulemaking." *Yale Law Journal* 88 (1979): 451–508.

Bruff, Harold H. "Separation of Powers under the Texas Constitution." *Texas Law Review* 68 (1990): 1337–67.

Bryson, John E., and Angus Macbeth. "Public Nuisance, the Restatement (Second) of Torts, and Environmental Law." *Ecology Law Quarterly* 2 (1972): 241–81.

Bulow, Jeremy, and Paul Klemperer. "The Tobacco Deal." In *Brookings Papers on Economic Activity: Microeconomics*, ed. Marin Neil Baily, Peter C. Reiss, and Clifford Winston, 323–94. Washington, DC: Brookings Institution, 1998.

Burney, L. E. "Smoking and Lung Cancer: A Statement of the Public Health Service." *Journal of the American Medical Association* 171 (1959): 1829–37.

Byers, Randolph K., and Elizabeth E. Lord. "Late Effects of Lead Poisoning on Mental Development." *American Journal of Diseases of Children* 66 (1943): 471–94.

Calabresi, Guido. "Concerning Cause and the Law of Torts: An Essay for Harry Kalven, Jr." *University of Chicago Law Review* 43 (1975): 69–108.

Calabresi, Guido. *The Costs of Accidents: A Legal and Economic Analysis*. New Haven, CT: Yale University Press, 1970.

Calabresi, Steven G., and Kevin H. Rhodes. "The Structural Constitution: Unitary Executive, Plural Judiciary." *Harvard Law Review* 105 (1992): 1153–1216.

Cardozo, Benjamin N. *The Nature of the Judicial Process*. New Haven, CT: Yale University Press, 1921.

Carroll, Stephen J., Deborah Hensler, Allan Abrahamse, Jennifer Gross, Michelle White, Scott Ashwood, and Elizabeth Sloss. *Asbestos Litigation Costs and Compensation: An Interim Report*. Doc. No. DB-397-ICS. Santa Monica, CA: RAND Institute for Civil Justice, 2002. http://www.rand.org/pubs/documented_briefings/DB397/DB397.pdf.

Carroll, Stephen J., Deborah Hensler, Jennifer Gross, Elizabeth M. Sloss, Matthias Schonlau, Allan Abrahamse, and J. Scott Ashwood. *Asbestos Litigation*, Doc. No. MG-162-ICJ. Santa Monica, CA: RAND Institute for Civil Justice, 2005. http://www.rand.org/pubs/monographs/2005/RAND_MG162.pdf.

Carson, Rachel. *Silent Spring*. Boston: Houghton Mifflin; Cambridge, MA: Riverside Press, 1962.

Castleman, Barry I. *Asbestos: Medical and Legal Aspects*. 5th ed. New York: Aspen, 2005.

Center for Legal Policy at the Manhattan Institute. *Regulation by Litigation: The New Wave of Government-Sponsored Litigation*. New York: Center for Legal Policy at the Manhattan Institute, 1999. Available at http://www.manhattan-institute.org/pdf/mics1.pdf.

Chisolm, Julian, Jr. "Lead Poisoning." *Scientific American* 224, no. 2 (Feb. 1971): 15–23.

Coleman, Jules L. *Risks and Wrongs*. New York: Cambridge University Press, 1992.

Common Cause/NY. *Lead Poisoning Legislation and the Political Power of Real Estate in New York City*. New York: Common Cause/NY, 2003.

Curriden, Mark. "Tobacco Fees Give Plaintiffs' Lawyers New Muscle for Other Litigation." *Dallas Morning News*, Oct. 31, 1999, H1.

Curriden, Mark. "Up in Smoke: How Greed, Hubris, and High-Stakes Lobbying Laid Waste to the $246 Billion Tobacco Settlement." *American Bar Association Journal* 93, no. 3 (2007): 27–32.

Derthick, Martha A. *Up in Smoke: From Legislation to Litigation in Tobacco Politics*. 2nd ed. Washington, DC: CQ Press, 2004.

Diaz, Tom. "The American Gun Industry: Designing and Marketing Increasingly Lethal Weapons." In Lytton, *Suing the Gun Industry*, 84–104.

Diver, Colin S. "The Judge as Political Powerbroker: Superintending Structural Change in Public Institutions." *Virginia Law Review* 65 (1979): 43–106.

Doll, Richard. "Conversation with Sir Richard Doll." *British Journal of Addiction* 86 (1991): 365–77.

Doll, Richard, and A. Bradford Hill. "Smoking and Carcinoma of the Lung: Preliminary Report." *British Medical Journal* 2 (1950): 739–48.

Dorf, Michael C. "The Relevance of Federal Norms for State Separation of Powers." *Roger Williams University Law Review* 4 (1998): 51–77.

Dryer, Trevor D. "Gaining Access: A State Lobbying Case Study." *Journal of Law and Politics* 23 (2007): 283–329.

Eisenberg, Melvin Aron. *The Nature of the Common Law*. Cambridge, MA: Harvard University Press, 1988.

English, Peter C. *Old Paint: A Medical History of Childhood Lead-Paint Poisoning in the United States to 1980*. New Brunswick, NJ. Rutgers University Press, 2001.

Erichson, Howard M. "Informal Aggregation: Procedural and Ethical Implications of Coordination among Counsel in Related Lawsuits." *Duke Law Journal* 50 (2000): 381–471.

Erichson, Howard M. "Private Lawyers, Public Lawsuits: Plaintiffs' Attorneys in Municipal Gun Litigation." In Lytton, *Suing the Gun Industry*, 129–51.

Felstiner, W. L. F., and Peter Siegelman. "Neoclassical Difficulties: Tort Deterrence for Latent Injuries." *Law and Policy* 11 (1989): 309–29.

Finkelstein, Claire. "Is Risk a Harm?" *University of Pennsylvania Law Review* 151 (2003): 963–1001.

Fiorina, Morris P., and Paul E. Peterson. *The New American Democracy*. Boston: Allyn and Bacon, 1998.

Ford, Deborah Ann, and Michele Gilligan. "The Effect of Lead Paint Abatement Laws on Rental Property Values." *American Real Estate and Urban Economics Association Journal* 16 (1988): 84–94.

Gangarosa, Raymond E., Frank J. Vandall, and Brian M. Willis. "Suits by Public Hospitals to Recover Expenditures for the Treatment of Disease, Injury and

Disability Caused by Tobacco and Alcohol." *Fordham Urban Law Journal* 22 (1994): 81–139.

Gifford, Donald G. "The Challenge to the Individual Causation Requirement in Mass Products Torts." *Washington and Lee Law Review* 62 (2005): 873–935.

Gifford, Donald G. "The Death of Causation: Mass Products Torts' Incomplete Incorporation of Social Welfare Principles." *Wake Forest Law Review* 41 (2006): 943–1002.

Gifford, Donald G. "Impersonating the Legislature: State Attorneys General and *Parens Patriae* Products Litigation," *Boston College Law Review* 49 (2008): 913–69.

Gifford, Donald G. "Public Nuisance as a Mass Products Liability Tort." *University of Cincinnati Law Review* 71 (2003): 741–837.

Gillette, Clayton P., and James E. Krier. "Risks, Courts, and Agencies." *University of Pennsylvania Law Review* 138 (1990): 1027–1109.

Glantz, Stanton A., and Michael E. Begay. "Tobacco Industry Campaign Contributions Are Affecting Tobacco Control Policymaking in California." *Journal of the American Medical Association* 272 (1994): 1176–82.

Goldberg, John C. P., and Benjamin C. Zipursky. "Accidents of the Great Society." *Maryland Law Review* 64 (2005): 364–408.

Gordon, John E. "The Epidemiology of Accidents." *American Journal of Public Health* 39 (1949): 504–15.

Green, Michael D. "The Inability of Offensive Collateral Estoppel to Fulfill Its Promise: An Examination of Estoppel in Asbestos Litigation." *Iowa Law Review* 70 (1984): 141–235.

Grob, Gerald N. *The Deadly Truth: A History of Disease in America.* Cambridge, MA: Harvard University Press, 2002.

Haddon, William, Jr. "The Changing Approach to the Epidemiology, Prevention, and Amelioration of Trauma: The Transition to Approaches Etiologically Rather Than Descriptively Based." *American Journal of Public Health* 58 (1968): 1431–38.

Haltom, William, and Michael McCann. *Distorting the Law: Politics, Media, and the Litigation Crisis.* Chicago: University of Chicago Press, 2004.

Hammond, E. Cuyler, and David Horn. "The Relationship between Human Smoking Habits and Death Rates." *Journal of the American Medical Association* 155 (1954): 1316–28.

Hanson, Jon D., Kyle D. Logue, and Michael S. Zamore. "Smokers' Compensation: Toward a Blueprint for Federal Regulation of Cigarette Manufacturers." *Southern Illinois University Law Journal* 22 (1998): 519–600.

Hart, Henry M., Jr., and Albert M. Sacks. *The Legal Process: Basic Problems in the Making and Application of Law.* Ed. William N. Eskridge Jr. and Philip P. Frickey. Westbury, NY: Foundation Press, 1994.

Hearst, William Randolph, Jr. "The Traffic Accident Problem and the U.S. President's Committee for Traffic Safety." *Journal of Criminal Law, Criminology, and Police Science* 51 (1960): 90–92.

Henderson, James A., Jr. "The Lawlessness of Aggregative Torts." *Hofstra Law Review* 34 (2005): 329–43.

Henderson, James A., Jr. "Product Liability and the Passage of Time: The Imprisonment of Corporate Rationality." *New York University Law Review* 58 (1983): 765–95.

Hirayama, Takeshi. "Non-Smoking Wives of Heavy Smokers Have a Higher Risk of Lung Cancer: A Study from Japan." *British Medical Journal* 282 (1981): 183–85.

Holmes, Oliver Wendell, Jr. *The Common Law.* Ed. Mark deWolfe Howe. 1881. Reprint, Cambridge, MA: Harvard University Press, Belknap Press, 1963.

Holmes, Oliver Wendell, Jr. "The Theory of Torts." *American Law Review* 7 (1873): 652–63.

Holt, L. Emmett. *The Diseases of Infancy and Childhood, for the Use of Students and Practitioners of Medicine.* 8th ed. New York: D. Appleton, 1922.

Hylton, Keith N. "Calabresi and the Intellectual History of Law and Economics." *Maryland Law Review* 64 (2005): 85–107.

Hylton, Keith N. "The Theory of Tort Doctrine and the Restatement (Third) of Torts." *Vanderbilt Law Review* 54 (2001): 1413–38.

Ibbetson, David. *A Historical Introduction to the Law of Obligations.* New York: Oxford University Press, 1999.

Ieyoub, Richard P., and Theodore Eisenberg. "State Attorney General Actions, the Tobacco Litigation, and the Doctrine of *Parens Patriae.*" *Tulane Law Review* 74 (2000): 1859–83.

Jacobs, David E., Robert P. Clickner, Joey Y. Zhou, Susan M. Viet, David A. Marker, John W. Rogers, Darryl C. Zeldin, Pamela Broene, and Warren Friedman. "The Prevalence of Lead-Based Paint Hazards in U.S. Housing." *Environmental Health Perspectives* 110 (2002): A599–A606.

Jacobson, Peter D., and Kenneth E. Warner. "Litigation and Public Health Policy Making: The Case of Tobacco Control." *Journal of Health Politics, Policy, and Law* 24 (1999): 769–804.

Johnson, Barry L. *Environmental Policy and Public Health.* Boca Raton, FL: CRC Press, 2007.

Kagan, Robert A. *Adversarial Legalism: The American Way of Life.* Cambridge, MA: Harvard University Press, 2001.

Kagan, Robert A., and William P. Nelson. "The Politics of Tobacco Regulation in the United States." In Rabin and Sugarman, *Regulating Tobacco*, 11–38.

Kairys, David. "The Governmental Handgun Cases and the Elements and Underlying Policies of Public Nuisance Law." *Connecticut Law Review* 32 (2000): 1175–87.

Kairys, David. "The Origin and Development of the Governmental Handgun Cases." *Connecticut Law Review* 32 (2000): 1163–74.

Kelder, Graham E., Jr., and Richard A. Daynard. "The Role of Litigation in the Effective Control of the Sale and Use of Tobacco." *Stanford Law and Policy Review* 8 (1997): 63–98.

Kennedy, Edward M. "The Need for FDA Regulation of Tobacco Products." *Yale Journal of Health Policy, Law, and Ethics* 3 (2002): 101–8.

Kessler, David A. Letter to Scott D. Ballin, Chairman, Coalition on Smoking and Health, Feb. 25, 1994. Available at http://tobaccodocuments.org/batco/500821335-1337.html.

Kessler, David. *A Question of Intent: A Great American Battle with a Deadly Industry.* New York: Public Affairs, 2001.

Klein, David, and Julian A. Waller. *Causation, Culpability, and Deterrence in Highway Crashes, Department of Transportation Automobile Insurance and Compensation Study.* Washington, DC: U.S. Department of Transportation, 1970.

Kluger, Richard. *Ashes to Ashes: America's Hundred-Year Cigarette War, the Public Health, and the Unabashed Triumph of Philip Morris.* New York: Vintage Books, 1997.

Kovarik, William. "Ethyl-Leaded Gasoline: How a Classic Occupational Disease Became an International Public Health Disaster." *International Journal of Occupational and Environmental Health* 11 (2005): 384–97.

Landes, William M., and Richard A. Posner. "Causation in Tort Law: An Economic Approach." *Journal of Legal Studies* 12 (1983): 109–34.

Landes, William M., and Richard A. Posner. "Joint and Multiple Tortfeasors: An Economic Analysis." *Journal of Legal Studies* 9 (1980): 517–55.

Landrigan, Philip J. "Pediatric Lead Poisoning: Is There a Threshold?" *Public Health Reports* 115 (2000): 530–31.

Lanphear, Bruce P., Kim Dietrich, Peggy Auinger, and Christopher Cox. "Cognitive Deficits Associated with Blood Lead Concentrations <10 μg/dL in US Children and Adolescents." *Public Health Reports* 115 (2000): 521–29.

Levinson, Sanford. "The Rhetoric of the Judicial Opinion." In *Law's Stories: Narrative and Rhetoric in the Law*, ed. Peter Brooks and Paul Gewirtz, 187–205. New Haven: Yale University Press, 1996.

Llewellyn, Karl. *The Case Law System in America.* Ed. Paul Gewirtz. Trans. Michael Ansaldi. Chicago: University of Chicago Press, 1989.

Lytton, Timothy D. "The NRA, the Brady Campaign, and the Politics of Gun Litigation." In Lytton, *Suing the Gun Industry,* 152–75.

Lytton, Timothy D., ed. *Suing the Gun Industry: A Battle at the Crossroads of Gun Control and Mass Torts.* Ann Arbor: University of Michigan Press, 2005.

Lytton, Timothy D. "Tort Claims against Gun Manufacturers for Crime-Related Injuries: Defining a Suitable Role for the Tort System in Regulating the Firearms Industry." *Missouri Law Review* 65 (2000): 1–81.

Lytton, Timothy D. "Using Litigation to Make Public Health Policy: Theoretical and Empirical Challenges in Assessing Product Liability, Tobacco, and Gun Litigation." *Journal of Law, Medicine, and Ethics* 32 (2004): 556–64.

Lytton, Timothy D. "Using Tort Litigation to Enhance Regulatory Policy Making: Evaluating Climate-Change Litigation in Light of Lessons from Gun-Industry and Clergy-Sexual-Abuse Lawsuits." *Texas Law Review* 86 (2008): 1837–76.

Magill, M. Elizabeth. "The Real Separation in Separation of Powers Law." *Virginia Law Review* 86 (2000): 1127–98.

Mair, Julie Samia, Stephen Teret, and Shannon Frattaroli. "A Public Health Perspective on Gun Violence Prevention." In Lytton, *Suing the Gun Industry,* 39–61.

Markowitz, Gerald, and David Rosner. *Deceit and Denial: The Deadly Politics of Industrial Pollution.* Berkeley: University of California Press, 2002.

Mashaw, Jerry L., and David L. Harfst. *The Struggle for Auto Safety.* Cambridge, MA: Harvard University Press, 1990.

Mather, Lynn. "Theorizing about Trial Courts: Lawyers, Policymaking, and Tobacco Litigation." *Law and Social Inquiry* 23 (1998): 897–940.

McCann, Michael, and William Haltom. "Seeing through the Smoke: Adversarial Legalism and U.S. Tobacco Politics." Unpublished conference paper made available by the authors.

Merrill, Thomas W. "Agency Capture Theory and the Courts: 1967–1983." *Chicago-Kent Law Review* 72 (1997): 1039–1117.

Mollenkamp, Carrick, Adam Levy, Joseph Menn, and Jeffrey Rothfeder. *The People vs. Big Tobacco: How the States Took on the Cigarette Giants.* Princeton, NJ: Bloomberg Press, 1998.

Moore, Stephen, Deborah Lindes, Sidney M. Wolfe, and Cliff Douglas. *Contributing to Death: The Influence of Tobacco Money on the U.S. Congress.* Washington, DC: Public Citizen's Health Research Group, 1993.

Moore, Stephen, Sidney M. Wolfe, Deborah Lindes, and Clifford E. Douglas. "Epidemiology of Failed Tobacco Control Legislation." *Journal of the American Medical Association* 272 (1994): 1171–75.

Moynihan, Daniel P. "Public Health and Traffic Safety." *Journal of Criminal Law, Criminology, and Police Science* 51 (1960): 93–98.

Müller, Franz Hermann. "Abuse of Tobacco and Carcinoma of Lungs." Abstract. *Journal of the American Medical Association* 113 (1939): 1372.

Nader, Ralph. *Unsafe at Any Speed: The Designed-In Dangers of the American Automobile.* New York: Grossman, 1972.

Nagareda, Richard A. *Mass Torts in a World of Settlement.* Chicago: University of Chicago Press, 2007.

National Research Council. Committee on Passive Smoking. *Environmental Tobacco Smoke: Measuring Exposures and Assessing Health Effects.* Washington, DC: National Academy Press, 1986.

Noah, Lars, and Barbara A. Noah. "Nicotine Withdrawal: Assessing the FDA's Effort to Regulate Tobacco Products." *Alabama Law Review* 48 (1996): 1–63.

Owen, David G. *Products Liability Law.* 2nd ed. St. Paul, MN: West, 2008.

Parker-Pope, Tara. *Cigarettes: Anatomy of an Industry from Seed to Smoke.* New York: New Press, 2001.

Percival, Robert V. "The Clean Water Act and the Demise of the Federal Common Law of Interstate Nuisance." *Alabama Law Review* 55 (2004): 717–74.

Percival, Robert V., Christopher H. Schroeder, Robert M. Miller, and James P. Leape. *Environmental Regulation: Law, Science, and Policy.* 5th ed. New York: Aspen, 2006.

Plumer, Christy. "Setting Priorities for Prevention of Childhood Lead Poisoning in Providence." Master's thesis abstract, Brown University, 2000. http://envstud ies.brown.edu/oldsite/dept/thesis/master9900/christy_plumer.htm.

Posner, Richard A. "Guido Calabresi's *The Costs of Accidents:* A Reassessment." *Maryland Law Review* 64 (2005): 12–23.

President's Task Force on Environmental Health Risks and Safety Risks to Children. *Eliminating Childhood Lead Poisoning: A Federal Strategy Targeting Lead Paint Hazards.* Washington, DC: Government Printing Office, 2000. Available at http://www.cdc.gov/nceh/Lead/about/fedstrategy2000.pdf.

Pringle, Peter. *Cornered: Big Tobacco at the Bar of Justice.* New York: Henry Holt, 1998.

Prosser, William L. "Nuisance without Fault." *Texas Law Review* 20 (1942): 399–426.

Prosser, William L. "Private Action for Public Nuisance." *Virginia Law Review* 52 (1966): 997–1027.

Rabin, Robert L. "Essay: A Sociolegal History of the Tobacco Tort Litigation." *Stanford Law Review* 44 (1992): 853–78.

Rabin, Robert L. "Some Thoughts on the Efficacy of a Mass Toxics Administrative Compensation Scheme." *Maryland Law Review* 52 (1993): 951–82.

Rabin, Robert L. "The Tobacco Litigation: A Tentative Assessment." *DePaul Law Review* 51 (2001): 331–57.

Rabin, Robert L., and Stephen D. Sugarman, eds. *Regulating Tobacco.* New York: Oxford University Press, 2001.

Redhead, C. Stephen. *Tobacco Master Settlement Agreement (1998): Overview, Implementation by States, and Congressional Issues.* Washington, DC: Congressional Research Service, 1999.

Redish, Martin H. *The Constitution as Political Structure.* New York: Oxford University Press, 1995.

Rheingold, Paul D. "The MER/29 Story—An Instance of Successful Mass Disaster Litigation." *California Law Review* 56 (1968): 116–48.

Robinson, Glen O. "Multiple Causation in Tort Law: Reflections on the DES Cases." *Virginia Law Review* 68 (1982): 713–69.

Roscoe, Douglas D., and Shannon Jenkins. "A Meta-Analysis of Campaign Contributions' Impact on Roll Call Voting." *Social Science Quarterly* 86 (2005): 52–68.

Rosenberg, David. "The Causal Connection in Mass Exposure Cases: A 'Public Law' Vision of the Tort System." *Harvard Law Review* 97 (1984): 849–929.

Rosenberg, Gerald N. *The Hollow Hope: Can Courts Bring About Social Change?* 2nd ed. Chicago: University of Chicago Press, 2008.

Rossi, Jim. "Institutional Design and the Lingering Legacy of Antifederal Separation of Powers Ideals in the States." *Vanderbilt Law Review* 52 (1999): 1167–1240.

Sandler, Ross, and David Schoenbrod. *Democracy by Decree: What Happens When Courts Run Government.* New Haven: Yale University Press, 2003.

Sargent, James D., Madeline Dalton, Eugene Demidenko, Peter Simon, and Robert Z. Klein. "The Association between State Housing Policy and Lead Poisoning in Children." *American Journal of Public Health* 89 (1999): 1690–95.

Schroeder, Steven A. "Tobacco Control in the Wake of the 1998 Master Settlement Agreement." *New England Journal of Medicine* 350 (2004): 293–301.

Schuck, Peter H. *The Limits of Law: Essays on Democratic Governance.* Boulder: Westview Press, 2000.

Schwartz, Gary T. "Considering the Proper Federal Role in American Tort Law." *Arizona Law Review* 38 (1996): 917–51.

Sive, David. "The Litigation Process in the Development of Environmental Law." *Pace Environmental Law Review* 13 (1995): 1–32.

Sloan, F. A., C. A. Mathews, and J. G. Trogdon. "Impacts of the Master Settlement Agreement on the Tobacco Industry." *Tobacco Control* 13 (2004): 356–61.

Sobel, Robert. *They Satisfy: The Cigarette in American Life.* Garden City, NY: Anchor Books, 1978.

Spencer, J. R. "Public Nuisance—A Critical Examination." *Cambridge Law Journal* 48 (1989): 55–84.

Spengler, John D., and Ken Sexton. "Indoor Air Pollution: A Public Health Perspective." *Science* 221 (1983): 9–17.

Stein, Marcia L. "Cigarette Products Liability Law in Transition." *Tennessee Law Review* 54 (1987): 631–70.

Talbott, John H. "Smoking and Lung Cancer." *Journal of the American Medical Association* 171 (1959): 2104.

Tarr, G. Alan. *Understanding State Constitutions.* Princeton: Princeton University Press, 1998.

Teret, Stephen P. "Litigating for the Public's Health." *American Journal of Public Health* 76 (1986): 1027–29.

Thomson, Judith Jarvis. "The Decline of Cause." *Georgetown Law Journal* 76 (1987): 137–50.

Tobacco Industry Research Committee. "Confidential Report—Tobacco Industry Research Committee Meeting." Oct. 19, 1954. Bates Nos. CTRMN007295–97. Available at http://tobaccodocuments.org/ctr/CTRMN007295-7297.html.

Tobacco Industry Research Committee. "A Frank Statement to Cigarette Smokers." Jan. 4, 1954. Available at http://tobaccodocuments.org/ness/10245.html?pattern=frank%5Ba-z%5D%2A%5CW%2Bstatem%5Ba-z%5D%2A&#p4.

Todd, J. S., D. Rennie, R. E. McAfee, L. R. Bristow, J. T. Painter, T. R. Reardon, D. H. Johnson Jr., et al. "The Brown and Williamson Documents: Where Do We Go from Here?" *Journal of the American Medical Association* 274 (1995): 256–58.

Tribe, Laurence. *American Constitutional Law.* 3rd ed. Vol. 1. New York: Foundation Press, 2000.

Trichopoulos, Dimitrios, Anna Kalandidi, Loukas Sparros, and Brian MacMahon. "Lung Cancer and Passive Smoking." *International Journal of Cancer* 27 (1981): 1–4.

U.S. Department of Commerce. *Interagency Task Force on Product Liability: Final Report.* Washington, DC: Department of Commerce, 1978.

U.S. Department of Commerce. *United States Government Master Specification for Paint, White, and Tinted Paints Made on a White Base, Semipaste, and Ready Mixed, Federal Specifications Board Stand. and Spec. No. 10B.* 3rd ed. Circular of the Bureau of Standards, no. 89. Washington, DC: Government Printing Office, 1927.

U.S. Department of Health and Human Services. *The Health Consequences of Involuntary Smoking: A Report of the Surgeon General.* Washington, DC: Government Printing Office, 1986.

U.S. Department of Health and Human Services. *The Health Consequences of Smoking: Chronic Obstructive Lung Disease; A Report of the Surgeon General.* Washington, DC: Government Printing Office, 1984. Available at http://profiles.nlm.nih.gov/NN/B/C/C/S/_/nnbccs.pdf.

U.S. Department of Health and Human Services. *The Health Consequences of Smoking: Nicotine Addiction; A Report of the Surgeon General.* Washington, DC: Government Printing Office, 1988. Available at http://profiles.nlm.nih.gov/NN/B/B/Z/D/_/nnbbzd.pdf.

U.S. Department of Health, Education, and Welfare. *Smoking and Health: Report of the Advisory Committee to the Surgeon General of the Public Health Service.* Washington, DC: Government Printing Office, 1964.

Vandall, Frank J. "The Legal Theory and the Visionaries that Led to the Proposed $368.5 Billion Tobacco Settlement." *Southwestern University Law Review* 27 (1998): 473–85.

Vernick, Jon S., Julie Samia Mair, Stephen P. Teret, and Jason W. Sapsin. "Role of Litigation in Preventing Product-Related Injuries." *Epidemiologic Reviews* 25 (2003): 90–98.

Viscusi, W. Kip. *Smoke-Filled Rooms: A Postmortem on the Tobacco Deal.* Chicago: University of Chicago Press, 2002.

Wade, John W. "Environmental Protection, the Common Law of Nuisance, and the Restatement of Torts." *Forum* 8 (1972): 165–74.

Wagner, Wendy E. "Rough Justice and the Attorney General Litigation." *Georgia Law Review* 33 (1999): 935–77.

Wagner, Wendy E. "When All Else Fails: Regulating Risky Products through Tort Litigation." *Georgetown Law Journal* 95 (2007): 693–732.

Walker, Laurens, and John Monahan. "Sampling Liability." *Virginia Law Review* 85 (1999): 329–51.

Warren, Christian. *Brush with Death: A Social History of Lead Poisoning.* Baltimore: Johns Hopkins University Press, 2000.

Weinrib, Ernest J. "Corrective Justice." *Iowa Law Review* 77 (1992): 403–26.

Weinrib, Ernest J. *The Idea of Private Law.* Cambridge, MA: Harvard University Press, 1995.

Weinstein, Jack B. "Revision of Procedure: Some Problems in Class Actions." *Buffalo Law Review* 9 (1960): 433–70.

Wilson, Jonathan, Tim Pivetz, Peter Ashley, David Jacobs, Warren Strauss, John Menkedick, Sherry Dixon, et al. "Evaluation of HUD-funded Lead Hazard Control Treatments at 6 Years Post-Intervention." *Environmental Research* 102 (2006): 237–48.

Wilson, Joy Johnson. *Summary of the Attorneys General Master Tobacco Settlement Agreement.* National Conference of State Legislatures, Mar. 1999. http://www.tobacco.org/News/settlementresources.html.

Witt, John Fabian. "Toward a New History of American Accident Law: Classical Tort Law and the Cooperative First-Party Insurance Movement." *Harvard Law Review* 114 (2001): 690–841.

Woolhandler, Ann, and Michael G. Collins. "State Standing." *Virginia Law Review* 81 (1995): 387–520.

Wright, John R. "Campaign Contributions and Congressional Voting on Tobacco Policy, 1980–2000." *Business and Politics* 6, no. 3 (2004): 1–26.

Wright, John R. *Interest Groups and Congress.* Boston: Allyn and Bacon, 1996.

Wright, Richard W. "Causation, Responsibility, Risk, Probability, Naked Statistics, and Proof: Pruning the Bramble Bush by Clarifying the Concepts." *Iowa Law Review* 73 (1988): 1001–77.

Wynder, Ernst L., and Evarts A. Graham. "Tobacco Smoking as a Possible Etiologic Factor in Bronchiogenic Carcinoma: A Study of 684 Proved Cases." *Journal of the American Medical Association* 143 (1950): 329–36.

Zegart, Dan. *Civil Warriors: The Legal Siege on the Tobacco Industry.* New York: Delacorte Press, 2000.

Selected Judicial Opinions

Abel v. Eli Lilly & Co., 343 N.W.2d 164 (Mich. 1984).

Action on Smoking and Health v. Harris, 655 F.2d 236 (D.C. Cir. 1980).

Agency for Health Care Admin. v. Associated Indus. of Fla., Inc., 678 So. 2d 1239 (Fla. 1996).

Alfred L. Snapp & Son, Inc. v. Puerto Rico, 458 U.S. 592 (1982).

Amchem Prods., Inc. v. Windsor, 521 U.S. 591 (1997).

Baker v. Carr, 369 U.S. 186 (1962).

Blankenship v. Gen. Motors Corp., 406 S.E.2d 781 (W. Va. 1991).

Borel v. Fibreboard Paper Prods. Corp., 493 F.2d 1076 (5th Cir. 1973), *reh'g denied,* 493 F.2d 1103 (5th Cir. 1974), *cert. denied,* 419 U.S. 869 (1974).

Brenner v. Am. Cyanamid Co., 699 N.Y.S.2d 848 (App. Div. 1999).

Brenner v. Am. Cyanamid Co., 732 N.Y.S.2d 799 (App. Div. 2001).

Brooks v. Lewin Realty III, Inc., 835 A.2d 616 (Md. 2003).

Brown & Williamson Tobacco Corp. v. FDA, 153 F.3d 155 (4th Cir. 1998).

California v. Gen. Motors Corp., No. C06-05755 MJJ, 2007 U.S. Dist. LEXIS 68547 (N.D. Cal. Sept. 17, 2007).

Camden County Bd. of Chosen Freeholders v. Beretta, U.S.A. Corp., 273 F.3d 536 (3d Cir. 2001).

Castano v. Am. Tobacco Co., 160 F.R.D. 544 (E.D. La. 1995).

Castano v. Am. Tobacco Co., 84 F.3d 734 (5th Cir. 1996).

Cent. Valley Chrysler-Jeep, Inc. v. Witherspoon, No. CV F 04-6663 AWI LJO, 2007 U.S. Dist. LEXIS 3002 (E.D. Cal. Jan. 16, 2007).

Cimino v. Raymark Indus., 751 F. Supp. 649 (E.D. Tex. 1990).

Cimino v. Raymark Indus., 151 F.3d 297 (5th Cir. 1998).

Cipollone v. Liggett Group, 505 U.S. 504 (1992).

City of Boston v. Smith & Wesson Corp., No. 1999-02590, 2000 Mass. Super. LEXIS 352 (Mass. Super. Ct. July 13, 2000).

City of Chicago v. Am. Cyanamid Co., 823 N.E.2d 126 (Ill. 2005), *appeal denied*, 833 N.E.2d 1 (Ill. 2005).

City of Chicago v. Beretta U.S.A. Corp., 821 N.E.2d 1099 (Ill. 2004).

City of Cincinnati v. Beretta U.S.A. Corp., 708 N.E.2d 1136 (Ohio 2002).

City of Milwaukee v. Illinois, 451 U.S. 304 (1981).

City of Milwaukee v. NL Indus., Inc., 691 N.W.2d 888 (Wis. Ct. App. 2004).

City of Philadelphia v. Beretta U.S.A. Corp., 277 F.3d 415 (3d Cir. 2002).

City of Philadelphia v. Lead Indus. Ass'n, 994 F.2d 112 (3d Cir. 1993).

City of San Francisco v. Philip Morris, Inc., 957 F. Supp. 1130 (N.D. Cal. 1997).

City of St. Louis v. Benjamin Moore & Co., 226 S.W.3d 110 (Mo. 2007).

Collins v. Eli Lilly Co., 342 N.W.2d 37 (Wis. 1984).

Connecticut v. Am. Elec. Power Co., 406 F. Supp. 2d 265 (S.D.N.Y. 2005).

County of Santa Clara v. Atl. Richfield Co., 40 Cal. Rptr. 3d 313 (Ct. App. 2006).

County of Santa Clara v. Super. Ct. of Santa Clara County, 74 Cal. Rptr. 3d 842 (Ct. App. 2008).

Dames & Moore v. Regan, 453 U.S. 654 (1981).

District of Columbia v. Beretta, U.S.A., Corp., 872 A.2d 633 (D.C. 2005).

Escola v. Coca Cola Bottling Co., 150 P.2d 436 (Cal. 1944).

Evans v. Gen. Motors Corp., 359 F.2d 822 (7th Cir. 1966).

Falise v. Am. Tobacco Co., 94 F. Supp. 2d 316 (E.D.N.Y. 2000).

FDA v. Brown & Williamson Tobacco Corp., 529 U.S. 120 (2000).

Fischer v. Johns-Manville Corp., 512 A.2d 466 (N.J. 1986).

Floyd v. Thompson, 227 F.3d 1029 (7th Cir. 2000).

Ganim v. Smith & Wesson Corp., 780 A.2d 98 (Conn. 2001).

Georgia v. Tenn. Copper Co., 206 U.S. 230 (1907).

Georgine v. Amchem Prods., Inc., 83 F.3d 610 (3d Cir. 1996).

Green v. Am. Tobacco Co., 154 So. 2d 169 (Fla. 1963).

Greenman v. Yuba Power Prods., Inc., 377 P.2d 897 (Cal. 1963).

Hall v. E. I. Du Pont de Nemours & Co., 345 F. Supp. 353 (E.D.N.Y. 1972).

Hamilton v. Accu-Tek, 62 F. Supp. 2d 802 (E.D.N.Y. 1999).

Hamilton v. Beretta U.S.A. Corp., 750 N.E.2d 1055 (N.Y. 2001).

Hawaii v. Standard Oil Co. of Cal., 405 U.S. 251 (1972).

Horton v. Am. Tobacco Co., 667 So. 2d 1289 (Miss. 1995).

Hymowitz v. Eli Lilly & Co., 539 N.E.2d 1069 (N.Y. 1989).

Ileto v. Glock, Inc., 349 F.3d 1191 (9th Cir. 2003).

In re "Agent Orange" Prod. Liab. Litig., 597 F. Supp. 740 (E.D.N.Y. 1984).

In re "Agent Orange" Prod. Liab. Litig., 373 F. Supp. 2d 7 (E.D.N.Y. 2005).

In re Asbestos Prods. Liab. Litig. (No. VI), 771 F. Supp. 415 (J.P.M.L. 1991).

In re Brooklyn Navy Yard Asbestos Litig., 971 F.2d 831 (2d Cir. 1992).

In re E. & S. Dists. Asbestos Litig., 772 F. Supp. 1380 (E.D. & S.D.N.Y. 1991).

In re Fibreboard Corp., 893 F.2d 706 (5th Cir. 1990).

In re Lead Paint Litig., 924 A.2d 484 (N.J. 2007).

In re Methyl Tertiary Butyl Ether ("MTBE") Prods. Liab. Litig., 175 F. Supp. 2d 593 (S.D.N.Y. 2001).

In re Methyl Tertiary Butyl Ether ("MTBE") Prods. Liab. Litig., 438 F. Supp. 2d 291 (S.D.N.Y. 2006).

In re New York Asbestos Litig., 145 F.R.D. 644 (S.D.N.Y. 1993).

In re Related Asbestos Cases, 543 F. Supp. 1152 (N.D. Cal. 1982).

In re Rhone-Poulenc Rorer, Inc., 51 F.3d 1293 (7th Cir. 1995).

In re Sch. Asbestos Litig., 789 F.2d 996 (3d Cir. 1986).

In re Simon II Litig., 211 F.R.D. 86 (E.D.N.Y. 2002), *rev'd*, 407 F.3d 125 (2d Cir. 2005).

James v. Arms Tech., Inc., 820 A.2d 27 (N.J. Super. Ct. App. Div. 2003).

Kansas v. Colorado, 206 U.S. 46 (1907).

Karjala v. Johns-Manville Prods. Corp., 523 F.2d 155 (8th Cir. 1975).

Larsen v. Gen. Motors Corp., 391 F.2d 495 (8th Cir. 1968).

Lartigue v. R. J. Reynolds Tobacco Co., 317 F.2d 19 (5th Cir. 1963).

Louisiana v. Texas, 176 U.S. 1 (1900).

Malcolm v. Nat'l Gypsum Co., 995 F.2d 346 (2d Cir. 1993).

Massachusetts v. EPA, 549 U.S. 497 (2007).

McGraw ex rel. State v. Am. Tobacco Co., Civ. A. No. 94-C-1707, (W. Va. Cir. Ct. June 6, 1995), 1995 WL 569618.

Menne v. Celotex Corp., 861 F.2d 1453 (10th Cir. 1988).

Merrill v. Navegar, Inc., 89 Cal. Rptr. 2d 146 (Ct. App. 1999).

Merrill v. Navegar, Inc., 28 P.3d 116 (Cal. 2001).

Missouri v. Illinois, 180 U.S. 208 (1901).

Missouri v. Illinois, 200 U.S. 496 (1906).

Morial v. Smith & Wesson Corp., 785 So. 2d 1 (La. 2001).

New Jersey v. City of New York, 283 U.S. 473 (1931).

New York v. New Jersey, 256 U.S. 296, 298 (1921).

Or. Laborers-Employers Health & Welfare Trust Fund v. Philip Morris, Inc., 185 F.3d 957 (9th Cir. 1999).

Ortiz v. Fibreboard Corp., 527 U.S. 815 (1999).

Payton v. Abbott Labs., 437 N.E.2d 171 (Mass. 1982).

Pennsylvania v. West Virginia, 262 U.S. 553 (1923).

People v. Gold Run Ditch & Mining Co., 4 P. 1152 (Cal. 1884).

People v. Sturm, Ruger & Co., 761 N.Y.S.2d 192 (App. Div. 2003).

Pritchard v. Liggett & Myers Tobacco Co., 295 F.2d 292 (3d Cir. 1961).

Ross v. Philip Morris & Co., 328 F.2d 3 (8th Cir. 1964).

Rutherford v. Owens-Illinois, Inc., 941 P.2d 1203 (Cal. 1997).

Ryan v. Eli Lilly & Co., 514 F. Supp. 1004 (D.S.C. 1981).

Sabater v. Lead Indus. Ass'n, 704 N.Y.S.2d 800 (Sup. Ct. 2000).

Santiago v. Sherwin-Williams Co., 794 F. Supp. 29 (D. Mass. 1992).

Serv. Employees Int'l Union Health & Welfare Fund v. Philip Morris, Inc., 83 F. Supp. 2d 70 (D.D.C. 1999).

Sindell v. Abbott Labs., 607 P.2d 924 (Cal. 1980).

Skipworth v. Lead Indus. Ass'n, 690 A.2d 169 (Pa. 1997).

State v. Lead Indus. Ass'n, C.A. No. 99-5226, 2001 R.I. Super. LEXIS 37 (R.I. Super. Ct. Apr. 2, 2001).

State v. Lead Indus. Ass'n, C.A. No. 99-5226, 2005 R.I. Super. LEXIS 79 (R.I. Super. Ct. May 18, 2005).

State v. Lead Indus. Ass'n, C.A. No. 99-5226, 2005 R.I. Super. LEXIS 95 (R.I. Super. Ct. June 3, 2005).

State v. Lead Indus. Ass'n, C.A. No. PC 99-5226, 2007 R.I. Super. LEXIS 32 (R.I. Super. Ct. Feb. 26, 2007), *rev'd*, 951 A.2d 428 (R.I. 2008).

State v. Lead Indus. Ass'n, 951 A.2d 428 (R.I. 2008).

State ex rel Miller v. Philip Morris, Inc., 577 N.W.2d 401 (Iowa 1998).

Summers v. Tice, 199 P.2d 1 (Cal. 1948).

Station WCBS-TV, 8 F.C.C.2d 381 (1967), *aff'd sub nom. Banzhaf v. FCC*, 405 F.2d 1082 (D.C. Cir. 1968).

Texas v. Am. Tobacco Co., 14 F. Supp. 2d 956 (E.D. Tex. 1997).

Thomas v. Mallett, 701 N.W.2d 523 (Wis. 2005).

Tioga Pub. Sch. Dist. No. 15 v. U.S. Gypsum Co., 984 F.2d 915 (8th Cir. 1993).

United States v. Hooker Chems. & Plastics Corp., 680 F. Supp. 546 (W.D.N.Y. 1988).

United States v. Hooker Chems. & Plastics Corp., 722 F. Supp. 960 (W.D.N.Y. 1989).

United States v. Philip Morris USA Inc., 396 F.3d 1190 (D.C. Cir. 2005).

Urie v. Thompson, 337 U.S. 163 (1949).

Vieth v. Jubelirer, 541 U.S. 267 (2004).

Whitehouse v. Lead Indus. Ass'n, C.A. No. 99-5226, 2002 R.I. Super. LEXIS 43 (R.I. Super. Ct. Mar. 15, 2002).

Whitehouse v. Lead Indus. Ass'n, C.A. No. 99-5226, 2002 R.I. Super. LEXIS 90 (R.I. Super. Ct. July 3, 2002)

Wright v. Lead Indus. Ass'n, No. 94363042/CL190487, slip op. (Md. Cir. Ct. June 20, 1996).

Young v. Bryco Arms, 765 N.E.2d 1 (Ill. App. Ct. 2001), rev'd on other grounds, 821 N.E.2d 1078, 1091 (Ill. 2004).

Youngstown Sheet & Tube Co. v. Sawyer, 343 U.S. 579 (1952).

Selected Federal Statutes by Popular Name

Class Action Fairness Act of 2005, Public Law 109-2, U.S. Statutes at Large 119 (2005): 4 (codified at 28 U.S.C. §§ 1453, 1711–1715 (2005)).

Comprehensive Environmental Response, Compensation, and Liability Act, 42 U.S.C. §§ 9601–9675 (2006).

Consumer Product Safety Act, 15 U.S.C. §§ 2051–2089 (2006).

Federal Cigarette Labeling and Advertising Act, Public Law 89-92, U.S. Statutes at Large 79 (1965): 282 (codified as amended at 15 U.S.C. §§ 1331–1340 (2006)).

Lead-Based Paint Poisoning Prevention Act, Public Law 91-695, U.S. Statutes at Large 84 (1971): 2078 (codified as amended at 42 U.S.C. §§ 4821–4846 (2006)).

Protection of Lawful Commerce in Arms Act, Public Law 109-92, U.S. Statutes at Large 119 (2005): 2095 (codified at 15 U.S.C. §§ 7901–7903 (2006)).

Residential Lead-Based Paint Hazard Reduction Act, Public Law 102-550, U.S. Statutes at Large 106 (1992): 3897 (codified at 42 U.S.C. §§ 4851–4856 (2006)).

Universal Tobacco Settlement Act, S. 1415, 105th Cong., 1st sess. (Nov. 7, 1997).

Selected State Statutes and Legislative Materials

FLORIDA

Medicaid Third-Party Liability Act, 1994 Fla. Laws ch. 251, § 4 (codified at Fla. Stat. § 409.910 (9)(a) (repealed 1998))

MARYLAND

Md. Code Ann., Envir. §§ 6-801 to 6-852 (West 2002 & Supp. 2007).

NEW JERSEY

N.J. Stat. Ann. § 2A:58C-3(a)(2) (West 2000).

N.J. Stat. Ann. § 24:14A-5 (West 1997).

RHODE ISLAND

Lead Poisoning Prevention Act, R.I. Gen. Laws §§ 23-24.6-1 to 23-24.6-27 (1996 & Supp. 2006).

Lead Hazard Mitigation Act, R.I. Gen. Laws § 42-128.1-1 to 42-128.1-13 (2006).

Rhode Island Department of Administration. Housing Resources Commission. Rules and Regulations Governing Lead Hazard Mitigation (Jan. 2006). Available at http://www.hrc.ri.gov/documents/LHMR%2003.23.07%20Final.pdf.

Rhode Island Department of Health. *Rules and Regulations for Lead Poisoning Prevention*, R23-24.6-PB (Aug. 2007). Available at http://www2.sec.state.ri.us/dar/regdocs/released/pdf/DOH/4806.pdf.

2002 R.I. Pub. Laws ch. 187.

Selected Court Documents

Amended Complaint, *State v. Lead Indus. Ass'n*, C.A. No. 99-5226 (R.I. Super. Ct. Oct. 14, 1999).

Brief of Appellant NL Industries on Public Nuisance, *State v. Lead Indus. Ass'n*, No. SU-07-121-A (R.I. Jan. 31, 2008). Available as attachment to electronic version of "Sherwin-Williams, NL: Rhode Island Lawsuit Barred; Errors Require New Trial," *Mealey's Litigation Report: Lead* 17, no. 4 (Feb. 2008), 1.

Brief of Appellee on Public Nuisance Claim, *State v. Lead Indus. Ass'n*, No. SU-07-121-A (R.I. Mar. 17, 2008). Available as attachment to electronic version of "Rhode Island Says Nuisance Is a Valid Theory, Product Liability Not Applicable," *Mealey's Litigation Report: Lead* 17, no. 6 (Apr. 2008), 1.

Brief of Coalition for Public Nuisance Fairness and Property Casualty Insurers Association of America as Amici Curiae in Support of Appellants, *State v. Lead Indus. Ass'n*, No. 07-121A (R.I. Sup. Ct. Jan. 31, 2008). Available as attachment to electronic version of "Sherwin-Williams, NL: Rhode Island Lawsuit Barred; Errors Require New Trial," *Mealey's Litigation Report: Lead* 17, no. 4 (Feb. 2008), 1.

Brief of the State of Rhode Island 3, *State v. Lead Indus. Ass'n, Inc.*, 951 A.2d 428 (R.I. 2008) (No. 2004-63-M.P.), 2005 R.I. S. Ct. Briefs LEXIS 12.

Complaint, *City of Chicago v. Beretta U.S.A. Corp.*, No. 98-CH15596 (Ill. Cir. Ct. Nov. 12, 1998). Available at http://www.gunlawsuits.com/downloads/chicago.pdf.

Complaint, *Moore ex rel. State v. Am. Tobacco Co.*, No. 94-1429 (Miss. Ch. Ct. May 23, 1994), ¶ 2. Available at http://www.library.ucsf.edu/sites/all/files/ucsf_assets/ms_complaint.pdf.

Complaint, *State v. Lead Indus. Ass'n*, C.A. No. 99-5226 (R.I. Super. Ct. Oct. 12, 1999). Available at http://www.riag.ri.gov/documents/reports/lead/lead_complaint.pdf. Reprinted in *Mealey's Litigation Report: Lead* 9, no. 1 (Oct. 15, 1999), app. D.

Complaints filed by the states in the tobacco litigation. Galen Digital Library, University of California, San Francisco. http://www.library.ucsf.edu/tobacco/litigation/states.html.

Master Settlement Agreement (1998). Available at http://ag.ca.gov/tobacco/pdf/1msa.pdf.

State of Rhode Island and Providence Plantations. Department of Attorney General. *Rhode Island Lead Nuisance Abatement Plan* (Sept. 14, 2007). Available at http://www.riag.state.ri.us/documents/RILeadNuisanceAbatementPlan9-14-07.pdf and as attachment to LexisNexis electronic version of "Rhode Island Defendants to Respond to \$2.4B Abatement Plan by Nov. 15," *Mealey's Litigation Report: Lead* 16, no. 12 (Oct. 2007), 7, 9.

Supplemental Brief of Appellee on Procedural and Factual Background, *State v. Lead Indus. Ass'n,* No. 07-121A (R.I. Mar. 17, 2008). Available as attachment to electronic version of "Rhode Island Says Nuisance Is a Valid Theory," *Mealey's Litigation Report: Lead* 17, no. 6 (Apr. 2008), 1.

Third Party Complaint, *State v. Lead Indus. Ass'n,* C.A. No. 99-5226 (June 25, 2001). Reprinted in *Mealey's Litigation Report: Lead* 10, no. 19 (July 3, 2001), app. D.

INDEX

Aaronson, Roberta Hazen, 1
abatement
   remediation of lead-based paint hazards,
      113–15, 147, 149, 153, 186–89, 191,
      193, 195, 212, 223–25
   remedy for nuisance, 140–41, 154,
      172–73, 185, 186–87, 188, 196, 210,
      215, 228
Action on Smoking and Health (ASH), 107,
   109
   Action on Smoking and Health v. Harris,
      109
administrative cost avoidance. *See* litigation
   costs in mass products litigation
adversarial legalism, 2, 125–26, 140
Alfred L. Snapp & Son v. Puerto Rico,
   125–26, 140
*Allegheny General Hospital v. Philip
   Morris, Inc.*, 130
alternative compensation systems, 58, 72,
   76, 166
   lead-poisoned children compensation
      system (proposed), 220, 227–28
   tobacco disease compensation system
      (proposed), 220, 222–23
alternative liability, 55, 59, 60–62, 78
Amchem Products, Inc. v. Windsor, 72,
   73–74, 75, 76
American Bar Association, 196
American Cancer Society (ACS), 22, 25,
   106, 107, 175
American Law Institute (ALI), 35, 39, 91.
   See also *Restatement (Second) of Torts*
American Medical Association (AMA), 21,
   23, 102, 220
American Public Health Association, 102
Ankiewicz v. Kinder, 44
antitrust violations
   claims in mass products litigation, 5, 122
   Master Settlement Agreement, chal-
      lenges to, 181, 182
asbestos
   concealment of risks by manufacturers,
      49
   exposure to, 3, 14, 22, 45–50, 54, 59–61,

   67–69, 72–74, 76–78, 83–84, 101, 149,
      166, 215
   insulation and other uses, 45–50
   micronite cigarette filter, contained in,
      22
asbestos litigation, 3, 4, 14, 22, 34, 45–54,
   59–62, 66–69, 72–74, 76–78, 83–84,
   101, 149, 166, 215
asbestosis, 22, 45, 47, 49, 52, 72
ASH. *See* Action on Smoking and Health
assumption of risk, 5, 39–40, 121
attorney general (state)
   authority to file claims, 9, 197, 205, 207,
      208, 209, 213
   competence to resolve public health
      problems, 199–200, 208–9
   discretion in choosing targets of *parens
      patriae* actions, 196–97
   ethical rules governing, 196–97
   legitimacy in addressing product-caused
      public health problems, 200–202
   parens patriae litigation, role in, 4, 9,
      121, 127–28, 136, 171–76, 185–88,
      189, 192–213, 218–19; authority inde-
      pendent of legislature and governor,
      205; judicial reliance on in structuring
      remedial decree, 172, 186, 198, 201–2,
      218; legitimacy of role as regulator,
      194, 199, 200–202, 218
   partnership with plaintiffs' attorneys, 4,
      101, 103, 104–5, 138, 166, 174, 176,
      177, 185, 194, 209
   regulator, conceiving of role as, 192,
      193–94
   separation of powers and, 9, 153,
      192–96, 200, 201, 204–7, 210–11, 218
   settlement negotiations, role in, 131–32,
      166, 174–78, 185, 192, 194, 197–98,
      214, 216, 218, 220
automobile safety
   air bags, 97, 99
   crashworthiness litigation, 3–4, 98–100
   federal regulation, debate over, 83,
      97–99, 103
   ignition interlock system, 97

automobile safety (*continued*)
  and public health perspective, 3–4, 83, 96–100, 103, 163

Baker v. Carr, 202–3
Baltimore, childhood lead poisoning in, 28, 29, 30
Banzhaf, John F., III, 106, 107
Barrett, Don, 5, 121
Bates, Katharine Lee, 13
Battelle Institute, 114
Bichler v. Eli Lilly & Co., 67
Black, Hugo, 206–7
Bonsack, James, 15
Borel, Clarence. See *Borel v. Fibreboard Paper Products Corp.*
Borel v. Fibreboard Paper Products Corp., 45–54, 59, 68, 74, 83, 86, 87, 101
Brandt, Allan M., 15, 23, 105
Brenner v. American Cyanamid Co., 63
Breyer, Stephen, 212
Brown & Williamson Tobacco Corp. v. FDA, 176
Buckley, James L., 97
Bureau of Alcohol, Tobacco, Firearms and Explosives, 134
Burney, Leroy, 23
Bush, George W., 76, 202

Calabresi, Guido, 56–57, 98, 160, 163
Camden County Board of Chosen Free-holders v. Beretta, U.S.A., 134
campaign contributions
  attorney general elections, contributions of plaintiffs' firms specializing in *parens patriae* litigation, 200, 205, 212–13, 221
  campaign finance reform, 219–20
  effect on voting, 110–11
  by rental property owners, 9, 117–18
  by tobacco industry, 9, 104–5, 110–11, 177–78, 185
Campaign for Tobacco-Free Kids, 72
Carcieri, Donald L., 191
Cardozo, Benjamin, 34, 167
Carson, Rachel, 18–19, 31
Carter, Grady, 131
Castano Safe Gun Litigation Group, 133
Castano v. American Tobacco Co., 71, 131–32, 175
catalytic converter, 17
causation

cause in fact: critique of requirement, 56–57; requirement that claimant identify specific party causing harm, 2–3, 5, 41–43, 49, 54–58, 59–68, 70, 77–79, 86–87, 120, 122, 124, 137, 216
  proximate, 43, 49, 135–36, 150–52
  superseding, 42–43, 49–51
Centers for Disease Control and Prevention, 25, 31, 178, 183
  childhood lead poisoning and, 30–31
  tobacco-related illnesses and, 25, 178, 183
Cherensky, Steven, 161
child advocates for lead poisoned children, 1
childhood lead poisoning
  and blood lead level, 17, 27, 30, 31, 66, 113, 115, 152, 223
  causes, 25–31
  history, 25–31
  and racism, 140
  See also childhood lead poisoning prevention
childhood lead poisoning prevention
  abatement of lead paint hazards, 113–15, 147, 149, 153, 186–89, 191, 193, 195, 212, 223–25
  in Baltimore, 28–30
  education, 116, 223
  elevated blood lead levels, 30–31
  federal government, proposed role of, 225–28
  history of understanding consequences of exposure to lead, 25–31
  interim controls, 113–15, 118, 190–91, 196, 212, 223–28
  legislation (state): Maryland, 114–17, 219–20, 224; Massachusetts, 5, 114–15, 117, 219, 224; New Jersey, 193; Rhode Island, 115, 155
  loans to rental property owners, 117
  pica, 27–28
  proposed legislative solutions, 223–27
  and public health perspective, 9, 29–31, 65, 83–84, 104, 112–15, 118–19, 138, 139–40, 144, 152, 154, 155, 164–65, 171–72, 211, 217, 218, 227
  regulation of housing conditions, 43, 112–15, 117, 138–39, 147, 152, 163, 164, 167, 186–88, 219, 223, 224, 226
  remediation of lead-based paint hazards, 5, 9, 85, 113, 115–17, 119, 164, 166,

186, 194, 195, 208, 210, 217, 219, 223–27

Churchill, Winston, 228–29

cigarettes. *See* tobacco products (cigarettes and other)

City of Chicago v. American Cyanamid Co., 43, 147

City of Chicago v. Beretta U.S.A. Corp., 150

City of Cincinnati v. Beretta U.S.A. Corp., 156

City of Manchester v. National Gypsum Co., 149

City of Milwaukee v. Illinois, 90

civil conspiracy, 55, 59, 66–68, 78, 141

Class Action Fairness Act, 76

class action lawsuits
  certification of class actions: basic requirements, 68, 70–74; mass products torts action, inability to certify, 75–76
  collective form of plaintiff and, 55, 68–70
  due process concerns, 70
  global settlements, 72–74
  as means to overcome inability of mass products tort victim to identify manufacturer of product causing harm, 2, 55, 68, 141
  See also Class Action Fairness Act

Clean Air Act, 17, 20, 84, 126

Clean Water Act, 84

climate change litigation. *See* global warming litigation

Clinton, Bill, 109, 122, administration of, 132

Coale, John P., 2

Cobb, Ty, 20

Columbian Exposition, 13

Common Cause, 117

common knowledge doctrine, 5, 40, 121, 124, 158, 221

common law
  challenges to traditional tort law in mass products torts, 1, 32, 41, 119, 128, 204, 217, 221, 225, 227
  process, nature of, 7, 32, 94, 103, 141–43, 159–60, 167–68, 209, 216–17

Comprehensive Environmental Response, Compensation, and Liability Act (CERCLA), 84–87, 94, 150

concert of action, 55, 59, 66–68

concurrent causation resulting in indivisible harm, 54, 59–60, 78–79

Congress
  and administrative agencies, 8, 90, 104–6, 109, 199, 201, 204, 213, 215, 218, 220, 228
  and Global Settlement Agreement (tobacco), 132, 136, 174–75, 185
  influence of lobbyists on, 9, 118, 185, 194
  powers to regulate, tax, and spend, 225
  regulation: automobile safety, 97–98; handguns, 2, 6, 133, 136–37; lead paint, 8, 85, 112, 116, 228; tobacco, 2, 8, 9, 105–12, 118, 132, 173–77, 192, 218, 219, 220–21
  testimony of tobacco company executives, 24, 108–9

Connecticut v. American Electric Power Co., 203

consolidation, 55, 76–78, 174

Consumer Product Safety Commission, 30

contingent fees, 122, 176, 189, 210–13, 228
  private counsel partnering with state attorney general and effect on public policy, 210–13, 228

contributory negligence, 39–41, 48, 159

corrective justice
  as goal of tort law generally, 57–58, 160, 166, 217
  in mass products cases, 58, 62–63, 167

costs of litigating mass products torts. *See* litigation costs in mass products litigation

crashworthiness litigation. *See* automobile safety: crashworthiness litigation

Curriden, Mark, 179

Daley, Richard, 134

Dames & Moore v. Regan, 209

Daynard, Richard, 121, 171, 173, 179

DDT, 1, 18–19

Department of Defense, and prohibition of smoking, 107

Department of Housing and Urban Development (HUD)
  regulation of lead-based paint in homes, 114, 116, 219
  study of prevalence of lead-based paint, 112

Department of Transportation, automobile safety and standards, 1, 97

Destefanis, Kevin, 155

diethylstilbestrol (DES), 62

Doll, Richard, 22, 23
Durbin, Richard, 110

Eisenberg, Melvin, 7, 167
Elektro the Moto-Man, 13
English, Peter, 27, 30
enterprise liability. *See* industry-wide liability
environmental harm, products as a cause of, 3, 83–84, 120
environmental law
  emergence of modern environmental law, 84–87
  impact on mass product torts, 3, 9, 83, 84, 85–87
  and modern regulatory framework, 84–86
  and *parens patriae* litigation, 89–91, 125–26
  and public nuisance (*see* public nuisance)
Environmental Protection Agency, 17, 84, 113, 126–27
  Massachusetts v. EPA, 126–27
environmental tobacco smoke (ETS), 106–7, 111, 112, 146, 222
Escola v. Coca Cola Bottling Co., 35
Evans v. General Motors Corp., 99
excise taxes
  cigarettes, 108, 183–84, 222
  paint, 116, 225–26, 228

failure to warn. *See* warning, product
Family Smoking Prevention and Tobacco Control Act, 220
Federal Cigarette Labeling and Advertising Act of 1965, 38, 105, 248, 295
Federal Communications Commission (FCC), 106, 194
Federal Employers' Liability Act, 51
Federal Trade Commission (FTC), 22, 105, 106, 182, 194
Federal Water Pollution Control Act, 90
Felstiner, W. L. F., 161
Food, Drug, and Cosmetic Act (FDCA), 110
Food and Drug Administration, 108–10, 112, 118, 121, 131–32, 173–77, 189–90, 194, 216, 220–21
  proposed regulation of cigarettes during 1990s, 109–10
Ford, Henry, 204

Frank, John P., 91
Freedom Holdings, Inc. v. Spitzer, 181
Friedman, Lawrence M., 33

Gangarosa, Raymond E., 121
Ganim v. Smith & Wesson Corp., 125, 135
Gauther, Wendell, 71
Geier v. American Honda Motor Co., 99
Georgia v. Tennessee Copper Co., 89, 125, 127
Gillete, Clayton P., 162
Glantz, Stanley, 179
Global Settlement Agreement (tobacco), 132, 136, 173–74, 177, 182, 185. *See also* tobacco litigation: settlement by tobacco manufacturers
global settlement agreements (asbestos), 72–74, 76
global warming litigation, 6, 90–91, 128, 165, 203
Graham, Evarts, 21–22, 171
Gregoire, Christine, 175–76, 179
Green v. American Tobacco Co., 37
guns and gun litigation
  claims: negligence, 60, 121, 134–35; public nuisance, 134, 146–47, 159
  lawsuits against manufacturers, 5, 6, 8, 100–104, 120, 128, 133–37, 146, 156, 158, 164, 219; Chicago, action by City of, 134; New Orleans, action by City of, 133–34
  legislative failure to regulate, 2, 6, 83, 104, 133–34, 164, 167, 219
  legislative response, 136–37 (*see also* Protection of Lawful Commerce in Arms Act)
  lobbying (pro-gun), 134, 135, 167 (*see also* National Rifle Association)
  obstacles to recovery by state and municipal governments: proximate causation, lack of, 135–36; remoteness of harms, 125, 135–36; standing, lack of, 125, 135–36
  as public health problem, 83, 134, 135, 164, 167

Haddon, William, 96–97
Haltom, William, 188–90, 217
Hamilton, Alice, 19, 28
Hammond, Cuyler, 22
Harfst, David L., 100
Harshbarger, Scott, 6

Hart, Henry, 142
Henderson, James A., 200
Hill, A. Bradford, 22
Hippocrates, 25
Holmes, Oliver Wendell, Jr., 33, 89
Horn, Daniel, 22
Horton v. American Tobacco Co., 39, 40
Hymowitz v. Eli Lilly & Co., 63

In re Brooklyn Navy Yard Asbestos Litigation, 76–78
In re Lead Paint Litigation, 150, 193
indemnity, and *parens patriae* litigation, 5, 122, 128–30, 141
industry-wide liability, 59, 66, 226
instrumental theory underlying tort liability, 56–57, 58, 62, 64, 93, 95, 148, 149, 150, 160–65. *See also* loss distribution; loss minimization
interim controls of lead-based paint hazards, 113–15, 118, 191, 212, 223–26, 228

Jackson, Robert, *Youngstown Sheet & Tube Co. v. Sawyer* analytical framework, 206–10
Joe Camel, 14, 173, 176
Johnston, James W., 24
joint and several liability, 49, 57, 59, 60–61, 66, 87, 94
judges
    aptitude for deciding mass products public health cases, 7, 8, 167–68, 199–200
    influences on decision making, 4, 141–42, 167, 198
justiciability
    generally, 90, 165, 202–3
    and political question in environmental cases, 90–91, 203
    and political question in *parens patriae* products litigation, 203–4
    and standing, 4, 7, 89, 120–29, 133, 135, 140, 199

Kagan, Robert, 2, 100
Kairys, David, 133, 135, 164
Keating, Gregory C., 32
Kelder, Graham, 171, 173
Kennedy, John F., 23, 96
Kessler, David, 109, 175, 177, 190, 221
Kluger, Richard, 16, 21

Koop, C. Everett, 23–24, 175
Krier, James, 162

Landes, William, 57
Larsen v. General Motors Corp., 99
Lartigue v. R. J. Reynolds Tobacco Co., 37
lead (including lead pigment and lead paint)
    exposure to, 13, 16–18, 25–31
    and gasoline, 16–17, 26–27, 31, 113
    history, 16
    and occupational poisoning, 19, 26
    paint containing lead pigment: European regulation of in early twentieth century, 28; federal ban on, 30, 112, 116, 118, 163; federal regulations requiring minimum amount in paint, 28–29; prevalence of, 5, 8, 112, 139, 186, 217
    tetraethyl lead (TEL), 16–17, 26–27
    uses of, 13, 16–18
    See also childhood lead poisoning; lead pigment and lead paint industries; tetraethyl lead (TEL)
Lead Industry Association (LIA), 38
lead pigment and lead paint industries
    history of industry, 17–18, 28, 30
    Lead Industries Association, 30
    limits on lead in paint, establishing, 30
    lobbying and political process, 43, 116
    See also lead
lead pigment litigation
    causation, 5, 49, 64–66, 67
    conflict between proposed remedy in *parens patriae* litigation and existing statutory scheme, 152–53
    contribution from property owners, seeking, 154–55
    legal bases of claims: civil conspiracy, 55, 59, 66–68, 78, 141; fraudulent misrepresentation, 140; indemnity, 5, 122, 129–31, 141; negligence, 140; negligent misrepresentation, 140; public nuisance, 140–41, 143–45, 147, 148, 149, 150–55, 158, 213, 216; strict products liability, 140, unjust enrichment, 129, 141
    New Jersey litigation, 7, 43, 150, 153, 157, 168, 193, 208, 216, 225 (see also *In re Lead Paint Litigation*)

lead pigment litigation (*continued*)
  Rhode Island litigation, 1–3, 4, 7, 8, 123,
    129–30, 138–59, 163, 164, 167, 168,
    171–73, 185–91, 193, 195, 200, 201,
    203, 208, 210–14, 215–18, 225; attor-
    ney general's proposed abatement
    plan, 8, 172–73, 185–88, 194–96, 217
    (see also *State v. Lead Industries
    Ass'n*)
lead poisoning. *See* childhood lead poison-
    ing
lead-based paint hazards, remediation of,
    113
  abatement, 113–15, 147, 149, 153,
    186–89, 191, 193, 195, 212, 223–25
  interim controls, 113–15, 118, 191, 212,
    223–26, 228
  in owner-occupied residences, 224
  party responsible for bearing costs, 113,
    115–17, 148, 152–54, 191, 193, 203,
    224–27
  Rhode Island attorney general's proposal,
    172–73, 185–88, 194–96
legal process school, 142–43
legislation
  conflict with remedy sought by state in
    litigation, 152–53
  failure of legislation, 104–5, 215
  lead poisoning prevention, proposed,
    223–28; compensation for lead-poi-
    soned children, 227–28; federal excise
    tax, 228; remediation of lead-based
    paint hazards, 223–27
  lead poisoning prevention, state and lo-
    cal: Maryland, 114–17; Massachusetts,
    114–15, 117; New Jersey, 43, 153, 193;
    Rhode Island, 153, 190, 191, 193
  process, reform of legislative, 219–20
  tobacco control: Family Smoking Pre-
    vention and Tobacco Control Act, 220;
    federal, 38, 39, 40, 173, 220; proposed,
    220–23
Lenau, Gerald, 155
litigation costs in mass products litigation,
    166–67
Llewellyn, Karl, 142
lobbying
  of attorneys general, 211–13
  childhood lead poisoning: for lead poi-
    soned child, 28, 43, 220; for manufac-
    turers, 43, 118; for rental property
    owners, 9, 117–18

public interest, generally, 117, 220
  tobacco industry, 9, 40, 104–8, 111, 173,
    175, 176, 184, 185
  victims of tobacco-related diseases, on
    behalf of, 220
Lord, Elizabeth, 29
loss distribution, 35, 49, 56, 62, 160, 165,
    178, 179, 217
loss minimization, 49, 62, 98, 160–65
Lott, Trent, 131, 132, 177
Love Canal, 85, 93–95, 149
lung cancer
  asbestos exposure as a cause, 22, 59–60,
    72
  smoking as a cause, 5, 13, 21–25, 33, 38,
    40, 54, 106, 183, 222
Lynch, Patrick C., 172
Lytton, Timothy D., 188

MacPherson v. Buick Motor Co., 34
manufacturing defects, 38, 159
Marlboro Man, 14, 173
market share liability, 2, 55, 59, 62–66, 158.
  See also risk contribution theory
Maryland Reduction of Lead Risk in Hous-
    ing Act, 114–15
Mashaw, Jerry L., 180
Massachusetts lead poisoning prevention,
    114–15, 117
Massachusetts v. EPA, 126–27
mass products plaintiffs' attorneys
  distortion of public policy resulting from
    participation in *parens patriae* litiga-
    tion, 210–13
  fees from tobacco litigation, 184
  lobbying of state attorneys general,
    211–24
  as partners of state attorneys general in
    parens patriae litigation, 2, 210–13,
    215
  specialty, emergence as, 4, 101–2
Master Settlement Agreement (MSA), 8,
    132, 138, 166, 175–85, 189, 190, 194,
    201, 216, 218–20
  antitrust challenges, 181–82
  and cartelization of tobacco industry,
    179–82
  creating financial partnership between
    tobacco companies and states, 179
  efficacy in achieving goals of, 178–85,
    216; compensation and loss distribu-
    tion, 171, 182–84; reducing smoking,

171, 178–82; retribution, 171, 179–82;
transaction costs, minimizing, 184–85
enforcement of provisions by states,
181–82
financial success of industry since agreement, 179–80
impact on cigarette prices, 180, 183
and legal challenges, 181–82
securitization of proceeds, 179
See also tobacco litigation: settlement by
tobacco manufacturers
Mather, Lynn, 188
Mathews, C. A., 180
McCain, John, tobacco legislation in response to Global Settlement Agreement, 175, 182
McCann, Michael, 8, 189, 190, 217
McConnell, Jack, 138, 144
Medicaid
childhood lead poisoning, 5, 7, 123, 166,
171, 226
recovery of funds in *parens patriae* litigation, 122, 123, 124, 127, 171
tobacco-related disease, 5, 123, 166, 178,
182–84
Menne v. Celotex Corp., 60
Merrill, Thomas W., 112
Merryman, Walker, 24
mesothelioma, 22, 45, 49, 60, 61, 72
Michie v. Great Lakes Steel Division, 97
Missouri v. Illinois, 89
Mollenkamp, Carrick, 177
Moore, Michael C., 1, 121, 171, 175, 177
Morial, Marc, 134
Motley, Ron, 7, 138, 139, 171, 173, 178,
179
Moynihan, Daniel P., 97
Müller, Franz Hermann, 21
Myers, Matt, 72

Nader, Ralph, 97
Nagareda, Richard, 76
National Anti-Cigarette League, 20
National Center for Healthy Housing, 114
National Environmental Policy Act, 84
National Highway Safety Bureau, 96
National Resources Defense Council
(NRDC), 17, 92
National Rifle Association (NRA), 6, 134,
136
National Traffic and Motor Vehicle Safety
Act, 97

negligence per se for violation of interim
control standards, 227–28
New Jersey Lead Paint Act, 153, 193
New Jersey Supreme Court, 7, 43, 150,
153, 157, 168, 193, 208, 216
New York World's Fair (1939). *See* World
of Tomorrow World's Fair
Nicolet, Inc. v. Nutt, 67
no-fault compensation systems. *See* alternative compensation systems

Obama, Barack, 220
occupational health
history, 18–19
impact on mass product torts, 44, 51
and lead, 26, 28
O'Connor, Sandra Day, 110
Ortiz v. Fibreboard Corp., 72, 74, 75
Osler, William, 19
Owen, David G., 98
Oxycontin, lawsuits against manufacturer, 6

paint. *See* lead
parens patriae litigation, 1, 2, 3–8, 38, 45,
54–58, 61, 62, 89, 91, 104–5, 118,
120–41, 153, 157–60, 164–67, 171–72,
175, 178–85, 188–229
as alternative to legislation, 4, 9, 104–5,
118, 139, 153, 193–94 (*see also* attorney general: as regulator)
circumventing challenges facing individual plaintiffs, 3, 5, 38, 45, 54–58,
61–62, 123–28, 134–35, 140–41,
157–58
conflict with existing statutory framework, 152–53, 194
effect on political processes of reform,
188–91, 217–18
and environmental cases, 89–91,
125–126
history, 89–90, 120–23, 124–26
information within manufacturer's control related to product risk, release of,
189–91
political process, effect on, 188–91
and private attorneys partnered with
state attorneys general, 210–13,
215
and products, 1, 120–23, 127–31,
133–37, 138–41, 192–205
recognition by Supreme Court, 4,
89–90

parens patriae litigation (*continued*)
  separation of powers (*see* attorney general: separation of powers and; separation of powers)
  and settlements, 131–33, 175–78
  and standing, 4, 7, 89, 120, 121–22, 123–27, 128, 140–41, 199
  tobacco manufacturers, claims against, 120–28
  See also guns and gun litigation; lead pigment litigation; tobacco litigation
Parker, James M., 94
Parker, Robert M., 68–70, 73, 78
Payton v. Abbott Laboratories, 41
People v. General Motors Corp., 90
People v. Gold Run Ditch & Mining Co., 89
Perry v. American Tobacco Co., 129
Pershing, John "Black Jack," 14
pharmaceutical manufacturers, litigation involving, 6, 62
Philadelphia Children's Hospital, 27
Philip Morris, 21, 23, 24, 25, 32, 36, 72, 181
"pica," 27
Pitofsky, Robert, 182
plaintiffs' attorneys. *See* mass products plaintiffs' attorneys
political question. *See* justiciability
Posner, Richard, 57, 71, 98, 197
preemption, federal, 38, 204–5
President's Committee for Traffic Safety, 96
privity, 34–36
products liability, traditional, 3, 33–44
products litigation as public health measure, 2, 3, 102–3
Prosser, William, 41, 91, 92, 93, 149, 155
Protection of Lawful Commerce in Arms Act, 133, 136–37
Public Citizen, 110
  Health Research Group, 111
public health, history of movement, 1, 18–19
Public Health Cigarette Smoking Act of 1969, 38, 106
public nuisance, 1, 6, 7, 87–95, 116, 128–29, 131, 133, 134, 138–41, 143, 144–52, 155–67, 172, 186, 187, 192, 193, 195, 196, 199, 203, 208, 209, 210, 213, 216, 217, 227, 228
  crime as a, 88
  and environmental lawsuits, 88–91, 93–95
  evaluation of possible application in

parens patriae against product manufacturers, 159–67, 216–17; and corrective justice, 7, 166–67; and loss distribution, 165; and loss minimization, 160–65
  lead pigment litigation, 141, 144–52, 156–59
  origins of, 88–89, 145–46
  parens patriae lawsuits, as claim in, generally, 6, 96, 116, 140–41, 159–67
  private nuisance, distinguished from, 87
  products liability, potential overlap with, 157–59
  public health litigation and, 96, 116
  requirements for liability, 139; control of instrumentality, 148–50; defendant's conduct, nature of, 155–57; as land-based tort, 150; proximate causation, 150–52; public right, 144–48
  and *Restatement*, 91–93, 144–45, 149, 155–57
  statutes declaring activity to be public nuisance, 88
  statutes negating possible application in actions against product manufacturers, 193
  tobacco litigation, 6, 122, 128–29

Queensland study of effects of lead paint, 28

Rabin, Robert L., 36
Racketeer Influenced and Corrupt Organizations Act (RICO), 5, 122
Ramazzini, Bernadino, 25
RAND Corporation Institute for Civil Justice, 166
Reagan, Ronald, 23, 31, 112
Reich, Robert B., 8–9
remedial decrees in *parens patriae* products litigation, 8, 172, 185–88, 191, 194–95, 198, 199–201, 218
  constitutional implications of, 202, 204
  See also lead pigment litigation: Rhode Island litigation: attorney general's proposed abatement plan
Remediation of lead-based paint hazards. *See* lead-based paint hazards, remediation of
Residential Lead-Based Paint Hazard Reduction Act, 85

Resource Conservation and Recovery Act
of 1976, 84
Restatement (Second) of Torts
and the environmental movement, 91–93
and public nuisance provision, 91–93,
144, 146, 149, 155, 156
Section 402A, 35, 38, 40, 45, 46, 48, 53
tobacco products, provision specifically
addressing, 40
restitution. *See* unjust enrichment
Rhode Island lead pigment litigation. *See*
lead pigment litigation; *State v. Lead
Industries Ass'n*
Rhode Island lead poisoning prevention, 1,
7–8, 139–40, 153, 190, 191, 193
attorney general's proposed abatement
plan, 172–73, 185–88, 194–96
See also lead pigment litigation; *State v.
Lead Industries Ass'n*
Rice, Joe, 176
risk contribution theory, 64–66, 158. *See
also* market share liability; *Thomas v.
Mallett*
R. J. Reynolds, 21, 23, 24, 37
Roberts, John, 125
Rosenberg, David, 41–42, 54, 61
Ross v. Philip Morris & Co., 36
Rutherford v. Owens-Illinois, Inc., 59–60
Ryan v. Eli Lilly & Co., 66

Sacks, Albert, 142
Safe Drinking Water Act, 84–85
Santiago v. Sherwin-Williams Co., 67
Schuck, Peter H., 199–200
Schwartz, Gary, 194
Scruggs, Dickie, 131, 177
secondhand smoke. *See* environmental
tobacco smoke
securitization of tobacco settlement proceeds by states, 179
separation of powers, 153, 201, 202,
204–10
attorney general acting in legislative role,
8–9, 192–205, 218–19
federal structure, 206–10
federalism, intertwined with, 204–5
state constitutional requirements, 205
state systems, 205–10
U.S. Constitution, impact on state separation of powers, 206
See also justiciability; *Youngstown Sheet
& Tube Co. v. Sawyer*

Siegelman, Peter, 161
Silent Spring, 18–19, 31
Silverstein, Michael, 140, 143–44, 148, 151,
154, 157, 172
Sindell v. Abbott Laboratories, 62–64
Skipworth v. Lead Industries Association,
42, 63
Sloan, F. A., 180
Soper, Philip, 32
Spitzer, Eliott, 181–82, 197
Freedom Holdings, Inc. v. Spitzer,
181–82
State v. Lead Industries Ass'n, 1, 2, 129,
130, 138, 140, 141, 143, 144, 148, 154,
155, 157, 171–72, 175, 189, 216
State v. Schenectady Chemicals, Inc., 94,
95
statutes of limitations, 44, 51, 52, 69, 75,
124, 140, 141, 162
and latent diseases, 44, 51, 52, 69, 124
Stevens, John Paul, 127
strict products liability
common knowledge doctrine, 5, 40, 121,
124, 158, 221
defects, 38, 62, 69, 96, 98, 146, 159
historical development, 33–34, 35–36, 40
misuse of product as defense, 48
See also *Restatement (Second) of Torts:
Section 402A*
subrogation, 124, 157, 158
Summers v. Tice, 60–61
surgeon general reports on smoking
*The Health Consequences of Smoking:
Chronic Obstructive Lung Disease, A
Report of the Surgeon General 1984*,
23
*The Health Consequences of Smoking:
Nicotine Addiction; A Report of the
Surgeon General 1988*, 23–24
*Smoking and Health: Report of the Advi-
sory Committee to the Surgeon Gen-
eral of the Public Health 1964*, 23–35,
38, 39, 105, 106, 222
Synar, Mike, 107
Synar Amendment, 107

Teret, Stephen P., 182
tetraethyl lead (TEL), 16 17, 26–27
third parties, conduct of, in products cases,
42–43, 49–51
Thomas v. Mallett, 64–66
Tisch, Andrew, 24

tobacco industry
  congressional testimony by executives, 24
  history, 15–16, 21, 22–24
  lobbying and campaign contributions, 9, 40, 104–8, 110–12, 173, 175, 176, 184, 185
  manipulation of information by manufacturers, 22–25
Tobacco Industry Research Committee, 23
tobacco litigation, 1, 35–41
  antismoking activists, role of, 105, 106, 108, 112, 177, 179, 190, 221
  attorney fees, 122, 131, 166, 177, 184, 211
  claims: antitrust violations, 5, 122; common-law misrepresentation, 5, 122; deceptive advertising, 5, 122; indemnity, 5, 122, 129; public nuisance, 5, 122, 128–29, 131, 134, 146; Racketeer Influenced and Corrupt Organizations Act (RICO), 5, 122; unfair trade practices, 5; unjust enrichment, 5, 122
  federal government, role of, 131–32
  individuals, obstacles facing claims brought by, 1, 5, 35–41, 54, 124
  Mississippi litigation against manufacturers, 1, 5–6, 108, 118, 121–23, 129, 131, 171, 175, 177, 182, 190 (see also parens patriae litigation)
  "morality play," earlier litigation as, 35–41
  settlement by tobacco manufacturers, 6, 131–33, 173–84; as lawmaking process, 176–78 (see also Global Settlement Agreement; Master Settlement Agreement)
  state litigation, 120–28, 139 (see also parens patriae litigation; tobacco litigation: Mississippi litigation against manufacturers)
tobacco products (cigarettes and other)
  advertising and labeling, 5, 14, 15–16, 38, 105, 106, 110, 122, 171, 173, 175, 176, 178–80, 192, 194, 216, 220
  ammonia as additive, 24, 108
  cancer link, 5, 21–25, 36, 37, 38, 40, 54, 105, 107
  discovery of, 20–24
  diseases, related, and other health risks, 23, 106, 112, 146
  nicotine, 15, 22, 24, 71, 72, 108–9, 120,

131, 171, 174, 179, 189–90, 220, 222; addiction to, 15, 24; potency, manipulation of, 71, 108–9, 120, 131, 171, 189–90, 222
  promotion and marketing after MSA, 179
  and public health perspective, 20–25, 102
  smoking rates, 14, 16, 22, 25, 178
  and women, 14, 23
  and youth, 16, 173, 174, 176, 177, 178, 182, 185
tobacco regulation (other than litigation)
  Action on Smoking and Health (ASH), 107, 109
  and advocacy against smoking, 20, 109, 112, 175
  and age verification, 109
  and ban on airline flights, 106
  congressional testimony of manufacturers' executives, 108–9, 190
  federal, 105–10; exemption from broad-based regulatory statutes, 106, Federal Communications Commission, 106; Federal Trade Commission, 22, 105; Food and Drug Administration, proposed regulation during 1990s, 109–10; restrictions on sale and distribution of, 106–7, 109–10, 220–21; warnings, cigarette package, 105–6, 173, 220; youth, prohibition of sale to, 107, 109, 173, 176
  lobbying: by antitobacco activists, 106, 177, 179; by tobacco industry, 9, 40, 104–8, 111, 173, 175, 176, 184, 185
  proposed by Kessler, 221
  proposed by Viscusi, 220–21
  state and local regulation, 106–7
  taxation of, 108
tort law, traditional, 1–2, 32–34
  regulatory effect of, 193–94
Toxic Substances Control Act (TSCA) of 1976, 84
Trask, George, 20
Traynor, Roger, 35, 37, 52
Tribe, Laurence, 305
Trogdon, J. G., 180

unfair trade practices statutes, 5, 141
United States v. Hooker Chemicals & Plastics Corp., 93–95

unjust enrichment, 5, 122, 128–29, 131, 141

Urie v. Thompson, 51–52

Viscusi, W. Kip, 182–83, 184, 220–21

Wagner, Wendy, 189–90
warning, product
    asbestos, hazards of, 45–50, 60, 61, 62, 67
    duty owed to ultimate consumer or user, 38, 46, 47–48, 52–53, 62, 67, 160
    tobacco, hazards of, 38, 39, 40, 85, 105–6, 173, 220
Weiner, Charles, 73
Weinrib, Ernest, 57–58
Weinstein, Jack B., 66, 75, 77, 78

welfare economics, 56
Whitehouse, Sheldon, 45–53, 59, 86
Wigand, Jeffrey, 108
Williams, Huntington, 29
Williams, Merrell, 112, 190
Wisdom, John Minor, 45–53, 59, 86
Witt, John F., 39
workers' compensation, 76, 222–23
World of Tomorrow World's Fair, 13
Wright, John, 111
Wynder, Ernst L., 21–22

Youngstown Sheet & Tube Co. v. Sawyer, 206–7
    as analytical framework for *parens patriae* litigation, 207–10